I0044481

Systems Biology: Mathematical Modeling and Model Analysis

Systems Biology: Mathematical Modeling and Model Analysis

Editor: Lynda Feidan

www.callistoreference.com

Callisto Reference,
118-35 Queens Blvd., Suite 400,
Forest Hills, NY 11375, USA

Visit us on the World Wide Web at:
www.callistoreference.com

© Callisto Reference, 2019

This book contains information obtained from authentic and highly regarded sources. Copyright for all individual chapters remain with the respective authors as indicated. All chapters are published with permission under the Creative Commons Attribution License or equivalent. A wide variety of references are listed. Permission and sources are indicated; for detailed attributions, please refer to the permissions page and list of contributors. Reasonable efforts have been made to publish reliable data and information, but the authors, editors and publisher cannot assume any responsibility for the validity of all materials or the consequences of their use.

ISBN: 978-1-64116-220-3 (Hardback)

Trademark Notice: Registered trademark of products or corporate names are used only for explanation and identification without intent to infringe.

Cataloging-in-Publication Data

Systems biology : mathematical modeling and model analysis / edited by Lynda Feidan.
    p. cm.
Includes bibliographical references and index.
ISBN 978-1-64116-220-3
1. Systems biology--Mathematical models. 2. Bioinformatics. 3. Biological systems.
4. Computational biology. I. Feidan, Lynda.
QH324.2 .S97 2019
572.8--dc23

Table of Contents

Preface

Systems biology is the mathematical and computational modeling of complex biological systems. It is an interdisciplinary field of study concerned with complex interactions within biological systems. One of the primary objectives of systems biology is to discover and model emergent properties and explore the properties of cells, tissues and organisms functioning as a system. The foundations of systems biology are control theory and cybernetics, quantitative modeling of enzyme kinetics, synergetics, simulations for the study of neurophysiology and the mathematical modeling of population dynamics. The topics covered in this extensive book deal with the core aspects of mathematical modeling and model analysis in the discipline of systems biology. For all readers who are interested in this field, the case studies included in this book will serve as an excellent guide to develop a comprehensive understanding. It aims to equip students and experts with the advanced topics and upcoming concepts in this area of study.

Various studies have approached the subject by analyzing it with a single perspective, but the present book provides diverse methodologies and techniques to address this field. This book contains theories and applications needed for understanding the subject from different perspectives. The aim is to keep the readers informed about the progresses in the field; therefore, the contributions were carefully examined to compile novel researches by specialists from across the globe.

Indeed, the job of the editor is the most crucial and challenging in compiling all chapters into a single book. In the end, I would extend my sincere thanks to the chapter authors for their profound work. I am also thankful for the support provided by my family and colleagues during the compilation of this book.

Editor

Inferring extrinsic noise from single-cell gene expression data using approximate Bayesian computation

Oleg Lenive[1], Paul D. W. Kirk[2] and Michael P. H. Stumpf[3]*

Abstract

Background: Gene expression is known to be an intrinsically stochastic process which can involve single-digit numbers of mRNA molecules in a cell at any given time. The modelling of such processes calls for the use of exact stochastic simulation methods, most notably the Gillespie algorithm. However, this stochasticity, also termed "intrinsic noise", does not account for all the variability between genetically identical cells growing in a homogeneous environment.

Despite substantial experimental efforts, determining appropriate model parameters continues to be a challenge. Methods based on approximate Bayesian computation can be used to obtain posterior parameter distributions given the observed data. However, such inference procedures require large numbers of simulations of the model and exact stochastic simulation is computationally costly.

In this work we focus on the specific case of trying to infer model parameters describing reaction rates and extrinsic noise on the basis of measurements of molecule numbers in individual cells at a given time point.

Results: To make the problem computationally tractable we develop an exact, model-specific, stochastic simulation algorithm for the commonly used two-state model of gene expression. This algorithm relies on certain assumptions and favourable properties of the model to forgo the simulation of the whole temporal trajectory of protein numbers in the system, instead returning only the number of protein and mRNA molecules present in the system at a specified time point. The computational gain is proportional to the number of protein molecules created in the system and becomes significant for systems involving hundreds or thousands of protein molecules.

Conclusions: We employ this simulation algorithm with approximate Bayesian computation to jointly infer the model's rate and noise parameters from published gene expression data. Our analysis indicates that for most genes the *extrinsic* contributions to noise will be small to moderate but certainly are non-negligible.

Keywords: Stochastic simulation, Gene expression, Extrinsic noise, Approximate Bayesian computation

Background

Experiments have demonstrated the presence of considerable cell-to-cell variability in mRNA and protein numbers [1–5] and slow fluctuations on timescales similar to the cell cycle [6, 7]. Broadly speaking, there are two plausible causes of such variability. One is the inherent stochasticity of biochemical processes which are dependent on small numbers of molecules. The other relates to differences in numbers of protein, mRNA, metabolites and other molecules available for each reaction or process within a cell, as well as any heterogeneity in the physical environment of the cell population. These sources of variability have been dubbed as "intrinsic noise" and "extrinsic noise", respectively.

One of the earliest investigations into the relationship between intrinsic and extrinsic noise employed two copies of a protein with different fluorescent tags, expressed from identical promoters equidistant from the replication origin in *E. coli* [8]. By quantifying fluorescence for a range of expression levels and genetic backgrounds the authors

*Correspondence: m.stumpf@imperial.ac.uk
[3]Imperial College, London, Centre for Integrative Systems Biology and Bioinformatics, SW7 2AZ London, UK
Full list of author information is available at the end of the article

concluded that intrinsic noise decreases monotonically as transcription rate increases while extrinsic noise attains a maximum at intermediate expression levels. Other studies have considered extrinsic noise in the context of a range of cellular processes including the induction of apoptosis [9]; the distribution of mitochondria within cells [10]; and progression through the cell cycle [11]. From a computational perspective, extrinsic variability has been modelled by linking the perturbation of model parameters to perturbation of the model output using a range of methods, including the Unscented Transform [12] the method of moment closure [13], and density estimation [14].

Taniguchi et al. [7] carried out a high-throughput quantitative survey of gene expression in *E. coli*. By analysing images from fluorescent microscopy they obtained discrete counts of protein and mRNA molecules in individual *E. coli* cells. They provided both the measurements of average numbers of protein and mRNA molecules in a given cell, as well as measurements of cell-to-cell variability of molecule numbers. The depth and scale of their study revealed the influence of extrinsic noise on gene expression levels. The authors demonstrated that the measured protein number distributions can be described by Gamma distributions, the parameters of which can be related to the transcription rate and protein burst size [15]. To quantify extrinsic noise they consider the relationship between the means and the Fano factors of the observed protein distributions. They also illustrate how extrinsic noise in protein numbers may be attributed to fluctuations occurring on a timescale much longer than the cell cycle.

Here we aim to describe extrinsic noise at a more detailed, mechanistic, level using a stochastic model of gene expression. A relatively simple mechanistic model of gene expression may represent mRNA production as a zero order reaction with protein being produced from each mRNA via first order reactions. This can be described as the one-state model since the promoter is modelled as being constitutively active (Fig. 1). In the one-state model, mRNA production is represented by a homogeneous Poisson process and the Fano factor of the mRNA distribution at any time point will be one. However, experimental counts of mRNA molecules in single cells indicate that the Fano factor is often considerably higher than one [7].

Such a description calls for quantitative inference of the model's parameters. We achieve this by relying on the data made available by Taniguchi et al. and employing approximate Bayesian computation (ABC) [16, 17]. One difficulty that arises when trying to investigate the extent and effect of extrinsic noise is that it is difficult to separate it from intrinsic noise. To overcome this confounding effect, the parameters of our model come in two varieties. Firstly, reaction rate parameters describe the probability of events occurring per unit of time. These correspond to

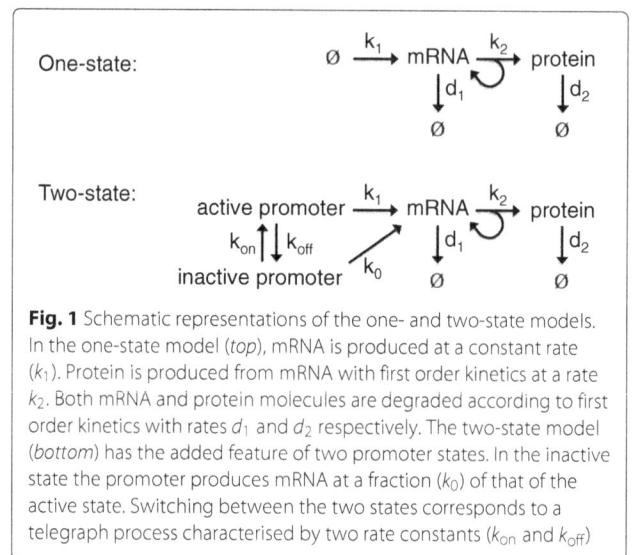

Fig. 1 Schematic representations of the one- and two-state models. In the one-state model (*top*), mRNA is produced at a constant rate (k_1). Protein is produced from mRNA with first order kinetics at a rate k_2. Both mRNA and protein molecules are degraded according to first order kinetics with rates d_1 and d_2 respectively. The two-state model (*bottom*) has the added feature of two promoter states. In the inactive state the promoter produces mRNA at a fraction (k_0) of that of the active state. Switching between the two states corresponds to a telegraph process characterised by two rate constants (k_{on} and k_{off})

the reaction rate parameters of a typical stochastic model which accounts for intrinsic noise. Secondly, noise parameters describe the variability in reaction rate parameters caused by the existence of extrinsic noise. In this model, extrinsic noise is represented by a perturbation of the model's rate parameters using a truncated Gaussian distribution. The magnitude of the perturbation of each rate parameter depends on the corresponding noise parameter, which is closely related to the standard deviation of the relevant Gaussian (see "Methods"). This approach allows us to simultaneously infer the rate parameters and the magnitude of extrinsic noise and may be thought of as an application of mixed effect modelling [18] in the context of exact stochastic simulation.

Stochastic simulation and ABC inference methods are both computationally costly endeavours. In this particular case, the experimental data corresponds to snapshots of the system at a single time point. The data are made available in the form of summary statistics, measures of central tendency (e.g. mean) and statistical dispersion (e.g. variance).

Thus, a complete temporal trajectory of the system is not necessary to carry out comparisons with the data. This allows us to make the problem computationally tractable. To this end, we develop a model-specific simulation method which takes advantage of the Poissonian relationship between the number of surviving protein molecules produced from a given mRNA molecule and its lifetime, under certain assumptions.

Results and discussion
Posterior distributions of parameters
We begin our analysis by examining the posterior distributions of parameters obtained for each gene using

the ABC Sequential Monte Carlo (ABC-SMC) inference procedure [16] A selection of distributions is shown in Fig. 3 and the Additional files 2, 3, 4 and 5 supplementary figures. The simulated summary statistics converged to within the desired threshold of the experimental measurements for 86 out of 87 genes. The inferred posterior for the one remaining gene converged relatively slowly and we chose to terminate the process after 30 days of CPU time.

Figure 2 shows a contour plot of the distribution of summary statistics and the mRNA degradation rate, obtained from particles in the final ABC-SMC population for a typical gene (*dnaK*).

We begin with a discussion of features of the posterior parameter distributions, that are common to most genes. Next, we examine the relationships between model parameters and summary statistics of the model outputs. Lastly, we carry out a sensitivity analysis on the inferred posteriors to assess the importance of each parameter in setting the overall levels of extrinsic noise.

In the two-state model, the switching of the promoter between active and inactive states is described by a telegraph process that can be parametrised either in terms of the switching reaction rates (k_{on} and k_{off}) or in terms of the on/off bias (k_r) and frequency of switching events (k_f)

(Fig. 1). The simulation algorithm takes parameters in the form of k_{on} and k_{off}. However, the effects of k_r and k_f on the observed mRNA distribution may be interpreted more directly and intuitively.

For the majority of genes the k_0 and k_r parameters are relatively small. This appears to be a prerequisite for a high Fano factor of the mRNA distribution and the mean marginal inferred values of these parameters are negatively correlated with Fano factors across all 86 genes as discussed below. A low switching rate combined with a low basal expression rate ensures that there are two distinct mRNA expression levels. This in turn produces a larger variance in measured mRNA counts and results in Fano factor values well above one. Conversely, genes for which mRNA production appears to be more Poissonian were inferred to have basal mRNA production rates close to one, i.e. similar to the active mRNA production rates. In other words, these genes appear to be constitutively active. Here again, we point out that the two-state promoter model provides a convenient abstraction and a hypothesis for explaining the super-Poissonian variance in mRNA copy number [5, 19]. However, based on these observations it is difficult to determine whether a model with more states or some other more elaborate regulatory model, would not be more appropriate. Our attempts

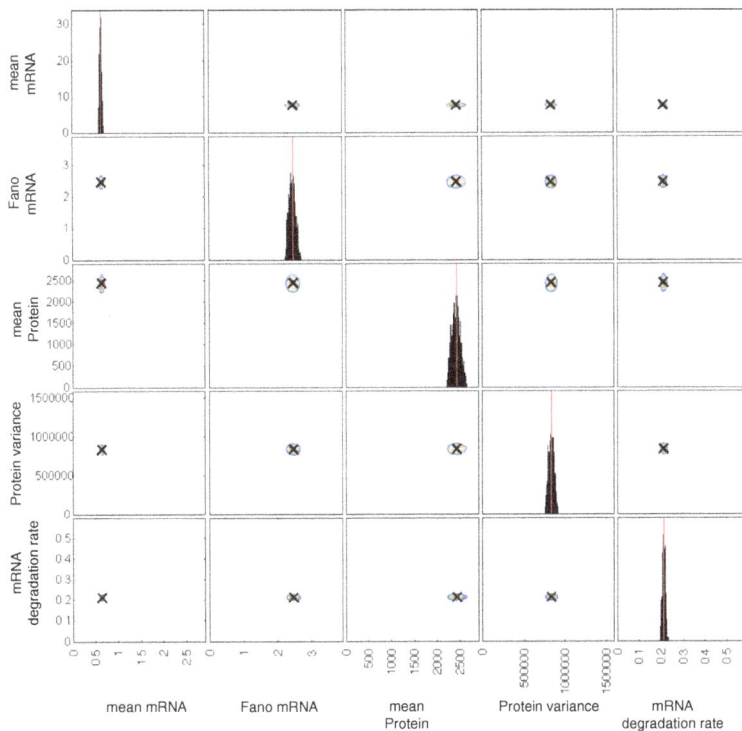

Fig. 2 Posterior distribution of summary statistics and the mRNA degradation rate for the gene *dnaK*. Contour plots indicating the density of points with the corresponding summary statistic for each particle in the final population. The summary statistics for each particle are calculated from 1000 simulation runs. The posterior distribution consists of 1000 particles

at carrying out the inference procedure with a one-state model indicate that extrinsic noise alone does not explain the observed mRNA distributions without also producing unacceptably high variability in protein numbers.

Our initial inference attempts used only the summary statistics from the data. We observed that the production and degradation rate parameters for mRNA (k_1 and d_1) and protein (k_2 and d_2) tended to be positively correlated in the posterior parameter distributions of many genes. This is due to limited identifiability of model parameters since different combinations of rates may produce similar steady state expression levels. We included the mRNA degradation rate in the inference procedure with the aim of overcoming the problem of unidentifiable parameters. However, this did not alleviate the problem entirely and there is still considerable uncertainty, or sloppiness, in the posterior with regard to some directions in parameter space. While this does make it difficult to pick precise parameter values it also illustrates how using ABC provides us with a way of measuring the model's sensitivity to changes in parameters. Our approach provides an indication of the possible range of extrinsic noise values that can account for the observed variability in mRNA and protein numbers (Fig. 3).

Although the posterior summary statistics (and mRNA degradation rate) are reasonably well constrained and distinct for each gene, the distributions of model parameters can still be relatively broad (Fig. 3). There are a number of reasons for this. Firstly, changes in parameters associated with active transcription and translation, as well as degradation rates, are more easily inferred than parameters describing switching between promoter states, basal transcription or extrinsic noise. In particular, when the production and degradation rates for the same species are subjected to different extrinsic noise parameters, the inference procedure struggles to resolve between the different source of extrinsic noise. This explains the correlation between the means of inferred extrinsic noise parameters (Fig. 4). Such correlations between extrinsic noise parameters are not observed in the posterior of each gene or when taking the single particle with the highest weight from the final population of each gene as in Fig. 5.

A comparison of Figs. 4 and 5 suggests that a certain level of extrinsic noise is expected for all genes. However, the extrinsic noise may affect various combinations of rate parameters and it may not be possible to discern if, for example, the production rate or the degradation rate is more affected by extrinsic variability. While our inference procedure does not indicate a distinctive lower boundary for the amount of extrinsic noise affecting each reaction rate, there is usually an upper limit to the inferred noise

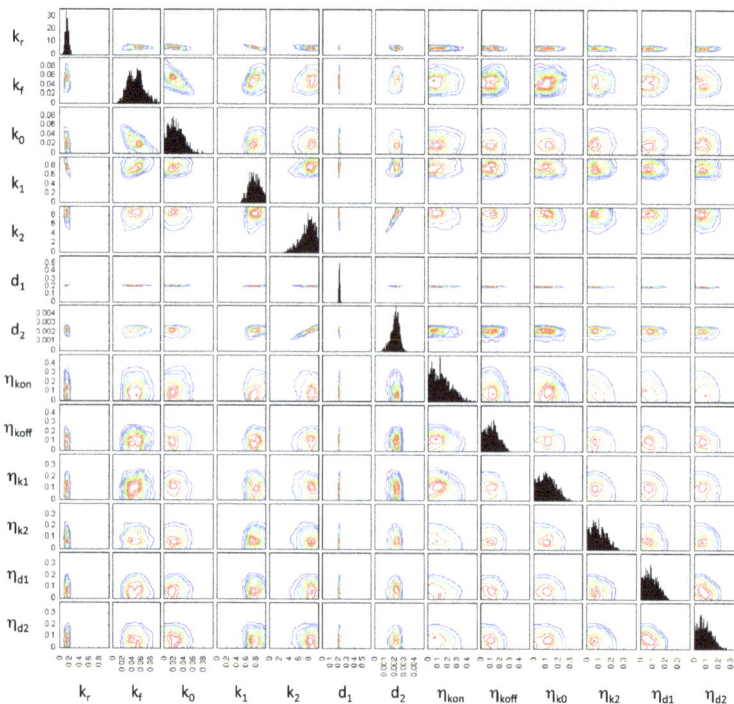

Fig. 3 Posterior distribution of model parameters for the gene *dnaK*. Contour plots indicating the density of points with the corresponding parameter values for each particle in the final population. The posterior distribution consists of 1000 particles

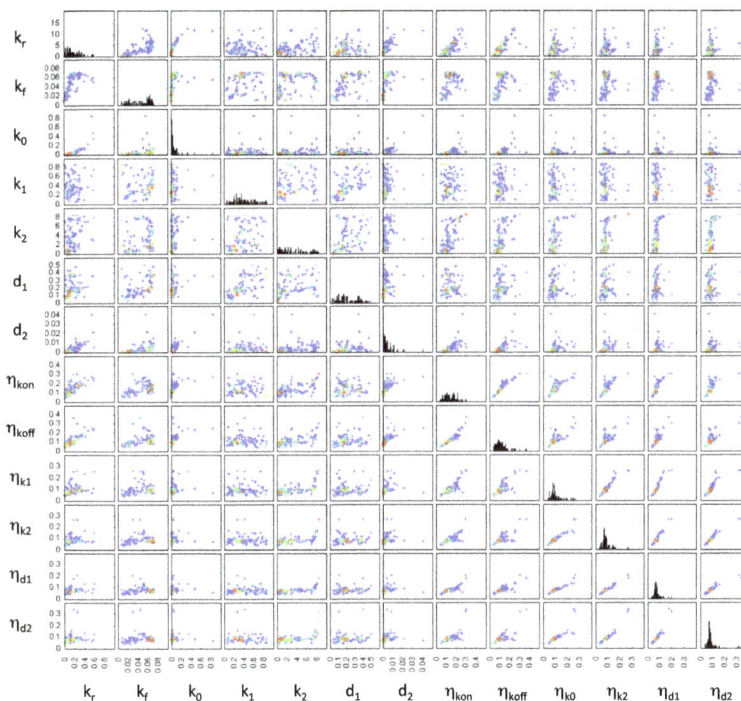

Fig. 4 Relationships between means of the marginal parameter posteriors. Scatter plots of the means of the marginal distributions of parameter posteriors are shown for all pairs of parameters. Each point corresponds to a gene. Warmer hues are used to indicate a higher density of data points

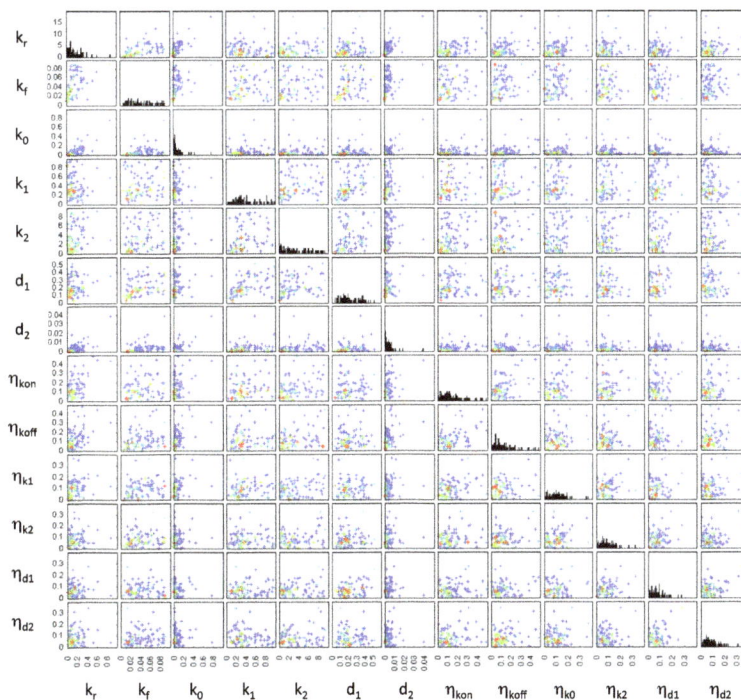

Fig. 5 Relationships between the heaviest particles. Scatter plots of the particles with the highest weight in the final ABC-SMC population, shown for all pairs of parameters. Each point corresponds to one particle from the inferred posterior of one gene. Warmer hues are used to indicate a higher density of data points

parameters ranges. The extrinsic noise parameters for most genes are below 0.2 in the units set here (Fig. 5); however, for some genes, $\eta_{k_{on}}$ and $\eta_{k_{off}}$ have relatively broad posterior marginal distributions.

To better understand the relationship between model parameters and observed patterns of gene expression, we look for correlations between means and variances of the inferred marginal parameters of each gene and the summary statistics used in the inference procedure (Fig. 6). As expected, the correlation between the measured mRNA degradation rate, calculated form mRNA lifetime, and the inferred mRNA degradation rate parameter of the model, is close to one.

The promoter switching rate parameters, k_{on} and k_{off}, display positive and negative correlation with the mean mRNA number, respectively (as may be expected). They have the opposite relationship with the Fano factor associated with the mRNA distribution. This is consistent with the idea that distinct levels of transcription are required to account for the observed mRNA Fano factors. The corresponding extrinsic noise parameters $\eta_{k_{on}}$ and $\eta_{k_{off}}$ are positively correlated with mRNA abundance. However, the means and variances of the marginal distributions of these parameters are negatively correlated with the Fano factor of the mRNA distribution. This indicates that when promoter switching is affected by higher extrinsic noise, the mRNA distribution becomes more Poissonian as the effect of the two distinct promoter states is averaged out.

Curiously, the mean and variance of the protein degradation rate (d_2) are positively correlated with mean mRNA number and negatively correlated with the mRNA Fano factor. Unlike the translation rate (k_2), it shows no significant correlation with the mean or variance of the protein number.

Parameter sensitivity

There are two complementary approaches to investigating the sensitivity of a modelled system to its parameters or inputs [20]. One approach is to consider a single point in parameter space and study how the model responds to infinitesimal changes in parameters. This local approach usually involves calculating the partial derivatives of the model output with respect to the parameters of interest. Alternatively, one may consider how the model behaviour varies within a region of parameter space by sampling parameters and observing model behaviour. Regardless of the method used, different linear combinations of parameters will affect the model output to varying degrees [21]. Gutenkunst et al. [22] coined the terms "stiff" and "sloppy" to describe these differences. They defined a Hessian matrix,

$$H_{i,j}^{\chi^2} \equiv \frac{d^2 \chi^2}{d \log \theta_i d \log \theta_j},$$

where χ^2 provides a measure of model behaviour, such as the average squared change in the species time course.

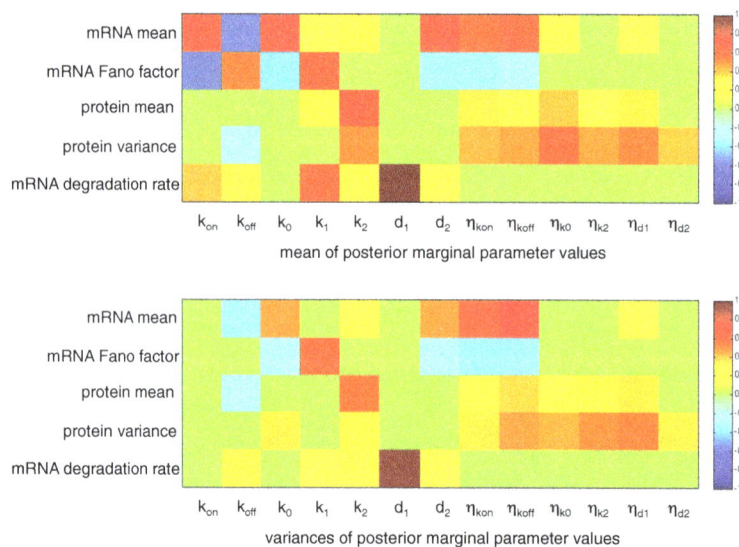

Fig. 6 Heat maps of correlation coefficients between parameters and summary statistics. Heat maps are of the correlation coefficients calculated between experimentally obtained summary statistics and the mean (*top*) or the variance (*bottom*) of the marginal posterior for each model parameter. Correlation coefficients for which the associated *p*-values are greater than 0.05, after correcting for multiple testing using the Benjamini-Hochberg method [43], are treated as zero for plotting purposes

By considering the eigenvalues of this Hessian, λ_i, the authors were able to quantify the (local) responsiveness of the system to a given change in parameters. Conceptually, moving along a stiff direction in parameter space causes a large change in model behaviour; conversely moving along a sloppy direction results in comparatively little effect on the output of the system.

Secrier et al. [23] later demonstrated how these ideas can be applied to the analysis of posterior distributions obtained by ABC methods [24]. Principal component analysis (PCA) may be used to approximate the log posterior density using a multivariate normal (MVN) distribution. They showed that the eigenvalues of the covariance matrix, s_i, of this MVN distribution are related to the eigenvalues of the Hessian as $\lambda_i = 1/s_i$.

To assess the the stiffness/sloppiness of the inferred parameters we carry out PCA of the covariance matrices of log posterior distributions for each gene. In interpreting the results of the PCA we assume that the posterior distribution is, in practice, unimodal. The principal components (eigenvectors), v, and the corresponding loadings (eigenvalues), s, provided by the PCA are then used to obtain the eigen-parameters, q, as

$$q_i = s_i v_i.$$

We calculate the projections of each parameter, θ_i, onto each eigen-parameter, q_j, as

$$c_{ij} = \theta_i \cdot q_j.$$

As a measure of the overall sloppiness of each parameter, l, we use the sum of the contributions of each parameter to the eigen-parameters, $l_i = \sum_j c_{ij}$. This can also be thought of as the sum of the projections of each principal component onto the parameter, weighted by the fraction of total variance explained by each of the principal components.

Having obtained a measure of the sloppiness of each parameter, for each gene, we carry out hierarchical clustering [25] of genes and parameters using a Euclidean distance metric for both (Fig. 7).

The majority of genes show a similar pattern of parameter stiffness/sloppiness. The most distinctive and the second most distinctive clusters consist of just two genes each, *yiiU* with *aceE* and *cspE* with *map*, respectively. These four genes are distinguished by unusually sloppy promoter activity ratio, k_r, and promoter switching frequency, k_f, parameters. The pair *yiiU* and *aceE* display a high ratio of protein variance to protein mean (Fano factor) and are stiff with regard to the protein degradation rate noise parameter η_{d_1}. *cspE* also has a high

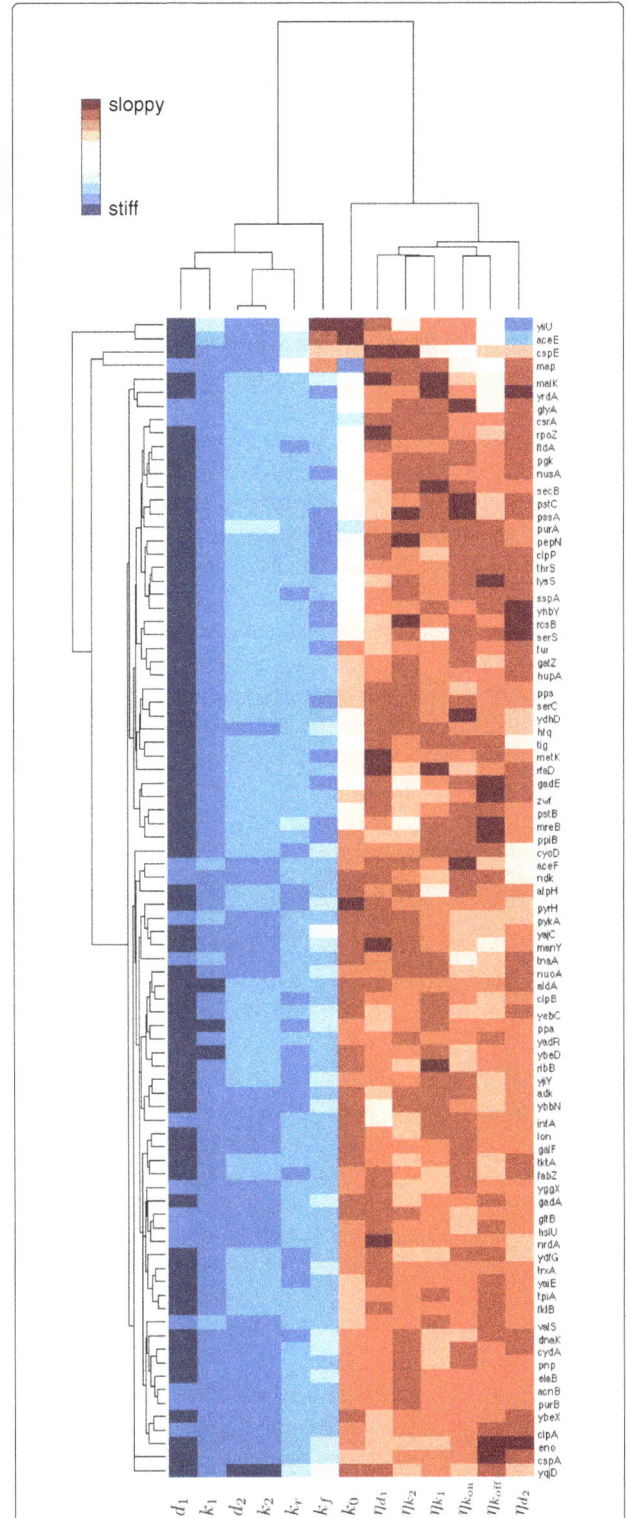

Fig. 7 Clustering of genes and inferred posteriors according to parameter sloppiness. The clustergram shows a heat map of parameter (*columns*) sloppiness for each gene (*rows*). Warmer hues indicate more sloppy parameters. Dendograms above and to the left of the heat map display the hierarchical tree obtained when clustering either the model parameters or the genes using a Euclidean distance metric

Fano factor of the protein distribution while *map* has an unusually low mRNA Fano factor. What these four genes appear to have in common is that the variability in their protein numbers is difficult to explain based solely on the mRNA variability. Thus, a higher level of extrinsic noise is inferred to account for the observed variability. Since these genes comprise a small minority, it may be that their expression is subject to regulatory mechanisms that are not well approximated by the two state model. The remaining majority of genes are broadly divided into two similar groups which differ mostly in the sloppiness of k_0.

The noise and rate parameters segregate into two clusters with the noise parameters generally being sloppier than the rate parameters (Fig. 7). The least sloppy parameter is the mRNA degradation rate (d_1). This is not surprising since it was used, together with the molecule number summary statistics, to infer the posterior distribution. Of the rate parameters, the basal transcription rate (k_0) is the sloppiest and often approaches the noise parameters in its sloppiness. Since this parameter is defined as a fraction of the active transcription rate (k_1), its relative sloppiness should not be equated to a lack of importance. For most genes the marginal posterior of k_0 is largely constrained to the lower half of its prior distribution, $U(0, 1)$. The only exception being the gene *map* for which the measured mRNA Fano factor was close to one and the marginal posterior of k_0 is in the top half of the prior range. The mean of the marginal posterior of k_0 is negatively correlated with the mRNA Fano factor across all genes (Fig. 6). The two other parameters that influence the mRNA Fano factor, k_r and k_f, are the next sloppiest rate parameters.

Conclusions

Cell-to-cell variability in genetically homogeneous populations of cells is a ubiquitous phenomenon [26–28]. Attempts to quantify it are complicated by the difficulty of assigning it to a single cellular process or any one experimentally measurable variable. It can also be difficult, for example, to distinguish between the intrinsic stochasticity of biochemical processes in the short term and longer term variations which may have been inherited from previous cell generations.

By including a representation of extrinsic noise in our model of gene expression we infer the extent to which the rates of biochemical processes can vary between cells while still producing the experimentally measured mRNA and protein variability. We demonstrate the usefulness of an efficient method for exact stochastic simulation of the two-state model of gene expression. The two-state model is necessary to explain the experimentally measured mRNA variation (Fano factor), and is capable of describing the majority of the observed data. The corresponding single-state model, with constant promoter

activity and extrinsic noise, does not produce mRNA Fano factors as high as those measured experimentally without leading to unacceptably high variability in the protein numbers. We show that the amount of extrinsic noise affecting most genes appears to be limited, but non-negligible.

The exact simulation method described here occupies a niche between those cases when only samples from the steady state mRNA distribution of the two-state model [3, 29, 30] are required, and cases when an approximation to the protein distribution [15, 31] is sufficient. The computational advantages of the simulation method described here are limited to specific conditions, such as, low numbers of mRNA molecules and higher numbers of protein molecules. The most limiting factor of this simulation method is that it is not applicable to models in which the protein products affect upstream processes such as promoter activity, transcription or translation. The addition of such interactions would mean that the assumptions used in deriving the Poissonian relationship between the number of surviving protein molecules produced form a given mRNA molecule and mRNA's lifetime would no longer be satisfied. Perhaps an approximate algorithm could be developed on the basis of Algorithm (1) to handle such situations. Alternatively, the tau-leaping algorithm [32], or moment expansion [33, 34], may be more appropriate for models involving these kinds of feedback interactions. Algorithm (1) could, however, be naturally extended to models involving regulatory interactions between non-coding RNAs as the simulation of that part of the model is equivalent to Gillespie's exact algorithm. Although here we use summary statistics of mRNA and protein number measurements, the simulation method is also applicable to cases where a direct comparison between sample distributions, for example using the Hellinger distance, is required.

Here we have worked under the assumption that experimental measurement error associated with individual mRNA or protein counts obtained by fluorescence microscopy are small relative to the combined effects of extrinsic and intrinsic noise. We deem this to be justifiable given the experimental method used by Tanaguchi et al. [7] and the results presented in their publication. More generally, such measurement errors would inflate estimates of the variances in molecule numbers and may skew the inferred extrinsic noise parameters. Other studies, which look directly at the interplay between intrinsic and extrinsic noise in single cells [35] — using time-resolved proteomics data — do also bear this out.

The inferred extrinsic noise parameters will also include the effects of regulatory mechanisms that are not well described by the two-state model. In this sense, our definition of noise becomes blurred with our ignorance about the regulatory interactions involved in the expression of each gene. Nonetheless, the biochemical mechanisms governing gene expression in a given species are shared between many genes. This is in agreement with our observation that, for most genes, inferred model parameters show similar patterns of sloppiness. If we are able to refine our understanding of the shared aspects of gene expression, we may be able to improve our understanding of both the nature of the noise affecting it, and the regulatory mechanisms controlling it. In practice this may mean finding a mechanistic explanation for the two-state model or further refining it to achieve a better agreement between simulations and experimental results.

The in silico approach used here not only relied on, but was inspired by the experimental work of Tanaguchi et al. [7]. As the resolution of high throughput experimental techniques and the quantity of data they generate continues to increase, more complete observations of cellular processes may begin to yield data amenable to statistical analysis and inference of extrinsic noise. These may in turn require other modelling, computational and theoretical approaches which would not rely on the assumptions and simplifications that we make in this work [36].

Methods
Modelling gene expression
A simple model of gene expression may represent the processes of transcription and translation using mass-action kinetics to describe production and degradation of various species as pseudo-first order reactions. Such a model may be simulated stochastically to take into account the intrinsic variability of processes involving low numbers of molecules. In the simplest version of this model, mRNA is produced from the promoter at a constant rate. However, such Poissonian mRNA production is often not sufficient to account for the variability in mRNA numbers measured experimentally in both prokaryotic and eukaryotic cells. In addition to this, for many genes, transcription appears to occur in bursts rather than at a constant rate. These characteristics of gene expression have been observed in organisms as diverse as bacteria [7], yeast [4], amoeba [2] and mammals [3]. One model of gene expression that takes this into account is the, so called, two-state model.

The two-state promoter model
In the two-state model of gene expression, a gene's promoter is represented as either active or inactive [5, 19]. Here we use a variant of the two-state model with the inactive state corresponding to a lower transcription rate rather than no transcription at all. For each state of the promoter, transcription events at that promoter are represented by a Poisson process with rate parameter corresponding to the transcription rate. Biochemical processes

such as transcription factor binding or reorganisation of chromatin structure may account for the existence of several distinct levels of promoter activity. However, which factors play a dominant role in the apparent switching, remains an unanswered question.

The Gillespie algorithm [37] may be used to simulate all the reactions represented by this model and obtain a complete trajectory of the system through time. However, in this case we are only interested in the number of molecules present at the time of measurement. We use a model-specific stochastic algorithm (Algorithm 1) which allows us to reduce the number of computational steps required to obtain a single realisation from the model.

The following reactions, represented using mass-action kinetics, comprise the two-state model:

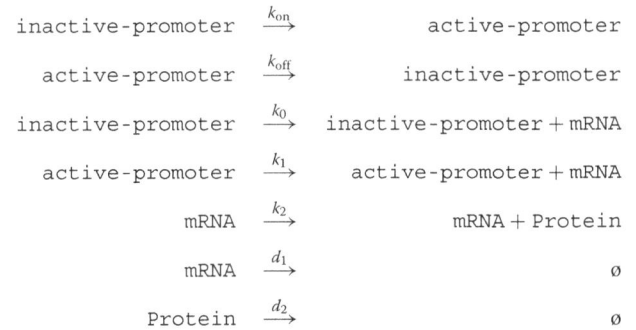

$$\texttt{inactive-promoter} \xrightarrow{k_{\mathrm{on}}} \texttt{active-promoter}$$
$$\texttt{active-promoter} \xrightarrow{k_{\mathrm{off}}} \texttt{inactive-promoter}$$
$$\texttt{inactive-promoter} \xrightarrow{k_0} \texttt{inactive-promoter} + \texttt{mRNA}$$
$$\texttt{active-promoter} \xrightarrow{k_1} \texttt{active-promoter} + \texttt{mRNA}$$
$$\texttt{mRNA} \xrightarrow{k_2} \texttt{mRNA} + \texttt{Protein}$$
$$\texttt{mRNA} \xrightarrow{d_1} \varnothing$$
$$\texttt{Protein} \xrightarrow{d_2} \varnothing$$

The propensity functions (hazards) for each of the above reactions are listed below:

$$h_0 = k_{\mathrm{on}}[\,\texttt{inactive-promoter}\,]$$
$$h_1 = k_{\mathrm{off}}[\,\texttt{active-promoter}\,]$$
$$h_2 = k_0[\,\texttt{inactive-promoter}\,]$$
$$h_3 = k_1[\,\texttt{active-promoter}\,]$$
$$h_4 = k_2[\,\texttt{mRNA}\,]$$
$$h_5 = d_1[\,\texttt{mRNA}\,]$$
$$h_6 = d_2[\,\texttt{Protein}\,]$$

Here the square brackets refer to the number of molecules of a species rather than its concentration.

The model presented here relies on a number of assumptions about the process of gene expression. Firstly, that the production of mRNA and protein can be described sufficiently well by pseudo-first order reactions. Secondly, that degradation of mRNA and protein can be described as an exponential decay. In a bacterial cell, mRNA molecules are degraded enzymatically and typically have a half-life on the scale of several minutes. The half-life of protein molecules usually exceeds the time required for cell growth and division during the exponential growth phase. Thus, dilution due to partitioning of protein molecules between daughter cells tends to be the dominant factor in decreasing the number of protein molecules. Here we do not build an explicit model of cell division, instead the decrease in protein numbers is approximated by an exponential decay. Finally, it is assumed that there is no feedback mechanism by which the number of mRNA or protein molecules produced by the gene affects its promoter switching, transcription or translation rates.

Representing extrinsic noise

We model extrinsic noise by perturbing the reaction rate parameters, using a Gaussian kernel, before each simulation of the model [35, 38]. The effect of extrinsic noise on each reaction is assumed to be independent. The reaction rates associated with a particular gene are termed nominal parameters (θ_n).

$$\theta_n = [\,k_{\mathrm{on}}, k_{\mathrm{off}}, k_0, k_1, k_2, d_1, d_2\,]$$

The values determining the magnitude of the perturbation are termed the noise parameters (η).

$$\eta = [\,\eta_{k_{\mathrm{on}}}, \eta_{k_{\mathrm{off}}}, \eta_{k_0}, \eta_{k_2}, \eta_{d_1}, \eta_{d_2}\,]$$

Together they comprise the full parameter set for the model $\theta = [\,\theta_n, \eta\,]$.

In the case of the two-state model of a single gene, each θ_n has a corresponding extrinsic noise parameter with the exception that the basal transcription rate (k_0') is defined as a fraction of the active transcription rate (k_1') so the two reaction rates are subject to the same perturbation (η_{k_0}) before each simulation. This is motivated by the idea that extrinsic factors affecting the transcription rate do not depend on the state of the promoter. The parameters used to generate a single realisation from the two-state model are obtained by sampling from $f(\mu, \sigma)$. Where f is a truncated normal distribution, restricted to non-negative values by rejection sampling, with μ and σ being the mean and standard deviation of the corresponding normal distribution.

$$k_{\mathrm{on}}' \sim f\left(k_{\mathrm{on}}, k_{\mathrm{on}}\eta_{k_{\mathrm{on}}}\right)$$
$$k_{\mathrm{off}}' \sim f\left(k_{\mathrm{off}}, k_{\mathrm{off}}\eta_{k_{\mathrm{off}}}\right)$$
$$k_1' \sim f\left(k_1, k_1\eta_{k_1}\right)$$
$$k_0' = k_0 k_1'$$
$$k_2' \sim f\left(k_2, k_2\eta_{k_2}\right)$$
$$d_1' \sim f\left(d_1, k_1\eta_{d_1}\right)$$
$$d_2' \sim f\left(d_2, k_2\eta_{d_2}\right)$$

The final time point of each simulation represents the number of mRNA and protein molecules in a single cell at the time of measurement.

Simulation procedure

In order to reduce the computational cost of each simulation, rather than using Gillespie's direct method to simulate the entire trajectory of mRNA and protein numbers, we employed Algorithm 1 to obtain samples of the numbers of mRNA and protein molecules at the time of measurement (t_m). First, a realisation of the telegraph process is used to obtain the birth and decay times of mRNA molecules. These are then used to sample the number of protein molecules that were produced from each mRNA molecule and survived until t_m. This procedure makes use of the Poisson relationship between the life time of an individual mRNA molecule and the number of surviving protein molecules that were produced from it. This relationship is derived in Additional file 1 and its use is illustrated in Fig. 8. The final result is the number of both mRNA (M) and protein (P) molecules present in the system at t_m.

Use of experimental data

Using an automated fluorescent imaging assay, Taniguchi et al. [7] were able to quantify the abundances of 1018 proteins from a yellow fluorescent protein fusion library. We focus on a subset of 87 genes from the published data set from [7]. These are all the genes for which, in addition to protein numbers, the experimental data include both fluorescence *in situ* hybridization measurements [39] of mRNA numbers and mRNA lifetimes measurements obtained using RNAseq [40]. We note that these genes are not a random sample from the set of all genes and exhibit higher than average expression levels.

To identify model parameters for which the two-state model, with extrinsic noise, is able to reproduce the experimental measurements, we carry out Bayesian inference using an ABC sequential Monte Carlo (SMC) algorithm that compares summary statistics from simulated and experimental data [41]. Specifically we used the following summary statistics: (1) the mean numbers of mRNA molecules; (2) the Fano factors of mRNA molecule distributions; (3) the mean numbers of protein molecules; (4) the variances of protein molecule numbers; and (5) mRNA lifetimes converted to expontial decay rate parameters. The distributions of these summary statistics are shown in Fig. 9. We assume that the summary statistics correspond to steady state expression levels for each gene. While there is no guarantee that this is the case for every gene, the majority of genes are unlikely to be undergoing major changes in their expression level given that the cells are in a relatively constant environment.

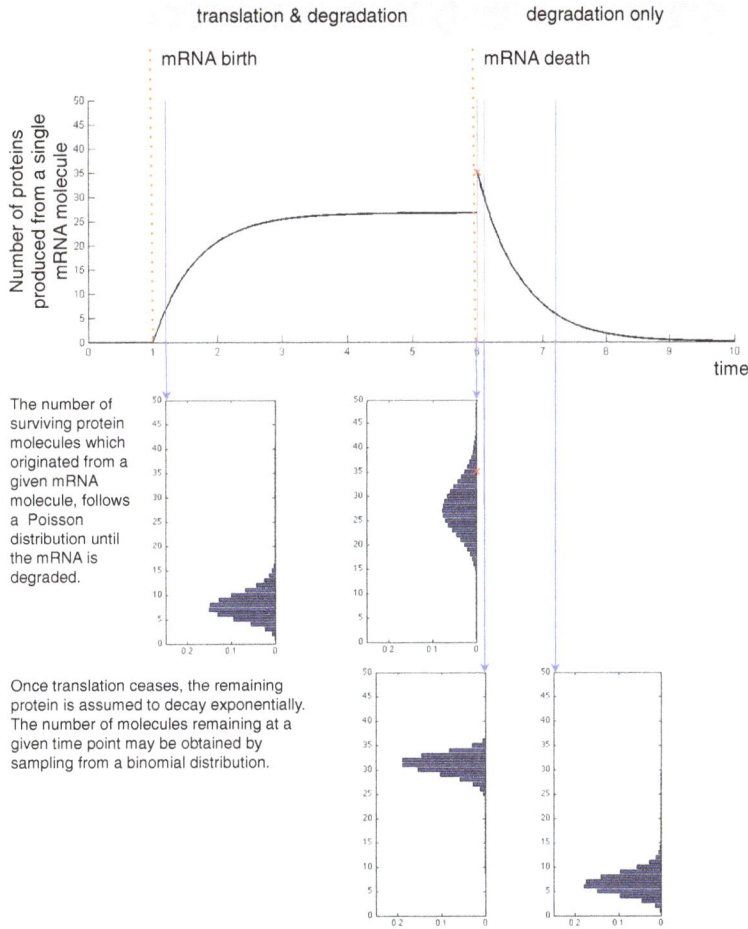

Fig. 8 Illustration of the principle behind Algorithm 1. An illustration of how the birth and death times of an mRNA molecule are used to obtain the number of proteins that were produced from it and then survived until the time at which mRNA and protein numbers were measured. According to the two-state model used here, the number of protein molecules that were translated using a given mRNA template and have not yet been degraded can be found by sampling from the corresponding Poisson distribution with a parameter which depends on the lifetime of the mRNA template. If the mRNA is degraded before the measurement time point, the remaining protein molecules are assumed to decay exponentially. Thus the number of protein molecules can be obtained by first sampling the number present at the point of mRNA decay and then sampling from the corresponding binomial distribution to determine the number of surviving molecules at the measurement time point

Taniguchi et al. [7] used images of about a thousand cells to obtain estimates of mean mRNA numbers, mRNA Fano factors, mean protein numbers and protein number variances. For this reason, we use 10^3 simulation runs when calculating summary statistics. The experimental measurements of mRNA lifetimes are compared directly to the mRNA degradation rate parameter (d_1) in the model by assuming that lifetimes correspond to the inverse of the decay rate.

Inference procedure

We use an ABC-SMC algorithm to infer plausible parameter sets for the two-state model based on the experimental data. The inference procedure is similar to that employed by [24, 41, 42], as described in Algorithm 2.

For the distance metric, d, we take the Euclidean distance between the logarithms of each type of experimental measurement (D_i) and the corresponding simulation results (x_i):

$$d(D, x) = \sqrt{\sum_{i=1}^{i=5} \left(\log D_i - \log x_i \right)^2}$$

$$D = \left[\mu_{mRNA}, \frac{\sigma_{mRNA}^2}{\mu_{mRNA}}, \mu_{prot}, \sigma_{prot}^2, \tau_{mRNA}^{-1} \right]$$

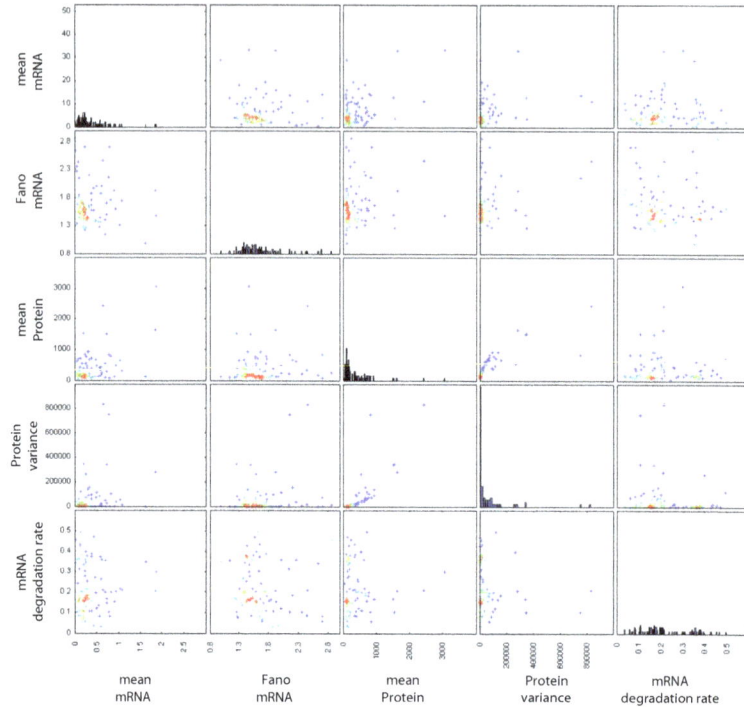

Fig. 9 Experimentally measured summary statistics. Each point on the scatter plots is an estimate of the corresponding summary statistic or mRNA degradation rate from experimental measurements. These data are taken from [7]. The mRNA degradation rates were taken to be the inverse of the mRNA lifetimes

Where μ_{mRNA} is the mean number of mRNA molecules; $\sigma^2_{mRNA}/\mu_{mRNA}$ is the Fano factor of the mRNA distribution; μ_{prot} is the mean number of protein molecules; σ^2_{prot} is the variance of the protein distribution and τ^{-1}_{mRNA} gives the exponential decay rate constant for mRNA degradation based on the measured mRNA lifetime (τ_{mRNA}).

$$x = \left[\mu_M, \frac{\sigma^2_M}{\mu_M}, \mu_P, \sigma^2_P, d_1 \right]$$

Where μ_M is the mean number of mRNA molecules; σ^2_M/μ_M is the Fano factor of the mRNA distribution; μ_P is the mean number of protein molecules; σ^2_P is the variance of the protein distribution and d_1 corresponds to the nominal mRNA degradation rate. The first sampled population of particles (population zero in Algorithm 2), provides a benchmark for the choice of ϵ values in the next population. Since we have no knowledge of the distribution of distances until a set of particles is sampled, all particles are accepted in the first population. For subsequent populations, ϵ values are chosen such that the probability of acceptance with the new ϵ value is equal to q_t. The vector q is chosen prior to the simulation. This allows for larger decreases in ϵ in the first few populations while keeping the actual epsilon values used, a function of the distances (g) in the previous population. New populations are sampled until the final epsilon value is reached $\epsilon_f = 0.1$. To obtain θ^* from θ we use a uniform perturbation kernel:

$$\theta^* \sim U(\theta - \mu_{t-1}, \theta + \mu_{t-1})$$

where μ_{t-1} is the vector of standard deviations of each parameter in the previous population.

Parameter prior

The telegraph process may be parametrized in terms of the ratio of probabilities of switching events (k_r) and the overall frequency with which events occur (k_f):

$$k_r = \frac{k_{\text{on}}}{k_{\text{on}} + k_{\text{off}}}$$

Algorithm 1 Simulation of the two-state model

Inputs: θ_n, η, t_m
Outputs: M, P

1: Obtain perturbed parameters using the nominal (θ_n) and noise (η) parameters.
   Stage one: simulate mRNA production subject to an underlying telegraph process.
2: $S \leftarrow 1$ with probability $k'_{on}/(k'_{on} + k'_{off})$), otherwise $S \leftarrow -1$
                                          $\triangleright$ Select the initial state of the telegraph process.
3: $t \leftarrow 0$ $\triangleright$ Initialise simulation time.
4: $M_b \leftarrow 0$ $\triangleright$ Initialise the number of mRNA molecules produced.
5: $i \leftarrow 1$ $\triangleright$ Initialise index of mRNA molecules.
6: **while** $t < t_m$ **do**
7:     **if** $S = -1$ **then**
8:         $k_S \leftarrow k'_{on}$
9:         $k_m \leftarrow k'_0$
10:     **else**
11:         $k_S \leftarrow k'_{off}$
12:         $k_m \leftarrow k'_1$
13:     **end if**
14:     $\tau \sim Exp(k_S)$ $\triangleright$ Sample the time until the next switching event.
15:     **if** $t + \tau > t_m$ **then** $\triangleright$ Ensure that $t + \tau$ does not exceed the final time point.
16:         $\tau \leftarrow t_m - t$
17:     **end if**
18:     $M_\tau \sim Poisson(\tau k_m)$ $\triangleright$ Sample the number of mRNA molecules produced.
19:     $M_b \leftarrow M_b + M_\tau$
20:     **while** $i \leq M_b$ **do**
21:         $u_i \sim Uniform(t, t + \tau)$ $\triangleright$ Sample birth times for each mRNA.
22:         $i \leftarrow i + 1$
23:     **end while**
24:     $t \leftarrow t + \tau$
25:     $S \leftarrow -S$
26: **end while**
   Stage two: simulate mRNA degradation; protein production and degradation.
27: $M \leftarrow 0$ $\triangleright$ Initialise the number of mRNA molecules at t_m.
28: $P \leftarrow 0$ $\triangleright$ Initialise the number of protein molecules at t_m.
29: $i \leftarrow 1$
30: **while** $i \leq M_b$ **do** $\triangleright$ For each mRNA molecule that was produced:
31:     $v \sim Exp(d'_1)$ $\triangleright$ Sample the time until mRNA decay.
32:     $T_l \leftarrow min(u_i + v; t_m) - u_i$ $\triangleright$ Calculate mRNA lifetime.
33:     $P_l \sim Poisson\left(\frac{k'_2}{d'_2}(1 - e^{-d'_2 T_l})\right)$
                             $\triangleright$ Sample the number of surviving proteins at time point $u_i + v$.
34:     $T_d \leftarrow t_m - min(u_i + v; t_m)$ $\triangleright$ Time since mRNA decay.
35:     **if** $T_d = 0$ **then**
36:         $M \leftarrow M + 1$ $\triangleright$ mRNA survived until t_m.
37:         $P \leftarrow P + P_l$
38:     **else**
39:         $P_s \sim Binomial(P_l, e^{-d'_2 T_d})$
                             $\triangleright$ Sample the number of surviving proteins at time t_m.
40:         $P \leftarrow P + P_s$
41:     **end if**
42:     $i \leftarrow i + 1$
43: **end while**

Algorithm 2 ABC-SMC with summary statistics

    Inputs: π, N, ϵ_f
    Outputs: Set of populations of N accepted particles
1: $i \leftarrow 1$
2: $t \leftarrow 0$
3: $q \leftarrow [\,0.01, 0.05, 0.25, 0.75, ..., 0.75\,]$
4: Initialise ϵ vector.
5: **while** $\epsilon > \epsilon_f$ **do**
6:     **if** $t = 0$ **then**
7:         **while** $i \leq N$ **do**
8:             Sample a new θ from π.
9:             Simulate from the model 10^3 times according to Algorithm 1.
10:           Calculate summary statistics, x, from the simulation outputs.
11:           **if** $d(D, x) < \epsilon$ **then**
12:             Accept particle.
13:             $\omega^{(i,t)} \leftarrow 1$
14:             $i \leftarrow i + 1$
15:           **end if**
16:         **end while**
17:     **else**
18:         **while** $i \leq N$ **do**
19:             Sample θ from $\{\theta^{(j,t-1)}\}_{1 \leq j \leq N}$ with probability $\{\omega^{(j,t-1)}\}_{1 \leq j \leq N}$.
20:             Perturb θ to obtain θ^*.
21:             Simulate from the model 10^3 times according to Algorithm 1.
22:           Calculate summary statistics, x, from the simulation outputs.
23:           **if** $d(D, x) < \epsilon$ **then**
24:             Accept particle
25:

$$\omega^{(i,t)} \leftarrow \frac{\pi(\theta^{(i,t)})}{\sum_{j=1}^{n} \omega^{(j,t)} K_t(\theta^{i,t} | \theta^{(j,t-1)})}$$

26:             $i \leftarrow i + 1$
27:           **end if**
28:         **end while**
29:     **end if**
30:     Normalise weights.
31:     $t \leftarrow t + 1$
32:     Set ϵ such that $Pr(g_i \leq \epsilon_i) = q_t$
33: **end while**

$$k_f = 2\frac{k_{\mathrm{on}} k_{\mathrm{off}}}{k_{\mathrm{on}} + k_{\mathrm{off}}}$$

To obtain θ, the vector of parameters used in the ABC-SMC inference procedure (Algorithm 2), rate and noise parameters are sampled from the following uniform priors,

$$k_r \sim U(0, 1)$$
$$k_f \sim U(0, 0.1)$$
$$k_0 \sim U(0, 1)$$
$$k_1 \sim U(0, 1)$$
$$k_2 \sim U(0, 10)$$
$$d_1 \sim U(0.01, 0.6)$$
$$d_2 \sim U(0.0005, 0.05)$$
$$\eta_{k_{\mathrm{on}}} \sim U(0, 0.5)$$
$$\eta_{k_{\mathrm{off}}} \sim U(0, 0.5)$$
$$\eta_{k_1} \sim U(0, 0.4)$$
$$\eta_{k_2} \sim U(0, 0.4)$$
$$\eta_{d_1} \sim U(0, 0.4)$$
$$\eta_{d_2} \sim U(0, 0.4).$$

The parameters for the telegraph process, sampled from the prior as k_r and k_f, are converted to k_{on} and k_{off} before being passed to the simulation algorithm (Algorithm 1) as follows,

$$k_{\mathrm{off}} = \frac{k_f}{2k_r}$$
$$k_{\mathrm{on}} = \frac{k_{\mathrm{off}} k_r}{1 - k_r}.$$

Rate parameters k_r and k_0 as well as the noise parameters (η) are unit-less. The remaining parameters have units $1s^{-1}$.

To ensure that M and P are from a distribution close to equilibrium, simulation duration is set depending on the nominal degradation rates for mRNA (d_1) and protein (d_2),

$$t_m = L\left(d_1^{-1} + d_2^{-1}\right)$$

where t_m is the final time point and L is a constant chosen arbitrarily to indicate the desired proximity to the steady state distribution. Here we use $L = 5$.

To confirm that our inference procedure is able to converge to the appropriate region of parameter space in an idealised case, we generate synthetic data by simulating 1000 times from the two-state model. We then calculate summary statistics from these data and carry out the inference procedure in the same manner as for the experimental data. Figures 10 and 11 show the resulting distributions of summary statistics and model parameters respectively.

To provide a comparison of the compute times required to simulate the two-state model using the Gillespie algorithm or our model-specific algorithm we take the final population of parameters obtained for the gene dnaK and run simulations on the same CPU using both methods. The extent of the improvement depends on the model parameters. In this case, the mean improvement is 26

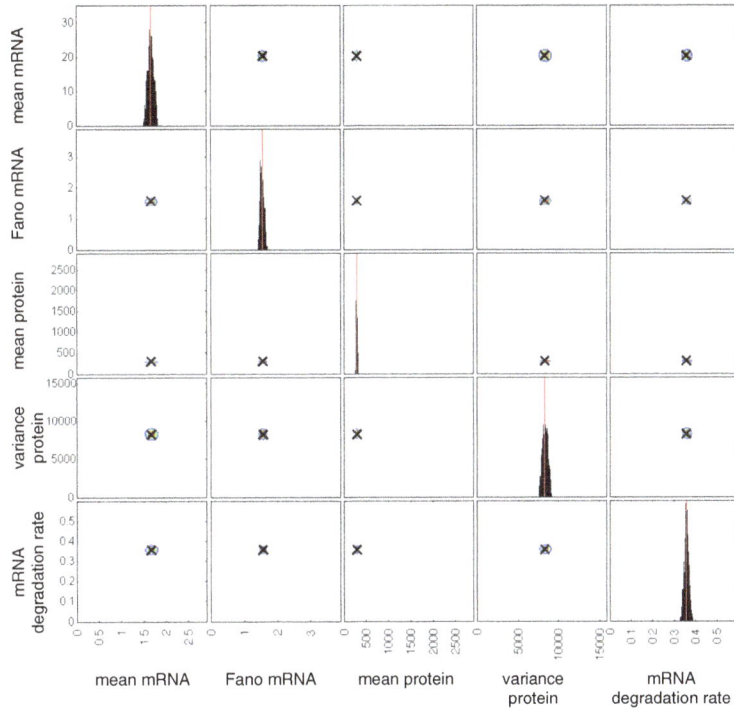

Fig. 10 Posterior distribution of summary statistics and the mRNA degradation rate for a test case where synthetic data were generated by simulating from a model with known parameters. Contour plots indicating the density of points with the corresponding summary statistic for each particle in the final population. The summary statistics for each particle are calculated from 1000 simulation runs. The posterior distribution consists of 1000 particles

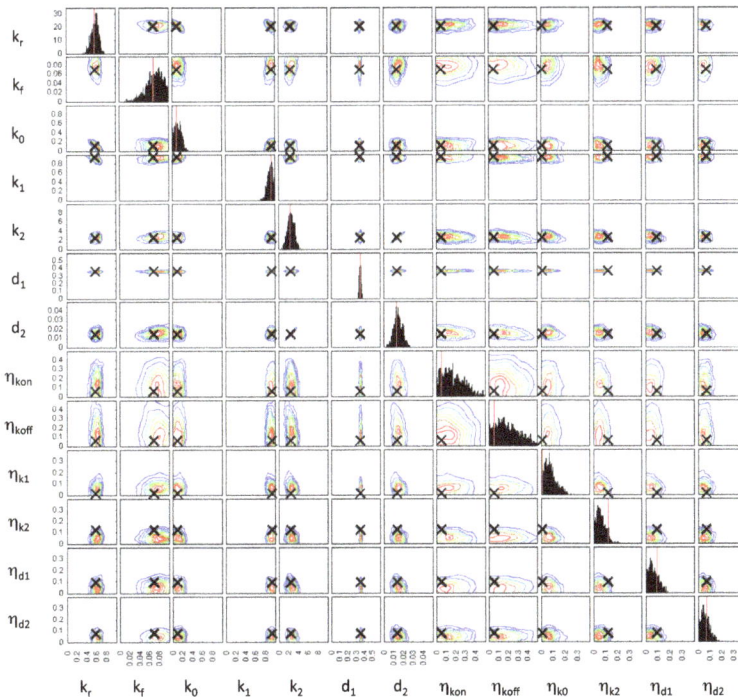

Fig. 11 Posterior distribution of model parameters for a test case where synthetic data were generated by simulating from a model with known parameters. Contour plots indicating the density of points with the corresponding parameter values for each particle in the final population. The posterior distribution consists of 1000 particles

fold with a variance of 12. The total times taken to simulate 1000 perturbed parameter samples from each of 1000 particles were 147 and 3786 s.

Additional files

Additional file 1: Derivation of the Poissonian relationship between the number of surviving protein molecules and mRNA lifetime.

Additional file 2: Parameter posteriors for the expression model of the *rcsB* gene.

Additional file 3: Parameter posteriors for the expression model of the *yiiU* gene.

Additional file 4: Parameter posteriors for the expression model of the *yebC* gene.

Additional file 5: Parameter posteriors for the expression model of the *eno* gene.

Abbreviations
ABC, approximate Bayesian computation; SMC, sequential Monte Carlo

Acknowledgements
We thank the members of the *Theoretical Systems Biology Group* at Imperial College London for helpful discussions and feedback.

Funding
The work was supported by a BBSRC Bioprocessing PhD studentship to O.L. and M.P.H.S. and P.D.W.K. was supported by the MRC (project reference MC_UP_0801/1).

Authors' contributions
OL, PK and MPHS designed the study. OL carried out the computational work. OL and MPHS wrote the paper. All authors read and reviewed the final paper.

Competing interests
The authors declare that they have no competing interests.

Author details
[1]ICR, SM2 5NG Sutton, UK. [2]MRC Biostatistics Unit, Cambridge Institute of Public Health, Cambridge, UK. [3]Imperial College, London, Centre for Integrative Systems Biology and Bioinformatics, SW7 2AZ London, UK.

References
1. Golding I, Paulsson J, Zawilski SM, Cox EC. Real-Time Kinetics of Gene Activity in Individual Bacteria. Cell. 2005;123(6):1025–36.
2. Chubb JR, Trcek T, Shenoy SM, Singer RH. Transcriptional Pulsing of a Developmental Gene. Curr Biol. 2006;16(10):1018–25.
3. Raj A, Peskin CS, Tranchina D, Vargas DY, Tyagi S. Stochastic mRNA Synthesis in Mammalian Cells. PLoS Bio. 2006;4(10):309.
4. Zenklusen D, Larson DR, Singer RH. Single-RNA counting reveals alternative modes of gene expression in yeast. Nat Struct Mol Biol. 2008;15(12):1263–71.
5. Tan RZ, van Oudenaarden A. Transcript counting in single cells reveals dynamics of rDNA transcription. Mol Syst Biol. 2010;6:358.
6. Rosenfeld N. Gene Regulation at the Single-Cell Level. Science. 2005;307(5717):1962–65.
7. Taniguchi Y, Choi PJ, Li GW, Chen H, Babu M, Hearn J, Emili A, Xie XS. Quantifying E. coli proteome and transcriptome with single-molecule sensitivity in single cells. Science. 2010;329(5991):533–8.
8. Elowitz MB, Levine AJ, Siggia ED, Swain PS. Stochastic gene expression in a single cell. Sci Adv. 2002;297(5584):1183–1186.
9. Spencer SL, Sorger PK, Gaudet S, Albeck JG, Burke JM. Non-genetic origins of cell-to-cell variability in TRAIL-induced apoptosis. 2009;459(7245):428–32.
10. Johnston IG, Gaal B, Neves RPd, Enver T, Iborra FJ, Jones NS. Mitochondrial variability as a source of extrinsic cellular noise. PLoS Comput Biol. 2012;8(3):1002416.
11. Kaufmann BB, Yang Q, Mettetal JT, van Oudenaarden A. Heritable stochastic switching revealed by single-cell genealogy. PLoS Biol. 2007;5(9):239.
12. Toni T, Tidor B. Combined model of intrinsic and extrinsic variability for computational network design with application to synthetic biology. PLoS Comput Biol. 2013;9(3):1002960.
13. Zechner C, Ruess J, Krenn P, Pelet S, Peter M, Lygeors J, Koeppl H. Moment-based inference predicts bimodality in transient gene expression. PNAS. 2012;109(21):8340–345.
14. Hasenauer J, Waldherr S, Doszczak M, Radde N, Scheurich P, Allgöwer F. Identification of models of heterogeneous cell populations from population snapshot data. BMC Bioinforma. 2011;12(1):1–15.
15. Cai L, Friedman N, Xie XS. Stochastic protein expression in individual cells at the single molecule level. Nature. 2006;440(7082):358–62.
16. Toni T, Welch D, Strelkowa N, Ipsen A, Stumpf MPH. Approximate Bayesian computation scheme for parameter inference and model selection in dynamical systems. J R Soc Interface. 2008;6(31):187–202.
17. Liepe J, Kirk P, Filippi S, Toni T, Barnes CP, Stumpf MPH. A framework for parameter estimation and model selection from experimental data in systems biology using approximate Bayesian computation. Nat Protoc. 2014;9(2):439–56.
18. Karlsson M, Janzen DLT, Durrieu L, Colman-Lerner A, Kjellsson MC, Cedersund G. Nonlinear mixed-effects modelling for single cell estimation: when, why, and how to use it. BMC Syst Biol. 2015;9:52.
19. Raj A, van Oudenaarden A. Nature, Nurture, or Chance: Stochastic Gene Expression and Its Consequences. Cell. 2008;135(2):216–26.
20. Nienałtowski K, Włodarczyk M, Lipniacki T, Komorowski M. Clustering reveals limits of parameter identifiability in multi-parameter models of biochemical dynamics. BMC Syst Biol. 2015;9(1):65.
21. Erguler K, Stumpf MPH. Practical limits for reverse engineering of dynamical systems: a statistical analysis of sensitivity and parameter inferability in systems biology models. Mol BioSyst. 2011;7(5):1593.
22. Gutenkunst RN, Waterfall JJ, Casey FP, Brown KS, Myers CR, Sethna JP. Universally sloppy parameter sensitivities in systems biology models. PLoS Comput Biol. 2007;3(10):1871–878.
23. Secrier M, Toni T, Stumpf MPH. The ABC of reverse engineering biological signalling systems. Mol BioSyst. 2009;5(12):1925.
24. Filippi S, Barnes CP, Cornebise J, Stumpf MPH. On optimality of kernels for approximate Bayesian computation using sequential Monte Carlo. Stat Appl Genet Mol Biol. 2013;12(1):87–107.
25. Bar-Joseph Z, Gifford DK, Jaakkola TS. Fast optimal leaf ordering for hierarchical clustering. Bioinformatics. 2001;17(Suppl 1):22–9.
26. Kacmar J, Zamamiri A, Carlson R, Abu-Absi NR, Srienc F. Single-cell variability in growing Saccharomyces cerevisiae cell populations measured with automated flow cytometry. J Biotechnol. 2004;109(3):239–54.
27. Yuan TL, Wulf G, Burga L, Cantley LC. Cell-to-Cell Variability in PI3K Protein Level Regulates PI3K-AKT Pathway Activity in Cell Populations. Curr Biol. 2011;21(3):173–83.
28. Li B, You L. Predictive power of cell-to-cell variability - Springer. Quant Biol. 2013;17(1):41–50.
29. Peccoud J, Ycart B. Markovian Modeling of Gene-Product Synthesis. Theor Popul Biol. 1995;48:222–34.
30. Stinchcombe AR, Peskin CS, Tranchina D. Population density approach for discrete mRNA distributions in generalized switching models for stochastic gene expression. Phys Rev E. 2012;85(6):061919.
31. Shahrezaei V, Swain PS. Analytical distributions for stochastic gene expression. Proc Natl Acad Sci. 2008;105(45):17256–17261.
32. Gillespie DT. Approximate accelerated stochastic simulation of chemically reacting systems. J Chem Phys. 2001;115(4):1716–1733.

33. Ale A, Kirk P, Stumpf MPH. A general moment expansion method for stochastic kinetic models. J Chem Phys. 2013;138(17):174101.

34. Lakatos E, Ale A, Kirk P, Stumpf MPH. Multivariate moment closure techniques for stochastic kinetic models. J Chem Phys. 2015;143(9): 094107.

35. Filippi S, Barnes CP, Kirk PDW, Kudo T, Kunida K, McMahon S, Tsuchiya T, Wada T, Kuroda S, Stumpf MPH. Robustness of the MEK-ERK core dynamics and origins of cell-to-cell variability. Cell Rep. 2016;15:2524–535.

36. Lillacci G, Khammash M. The signal within the noise: efficient inference of stochastic gene regulation models using fluorescence histograms and stochastic simulations. Bioinformatics. 2013;29(18):2311–319.

37. Gillespie DT. A general method for numerically simulating the stochastic time evolution of coupled chemical reactions. J Comput Phys. 1976;22: 403–34.

38. Mc Mahon SS, Lenive O, Filippi S, Stumpf MPH. Information processing by simple molecular motifs and susceptibility to noise. J R Soc Interface. 2015;12(110):20150597.

39. Levsky JM, Singer RH. Fluorescence in situ hybridization: past, present and future. J Cell Sci. 2003;116(Pt 14):2833–838.

40. Wang Z, Gerstein M, Snyder M. RNA-Seq: a revolutionary tool for transcriptomics. Nat Rev Genet. 2009;10(1):57–63.

41. Barnes CP, Filippi S, Stumpf M, Thorne T. Considerate approaches to constructing summary statistics for ABC model selection - Springer. Stat Comput. 2012;22(6):1181–197.

42. Toni T, Welch D, Strelkowa N, Ipsen A, Stumpf MPH. Approximate Bayesian computation scheme for parameter inference and model selection in dynamical systems. J R Soc Interf / R Soc. 2009;6(31):187–202.

43. Benjamini Y, Hochberg Y. Controlling the False Discovery Rate: a Practical and Powerful Approach to Multiple Testing. JSTOR: J R Stat Soc Ser B Methodol. 1995;57(1):289–300.

2

Sampling-based Bayesian approaches reveal the importance of quasi-bistable behavior in cellular decision processes on the example of the MAPK signaling pathway in PC-12 cell lines

Antje Jensch[†], Caterina Thomaseth[†] and Nicole E. Radde[*]

Abstract

Background: Positive and negative feedback loops are ubiquitous motifs in biochemical signaling pathways. The mitogen-activated protein kinase (MAPK) pathway module is part of many distinct signaling networks and comprises several of these motifs, whose functioning depends on the cell line at hand and on the particular context. The maintainance of specificity of the response of the MAPK module to distinct stimuli has become a key paradigm especially in PC-12 cells, where the same module leads to different cell fates, depending on the stimulating growth factor. This cell fate is regulated by differences in the ERK (MAPK) activation profile, which shows a transient response upon stimulation with EGF, while the response is sustained in case of NGF. This behavior was explained by different effective network topologies. It is widely believed that this sustained response requires a bistable system.

Results: In this study we present a sampling-based Bayesian model analysis on a dataset, in which PC-12 cells have been stimulated with different growth factors. This is combined with novel analysis methods to investigate the role of feedback interconnections to shape ERK response. Results strongly suggest that, besides bistability, an additional effect called quasi-bistability can contribute to explain the observed responses of the system to different stimuli. Quasi-bistability is the ability of a monostable system to maintain two distinct states over a long time period upon a transient signal, which is also related to positive feedback, but cannot be detected by standard steady state analysis methods.

Conclusions: Although applied on a specific example, our framework is generic enough to be also relevant for other regulatory network modeling studies that comprise positive feedback to explain cellular decision making processes. Overall, this study advices to focus not only on steady states, but also to take transient behavior into account in the analysis.

Keywords: Quasi-bistability, MAPK signaling pathway, Cellular decision making

Background

Feedback regulations are ubiquitous network motifs in all kinds of molecular interaction networks, such as for example metabolic networks, regulatory modules or signaling networks [1]. The role of single positive and negative feedback is well-characterized also from a theoretical

point of view. Negative feedback, which counteracts external perturbations, can cause oscillating behavior, but also has a stabilizing effect, implies robustness of cell states to internal and external perturbations [2], and plays a major role in maintaining homeostasis (see e.g. [3–5]). Furthermore, it can accelerate the response to a transient signal. By contrast, positive feedback amplifies an external perturbation or signal, which can cause multistability, hysteresis and memory effects or switch-like

*Correspondence: Nicole.radde@ist.uni-stuttgart.de
[†]Equal contributors
Institute for Systems Theory and Automatic Control, University of Stuttgart, Pfaffenwaldring 9, 70569 Stuttgart, Germany

behavior. Positive feedback is omnipresent in cellular decision processes, in which these phenomena arise. It can also produce ultrasensitivity and prolong the response to a transient external signal [5–7].

In this study we investigate the role of feedback regulation for proper signal processing by a case study on the well-known mitogen-activated protein kinase (MAPK) signaling pathway. This pathway is an evolutionary conserved signaling module, which is involved in many essential cellular processes such as proliferation, survival or differentiation [8–11]. It is de-regulated in various diseases and represents an important drug target [11]. The pathway module consists of a cascade of phosphorylation events, leading to the activation of ERK, which targets more than 80 substrates in the nucleus and the cytosol. It is integrated into multiple signaling pathways and shows a variety of different responses depending on the stimulus and the cell-type specific context [9, 11, 12]. Specificity of the cellular response is tightly related to distinct time courses of active ERK upon different stimuli, in particular amplitude and duration of the signal response [11–13]. A well-studied paradigm for such a context-specific response is the different behaviors of PC-12 cells upon stimulation with epidermal growth factors (EGF) and neural growth factors (NGF) [12, 14]. Cells stimulated with NGF show sustained activation of ERK, accompanied by a translocation of ERK into the nucleus, which eventually initiates cell differentiation. In contrast, ERK activity is transient and mainly restricted to the cytosol upon stimulation with EGF, which in turn triggers proliferation.

The pathway module is well-characterized experimentally and from a modeling point of view (for reviews see e.g. [9–11, 15–17]). Starting with the early work of Huang and Ferrell [18], many models of different complexity and with different foci have been suggested in the meantime [11, 19–22]. In particular, quite a number of studies focus on modeling and understanding the mechanisms behind the distinct responses upon EGF and NGF stimulation in PC-12 cells [12–14, 23–26].

It is commonly believed and well-described that a system which shows a sustained response to a transient signal, such as PC-12 cells upon NGF stimulation, is a bistable system [15, 19, 23, 25–30]. Hence modeling of this phenomenon usually focuses on the investigation of the bistability properties of respective models, and advanced methods have been developed tailored to the investigation of steady states in these models (see e.g. [23, 31, 32]).

In this study we turn our attention to a phenomenon called quasi-bistability and its role in the regulation of the MAPK module for cellular decision making. Quasi-bistability is the ability of a monostable system to maintain a second steady state for a long period of time upon a transient stimulus [33]. It is also related to positive feedback, but less well investigated and understood. Using a

dynamic modeling approach and a dataset of the MAPK module in PC-12 cell lines, the system is analyzed via Bayesian sampling techniques. Mechanisms behind sustained ERK responses are investigated by a combination of steady state analysis methods and novel methods that also allow to investigate time scales of transient behavior.

Methods
Experimental data used for model calibration
For our modeling study we used a dataset described in [12], where PC-12 cell lines were stimulated with EGF and NGF, and phosphorylation of the proteins in the cascade was measured via Western blotting and flow cytometry. For model calibration we used the data shown in Figs. 1 and S1b in [12]. This dataset contains data from control experiments, in which cells were stimulated with 100 ng/ml EGF or 50 ng/ml NGF, and measurements from RNA interference experiments.

In the control experiments the dynamic response of the system upon stimulation was measured in terms of phosphorylation levels of Raf (pRaf), MEK (ppMEK) and ERK (ppERK). In the following we will refer to the active states of the proteins by using the following variables:

$$v_1 = \text{pRaf}$$
$$v_2 = \text{ppMEK}$$
$$v_3 = \text{ppERK}.$$

We used data from flow cytometry experiments (Fig. S1b in [12]) as reference for model calibration, since all proteins were quantified in this experiment. Extracted values are illustrated in Fig. 1 and show a transient signal response in case of stimulation with EGF, and a sustained response after stimulation with NGF. The quantified values are a scaled version of the quantities v_i, and are defined as $\tilde{v}_i$.

In the siRNA experiments, Raf, MEK and ERK were consecutively downregulated. These data were used in [12] to analyze the network topology via Modular Response Analysis [34]. In this analysis, global response coefficients R_{ij}, $i, j = 1, 2, 3$ were calculated from the Western blot signals (Fig. S1c in [12]) via

$$R_{ij} = 2 \frac{\partial \ln(v_i)}{\partial \ln(p_j)} \approx 2 \frac{\bar{v}_i^{(s_j)} - \bar{v}_i^{(c)}}{\bar{v}_i^{(s_j)} + \bar{v}_i^{(c)}}. \tag{1}$$

The variables $\bar{v}_i^{(c)}$ and $\bar{v}_i^{(s_j)}$ denote the steady state concentrations of variable v_i before and after perturbation p_j, i.e. silencing of component j, respectively.

Equation (1) can be resolved for $\bar{v}_i^{(s_j)} / \bar{v}_i^{(c)}$,

$$\frac{\bar{v}_i^{(s_j)}}{\bar{v}_i^{(c)}} = \frac{2 + R_{ij}}{2 - R_{ij}}, \tag{2}$$

EGF stimulation				NGF stimulation			
Time [min]	$\tilde{v}_1(t)$	$\tilde{v}_2(t)$	$\tilde{v}_3(t)$	Time [min]	$\tilde{v}_1(t)$	$\tilde{v}_2(t)$	$\tilde{v}_3(t)$
5	0.79	0.76	0.76	5	0.78	0.75	0.80
10	0.47	0.4	0.33	10	0.31	0.34	0.55
15	0.34	0.3	0.10	15	0.39	0.42	0.30
30	0.15	0.14	0.03	30	0.26	0.35	0.30
60	0.09	0.09	0.02	60	0.35	0.30	0.20

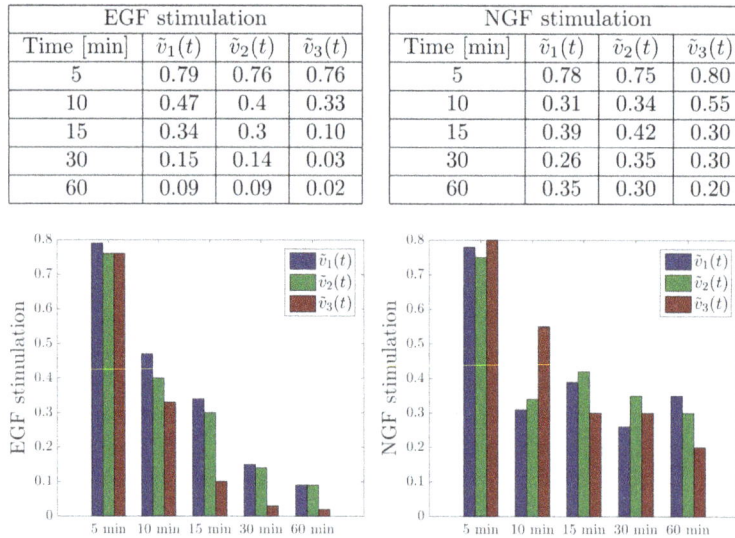

Fig. 1 Activities of Raf, MEK and ERK after stimulation. Scaled activities of Raf, MEK and ERK measured by polychromatic flow cytometry (by visual inspection from Fig. S1b in [12])

which gives the concentration change of component i relative to the control experiment in response to silencing of component j.

Values of the response coefficients of four replicates for silencing of each protein are provided in Table 1 in Fig. S1d in [12]. These data were used to calculate empirical means and standard deviations, as illustrated in Fig. 2, together with the respective relative changes of protein concentrations after silencing.

Time points were set to 5 min after EGF stimulation, the time about which the maximum of the signal response is reached in the control experiments, which is assumed to be close to a steady state condition. In case of NGF, global response coefficients are given at 5 and 15 min after stimulation. These two time points correspond to the times at which the maximum of the signal response was reached and at which the system seems to have reached the new steady state. In [12], these coefficients were used to extract the network structure based on the so-called local response coefficients. This analysis indicates a positive feedback from ERK to Raf upon NGF stimulation and a negative feedback when stimulated with EGF. This result will be taken into account in our modeling approach.

Sampling-based Bayesian approach for model calibration
Data-driven modeling approach
Based on the experimental data available for model calibration and on existing modeling studies for the MAPK module [10, 18, 23], we formulated a differential equation model based on mass action kinetics for the three-tiered phosphorylation cascade (Fig. 3).

In this cascade, both MEK and ERK require dual phosphorylation to become fully active. Double phosphorylation makes the cascade behave in an ultrasensitive way, which is advantageous for noise filtering [9, 11, 35]. MEK phosphorylation is processive, i.e. both sites are phosphorylated in a single step, whereas ERK phosphorylation is distributive and requires two interactions [9, 35]. We have taken this into account by modeling MEK double phosphorylation as a single reaction, while full activation of ERK is obtained in a two step reaction.

Furthermore, we exploited conservation of total protein concentrations,

$$\text{Raf}_{\text{TOT}} \cdot s_1 = \text{Raf} + \text{pRaf} \tag{3a}$$

$$\text{MEK}_{\text{TOT}} \cdot s_2 = \text{MEK} + \text{ppMEK} \tag{3b}$$

$$\text{ERK}_{\text{TOT}} \cdot s_3 = \text{ERK} + \text{pERK} + \text{ppERK} \tag{3c}$$

to end up with a four variable model, shown in Fig. 3b. Rate constants are denoted by $k_i^{+/-}$, $i = 1, \ldots, 4$, reduction of total protein amounts in the siRNA perturbation experiments are described by the silencing factors $s_i \in (0, 1]$ $(i = 1, 2, 3)$. These factors were extracted from quantification of the proteins in the control and the silencing experiments, as reported in Fig. 1c of [12]. Their values were set to $s_1 = 0.72$, $s_2 = 0.7$ and $s_3 = 0.65$ when simulating silencing of Raf, MEK or ERK, respectively.

The input $u(t)$, which mimics signal initiation after addition of growth factor and summarizes all upstream processes, was described via a sigmoidally decreasing

5 min after EGF stimulation						
j	$\widehat{\mathbb{E}}(R_{1j})$	$\widehat{\sigma}(R_{1j})$	$\widehat{\mathbb{E}}(R_{2j})$	$\widehat{\sigma}(R_{2j})$	$\widehat{\mathbb{E}}(R_{3j})$	$\widehat{\sigma}(R_{3j})$
1 (siRaf)	-0.6692	0.1913	-0.3312	0.3434	-0.4698	0.4684
2 (siMEK)	0.3727	0.3376	-0.4780	0.2923	-0.2985	0.2377
3 (siERK)	0.1525	0.1688	0.4970	0.3427	-0.7271	0.4068
5 min after NGF stimulation						
j	$\widehat{\mathbb{E}}(R_{1j})$	$\widehat{\sigma}(R_{1j})$	$\widehat{\mathbb{E}}(R_{2j})$	$\widehat{\sigma}(R_{2j})$	$\widehat{\mathbb{E}}(R_{3j})$	$\widehat{\sigma}(R_{3j})$
1 (siRaf)	-0.5600	0.0455	-0.3459	0.4273	-0.3869	0.4456
2 (siMEK)	-0.1314	0.1521	-0.2909	0.2268	-0.3295	0.3927
3 (siERK)	-0.1466	0.0500	0.2251	0.1456	-0.6345	0.2845
15 min after NGF stimulation						
j	$\widehat{\mathbb{E}}(R_{1j})$	$\widehat{\sigma}(R_{1j})$	$\widehat{\mathbb{E}}(R_{2j})$	$\widehat{\sigma}(R_{2j})$	$\widehat{\mathbb{E}}(R_{3j})$	$\widehat{\sigma}(R_{3j})$
1 (siRaf)	-0.7762	0.3307	-0.4829	0.4982	-0.8150	0.6924
2 (siMEK)	0.0787	0.2218	-0.2891	0.4811	-0.3451	0.4526
3 (siERK)	-0.4154	0.3704	0.4514	0.4218	-1.0215	0.5350

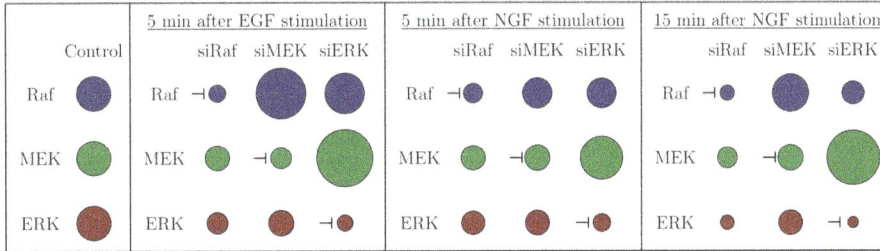

Fig. 2 Data from modular response analysis. Table. Means and standard deviations of the global response coefficients extracted from the silencing experiments via modular response analysis. These were calculated from replicates in Table S1d in [12]. Figure. Illustration of respective changes in protein concentrations in response to silencing relative to the control experiments (without silencing)

function, whose parameter K was also included in the optimization procedure,

$$u(t) = \begin{cases} 0 & t < 0 \\ 1 - \dfrac{t^3}{t^3 + K^3} & t \geq 0. \end{cases} \quad (4)$$

Thus, $u(t)$ jumps from 0 to 1 at time $t = 0$, which mimics addition of ligand, and subsequently decreases sigmoidally, reflecting observations of transient Ras activity, which is upstream of Raf and returns to its inactive Ras-GDP state within five minutes [19]. This implies that our

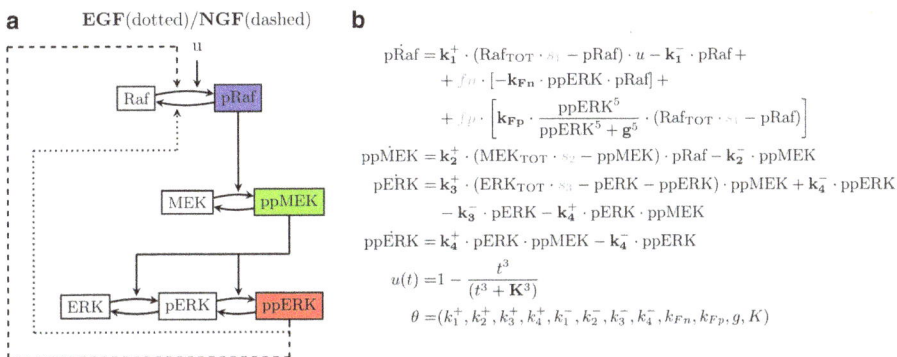

Fig. 3 Model structure of the MAPK module. **a** Reaction scheme of the MAPK module. Upon addition of growth factors, Raf, MEK and ERK are successively activated in a phosphorylation cascade. Different feedback topologies are assumed to shape context dependent ERK response: Effective negative feedback from ERK to Raf upon EGF stimulation (*dotted line* from ppERK to dephosphorylation of pRaf), and positive feedback in case of NGF stimulation (*dashed line* from ppERK to phosphorylation of Raf). **b** Differential equation model of the MAPK cascade. Bold parameters are the unknown constants, collected in the parameter vector θ, while gray parameters define the specific experimental condition for the simulation

model has a trivial steady state in which all variables are equal to 0 for $u = 0$, which is also a simplification, since proteins usually have minimal basal activities. However, since we do not have data for $t = 0$, which would reflect these basal activities, and since these are anyway assumed to be very low compared to the stimulated case [36], we consider this simplification not a crucial one.

In our model the input $u(t)$ is not directly coupled to the network structure, which is clearly a simplification, since EGF and NGF trigger different receptor systems. However, exactly the same model structure has been used in other studies as well (see e.g. [27]) and was shown to display a rich variety of different behaviors, including ultrasensitivity and bistability and, as we will demonstrate, is also sufficient to capture various observed responses. Moreover, we follow here the argumentation in [14], according to which the different ERK responses are unlikely to be caused by different receptor systems. The Boolean variables f_p and f_n account for the experimental condition and act as switches between the two network structures, depending on the growth factor.

The positive feedback from ERK to Raf that was postulated from the modular response analysis in [12] was described by a sigmoidal function in order to facilitate bistability. Although this feedback is not necessarily required for bistability in the MAPK signaling pathway [21, 28–30], it has been shown to enhance the range of bistable behavior and to make the occurrence of bistability less sensitive to stochastic fluctuations and parameter variations [30].

Model calibration procedure
In the next step we inferred the unknown model parameters

$$\theta = \left(k_1^+, k_2^+, k_3^+, k_4^+, k_1^-, k_2^-, k_3^-, k_4^-, k_{Fn}, k_{Fp}, g, K \right) \quad (5)$$

by using the described set of data y. For this model calibration procedure we used a sampling-based Bayesian approach, which provides a consistent statistical description for all quantities-of-interest. In a Bayesian approach, parameters θ and measurements y are interpreted as random variables that are characterized by probability distributions. Hence such an approach offers full information about uncertainties in terms of underlying distributions. A short explanation of the Bayesian idea is provided in Additional file 1.

In our Bayesian framework the ODE model is stochastically embedded by defining the underlying stochastic process from which the experimental data are assumed to be generated. This is sometimes also referred to as noise model (see Additional file 1 for more details). Here we exploit log-normal error models for protein concentrations, using the same standard deviation of 0.2 for

the logarithmic transformation of the experimental data, which by definition are normally distributed.

These are translated into respective error models for the global response coefficients via transformation of probability distributions. Altogether, this defines the likelihood function $l_y(\theta) = p(y|\theta)$, which is a measure of how likely it is to see the experimental data given a particular model.

In a Bayesian framework, the objective function of interest is the posterior distribution $p(\theta|y)$, which is a distribution of parameters conditional on the given dataset. According to the Bayes Theorem, the posterior distribution is proportional to the product of the prior distribution $p(\theta)$ of the parameters and of the likelihood function,

$$p(\theta|y) = \frac{p(y|\theta)p(\theta)}{p(y)}. \quad (6)$$

Since the light signals of the Western blot data require appropriate rescaling and normalization to a reference experiment for a comparison across different experimental conditions, the ODE model in Fig. 3b also had to be rescaled and normalized in order to enable a comparison with these data. This procedure is described in Additional file 2. Moreover, a detailed formulation of the posterior distribution is given in Additional file 3.

We investigate the posterior distribution by generating samples $\{\theta_i\}_{i=1,...,N}$ via Markov Chain Monte Carlo (MCMC) sampling. These samples are subsequently used for Monte Carlo estimates of other quantities-of-interest. For example, the posterior predictive distribution (PPD) to see new data $\tilde{y}$ in any experimental scenario is given by

$$
\begin{aligned}
p(\tilde{y}|y) &= \int_{\Theta} p(\theta, \tilde{y}|y)d\theta \quad \text{Marginalization} \\
&= \int_{\Theta} p(\tilde{y}|\theta, y)p(\theta|y)d\theta \quad \text{Factorization} \\
&= \int_{\Theta} p(\tilde{y}|\theta)p(\theta|y)d\theta \quad \tilde{y} \text{ is independent of } y \text{ given } \theta \\
&\approx \frac{1}{N} \sum_{i=1}^{N} p(\tilde{y}|\theta_i) \quad \theta_i \sim p(\theta|y) \quad \text{Monte Carlo estimate}
\end{aligned}
$$

If not stated otherwise, model predictions are consistently given in terms of these PPDs in this work.

Results
Calibrated model describes experimental data
We generated samples $\{\theta_i\}_{i=1,...,N}$ from the posterior distribution as described (see also Additional file 4 for implementation details). Kernel density estimates of the marginal parameter distributions and 2D scatter plots for the two-dimensional parameter marginals are shown in Additional files 5 and 6. Most of the parameters show a large variance. The only exceptions are the dephosphorylation rates of pRaf and ppMEK, which mainly determine

the speed of the decay of the signal. Moreover, the threshold parameter K of the input signal can be extracted from the data. There are also almost no correlations visible in the 2D scatter plots except a strong positive correlation between k_{Fp} and k_1^+.

Figure 4 shows the result of the Bayesian model calibration in the prediction space. Depicted are the Monte Carlo estimates of the PPDs in comparison with experimental data. Figure 4a and b show the dynamic responses of the observables pRaf, ppMEK and ppERK in the control experiments after stimulation with EGF (A) and NGF (B). The model captures the EGF scenario very well, with low variances in the PPDs. In case of NGF some data points are slightly overestimated, but the data are still within the predicted confidence intervals, which are larger here compared to the EGF scenario. The colors chosen for pRaf (blue), ppMEK (green) and ppERK (red) are maintained for all simulation results throughout the paper.

A comparison of the global response coefficients is depicted at the bottom (Fig. 4c). The sign structure is preserved for almost all silencing experiments, the only exception being MEK in the siERK experiments with NGF stimulation. This is due to the fact that we did not include the direct negative feedback from ERK to MEK that was postulated from the modular response analysis in [12] in our model, since the signal to noise ratio was rather low for this interaction, and we wanted to keep the model simple. At first glance the fits seem to be reasonable, which is however hard to judge solely from visual inspection, since error bars are large for most of these values. This is also mirrored by the variances of the PPDs. Thus we decided

to validate the model via predictions of further experiments with the same cell line that were not used for model calibration.

Model is able to predict various perturbation experiments
For model validation we decided to use the model to predict outcomes of a set of perturbation experiments that have not been used for model calibration. The result is shown in Fig. 5. In particular, the following experimental setups were considered:

Dose response profiles of ERK activation
We mimicked dose-response profiles of ERK activation to increasing EGF and NGF doses measured via flow cytometry (Fig. 2 in [12]). Since these datasets are single-cell measurements that represent a heterogeneous cell population, we interpreted our parameter samples to represent such a cell population, whose average is consistent with the data used for calibration, and whose distribution accounts for population heterogeneity. Increasing ligand concentration was reflected by multiplying the parameter k_1^+, which describes the input strength, by a factor k_u. Resulting mean values of ppERK are shown in Fig. 5a. We note here that this comparison can only be done in a qualitative way, since we lack a receptor model that directly relates growth factor concentrations to the input signal for Raf activation. Thus, it is here not possible to include the respective experimental data directly for comparison. However, model simulations capture the observed qualitative phenomena quite well: In case of stimulation with EGF the ERK activity profile is unimodal and raises

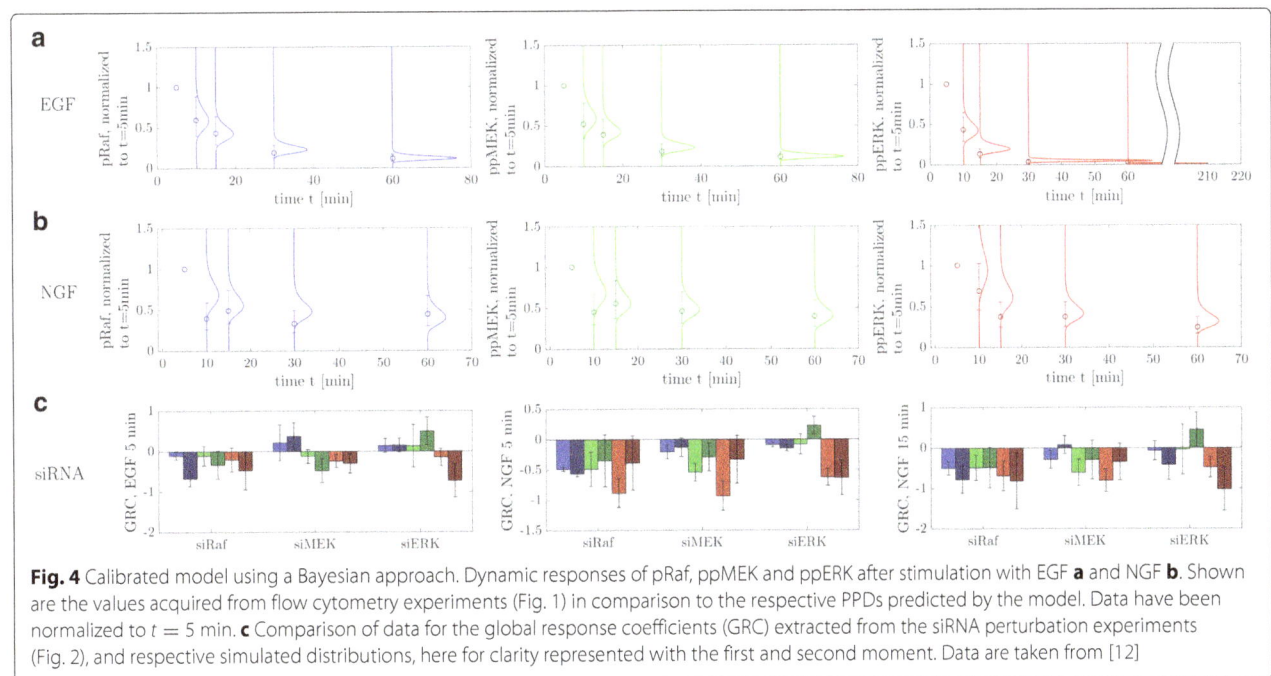

Fig. 4 Calibrated model using a Bayesian approach. Dynamic responses of pRaf, ppMEK and ppERK after stimulation with EGF **a** and NGF **b**. Shown are the values acquired from flow cytometry experiments (Fig. 1) in comparison to the respective PPDs predicted by the model. Data have been normalized to $t = 5$ min. **c** Comparison of data for the global response coefficients (GRC) extracted from the siRNA perturbation experiments (Fig. 2), and respective simulated distributions, here for clarity represented with the first and second moment. Data are taken from [12]

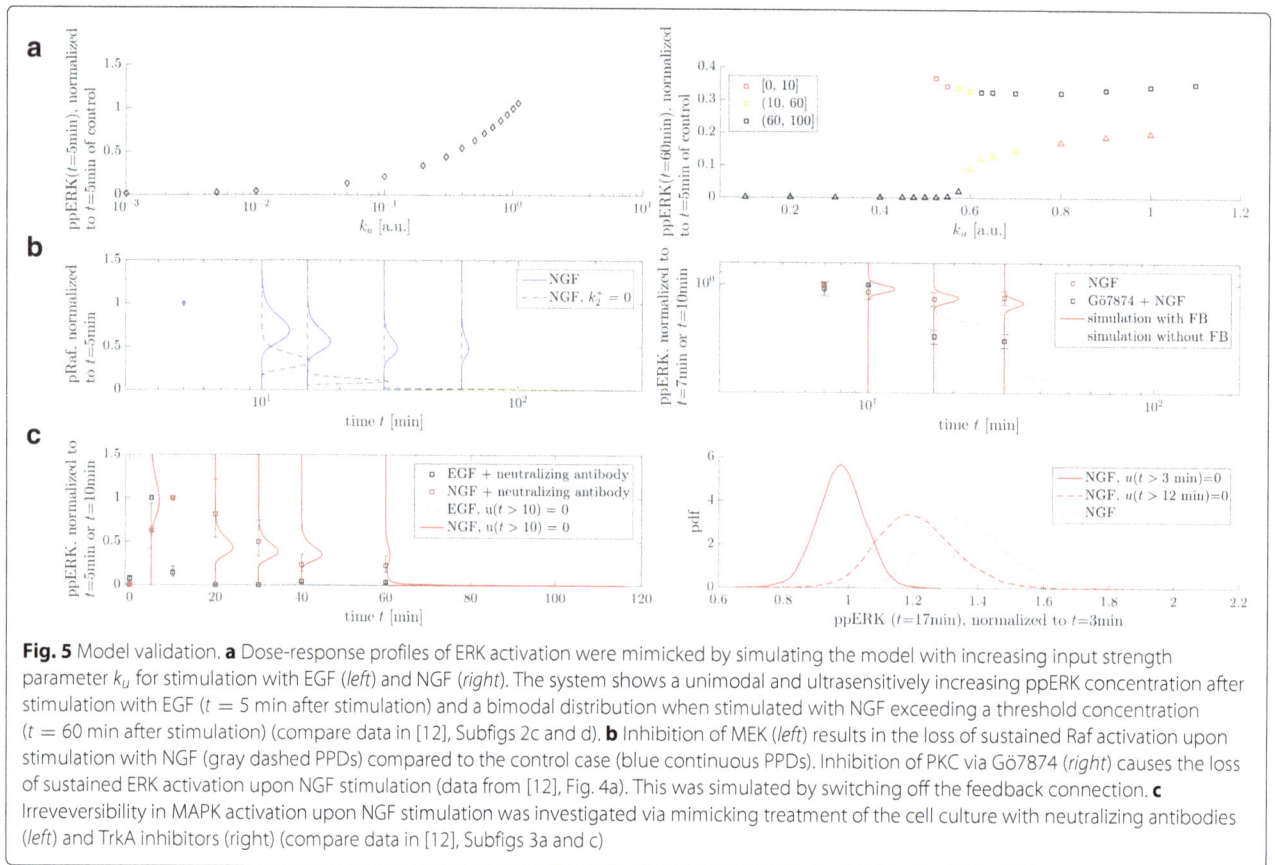

Fig. 5 Model validation. **a** Dose-response profiles of ERK activation were mimicked by simulating the model with increasing input strength parameter k_u for stimulation with EGF (*left*) and NGF (*right*). The system shows a unimodal and ultrasensitively increasing ppERK concentration after stimulation with EGF ($t = 5$ min after stimulation) and a bimodal distribution when stimulated with NGF exceeding a threshold concentration ($t = 60$ min after stimulation) (compare data in [12], Subfigs 2c and d). **b** Inhibition of MEK (*left*) results in the loss of sustained Raf activation upon stimulation with NGF (gray dashed PPDs) compared to the control case (blue continuous PPDs). Inhibition of PKC via Gö7874 (*right*) causes the loss of sustained ERK activation upon NGF stimulation (data from [12], Fig. 4a). This was simulated by switching off the feedback connection. **c** Irreversibility in MAPK activation upon NGF stimulation was investigated via mimicking treatment of the cell culture with neutralizing antibodies (*left*) and TrkA inhibitors (right) (compare data in [12], Subfigs 3a and c)

sigmoidally with increasing EGF concentration. In contrast, upon stimulation with NGF the profile becomes bimodal when NGF doses exceed a threshold. Moreover, with increasing NGF concentrations the fraction of cells with a sustained response as well as the mean ERK activities of both subpopulations increase.

Effect of feedback breaking via inhibition of MEK and PKC

We predicted the influence of MEK inhibition via the MEK inhibitor PD184352 on the temporal activity of Raf (Fig. S1e in [12]), by assuming that MEK activation is completely abolished. This was realized in our model by setting the MEK phosphorylation rate k_2^+ to zero, which destroys the feedback from ERK to Raf in the simulations, and inspection of Raf activity (Fig. 5b left). While the response is sustained in the control case (blue continuous PPDs), MEK inhibition results in the loss of sustained Raf activity, and pRaf follows the transient signal and rapidly drops within a few minutes (gray dashed PPDs). This result is in agreement with the observations in [12].

In addition, we mimicked the inhibition of PKC via Gö7874 during NGF stimulation (Fig. 4a in [12]). We considered the feedback to be completely eliminated as a result and realized this by removing the feedback connections from our model (Fig. 5b right). In the control case

(red continuous PPDs) activity of ppERK was sustained, whereas the feedback deletion caused a decrease in ERK activation (gray dotted PPDs), again in accordance with experimental findings.

Irreversibility in MAPK activation

Finally, we also compared our model to experimental data on the irreversibility in MAPK network activation upon NGF stimulation, which was investigated via terminating the signal by growth factor neutralizing antibodies and TrkA inhibitors (Subfigs 3a and c in [12]). Therefore, both perturbations, i.e. addition of neutralizing antibody and TrkA inhibitor after stimulation, were mimicked via abrupt signal termination at the respective time points. Results are shown in Fig. 5c. While in case of stimulation with EGF, ppERK was virtually zero shortly after addition of the neutralizing antibody (gray dotted PPDs in the left Figure), the NGF inhibition profile still showed some activity after 60 min (red PPDs).

For a further comparison we simulated ppERK time courses upon stimulation with NGF and addition of TrkA inhibitor at two different time points (Fig. 5c right). PPDs for ppERK are depicted at $t = 17$ min after stimulation when TrkA inhibitor was given at $t = 3$ min after stimulation (continuous curve) and $t = 12$ min after stimulation (dashed curve), compared to the control case (dotted curve). In agreement with experimental findings, results show that ERK activity rapidly drops in case that the stimulus terminated too early.

Overall, the results in Fig. 5 nicely demonstrate that our model is able to predict many important features of the signaling cascade quite accurately. Since these simulation scenarios capture the responses of the system to several treatments that are quite different from the experiments which have been used for fitting, the model is validated to have predictive power.

In the next step we decided to use the model to analyze mechanisms behind sustained ERK response in case of NGF stimulation.

Mechanism behind sustained response caused by NGF
Bifurcation analysis reveals that bistability is not sufficient to explain model outcomes upon NGF stimulation

In order to investigate the mechanisms behind sustained response to transient NGF signals, we combined our sampling-approach with the circuit-breaking algorithm (CBA) [32], which allows for an efficient calculation of steady states based on the topology of the signaling network, and for an automatic classification into mono- and bistable systems. Our approach is schematically illustrated

in Fig. 6. Figure 6a-c illustrate the steps of the CBA applied to our network model for a single parameter sample θ_i. The CBA operates on the topology of the interaction graph $G(V, E)$, which is a directed graph that shows dependencies between variables in the model (Fig. 6a). In the first step all feedback loops[1] are broken by deleting incoming edges for a suitably chosen subset $\tilde{V}$ of vertices and setting the respective variables to fixed values κ. The remaining vertices are collected in the set $\widehat{V}$. Here we set $\tilde{V} = \{x_4\}$, $x_4 = \kappa$ and $\widehat{V} = \{x_1, x_2, x_3\}$. The state variables $x_i, i = 1, \ldots, 4$, in the interaction graph refer to the rescaled states of our ODE model that we used for all simulations (see Additional file 2). Then we calculated the steady state coordinates of the variables in $\widehat{V}$ in dependence of the input κ, obtaining the set $\bar{x}_{\widehat{V}}(\kappa, \theta_i)$ (Fig. 6b). In the last step the circuits are released one after another by releasing vertices in the set $\tilde{V}$ (Fig. 6c). Mathematically, this translates here into the calculation of the zeros of the circuit-characteristic $c(\kappa, \theta_i)$, which is given by

$$c(\kappa, \theta_i) = f_{x_4}(x_4 = \kappa, x_{\widehat{V}} \in \{\bar{x}_{\widehat{V}}(\kappa, \theta_i)\}) = 0. \quad (7)$$

The obtained zeros $\bar{\kappa}$ of the circuit-characteristic correspond to the steady state coordinates of the state variable

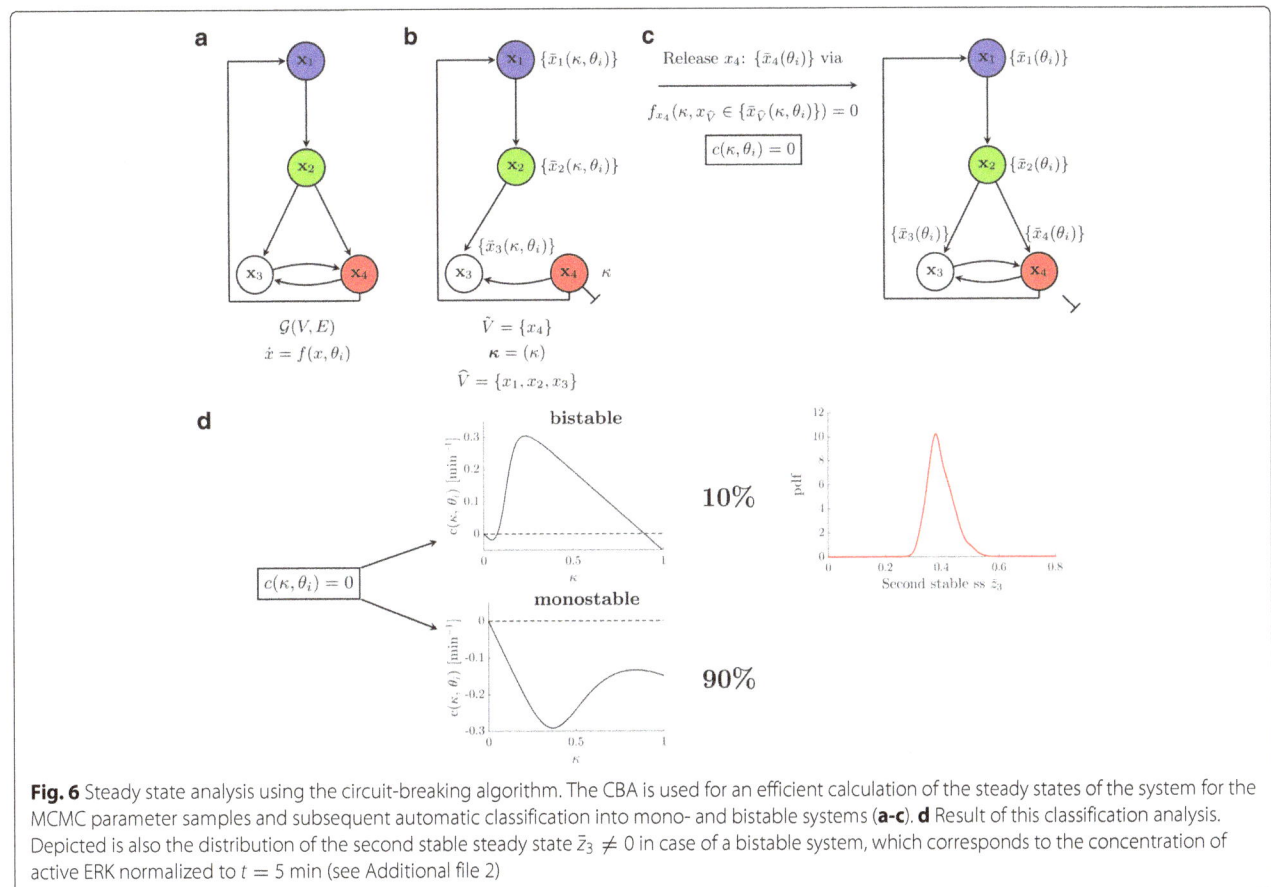

Fig. 6 Steady state analysis using the circuit-breaking algorithm. The CBA is used for an efficient calculation of the steady states of the system for the MCMC parameter samples and subsequent automatic classification into mono- and bistable systems (**a-c**). **d** Result of this classification analysis. Depicted is also the distribution of the second stable steady state $\bar{z}_3 \neq 0$ in case of a bistable system, which corresponds to the concentration of active ERK normalized to $t = 5$ min (see Additional file 2)

x_4, from which the set of steady states of the full system can be derived. All details about the calculation of the values for $\bar{\kappa}$ and of the expressions of the steady state coordinates for the other three state values $\bar{x}_{\widehat{V}}(\kappa, \theta_i)$, as functions of the parameter sample, are given in Additional file 7.

We applied the CBA to all parameters of the estimated posterior sample. The outcome was automatically classified, by using this analysis, according to the number of steady states of the system (Fig. 6d). Results show an overall probability of 10% for the system to be bistable. We found this a surprisingly small number, which indicates that bistability is probably not the main mechanism behind the observed sustained ERK activation. Even worse, our analysis only provides an upper bound in two respects: First, depending on the parameters θ_i, not all trajectories of bistable systems might be pushed to the basin of attraction of the second fixed point by the transient signal. Second, this set might also contain bistable systems in which the distance of the two steady states is rather small, such that the bistability will not be visible in any real experiment. Furthermore, we simulated the ODE model with the obtained subset of parameter samples θ_i leading to a bistable system, and we calculated the distribution of the second positive stable steady state $\bar{x}_4$. This is shown in Fig. 6d on the right, by considering the normalized state variable (see Additional file 2).

$$z_3(t) = x_4(t)/x_4(t = 5 \text{ min}).$$

Overall, this analysis suggests that bistability is not sufficient to explain the observed sustained activation of ERK after NGF stimulation.

Quasi-bistability can explain sustained ERK activation

We complemented our steady state analysis by a simulation-based classification of model trajectories after NGF stimulation, as illustrated in Fig. 7. A similar classification approach was used in [21], without explicitly investigating quasi-bistability.

We used the posterior sample to simulate model responses up to $t = 600$ min. These responses were automatically classified in a second step (Fig. 7a): Using ERK activity at $t = 5$ min as a reference value, samples were sorted according to the following classification scheme:

- Class 1 (Bistable systems):
  $\frac{\text{ppERK}(60 \text{ min})}{\text{ppERK}(5 \text{ min})} > 0.2$ and $\frac{\text{ppERK}(600 \text{ min})}{\text{ppERK}(5 \text{ min})} \geq 0.1$

- Class 2 (Quasi-bistable systems):
  $\frac{\text{ppERK}(60 \text{ min})}{\text{ppERK}(5 \text{ min})} > 0.2$ and $\frac{\text{ppERK}(600 \text{ min})}{\text{ppERK}(5 \text{ min})} < 0.1$

- Class 3 (Monostable systems):
  $\frac{\text{ppERK}(60 \text{ min})}{\text{ppERK}(5 \text{ min})} \leq 0.2$

The threshold 0.2 for $t = 60$ min was chosen such that all trajectories are above this value. This implies that class 3 is empty and no trajectory is classified to be simply monostable, as visualized in Fig. 7c. This implication is reasonable, since $u(t)$ is close to 0 already a few minutes after stimulation, and hence we expect simple monostable systems to follow the input with a delay that is much smaller than 1h. Figure 7c also shows that the classification result is rather insensitive to fine-tuning of the second threshold at $t = 600$ min: At this late time point the quasi-bistable and bistable trajectories are already well separated and there is a clear gap between the trajectories of classes 1 and 2.

The analysis revealed a fraction of 10% belonging to class 1. The estimated distribution of the second steady state equals that from the CBA analysis, which hints to the fact that trajectories of virtually all bistable systems detected via the CBA converge to the second steady state after stimulation with NGF. The rest of the samples, which are 90%, belong to class 2, which represent monostable systems that can show a sustained response for more than 60 min. after stimulation. However, trajectories in this class converge to their unique steady state at a later time point.

In order to understand the mechanism behind this highly prolonged response to a transient input signal, we filtered the parameter sample for monostable systems that belong to class 2 and investigated their behavior in more detail. Therefore, we used the input $u(t)$ as a bifurcation parameter and investigated the respective time-varying set $\{\bar{x}(u)\}$ of steady states of the system via the CBA. Figure 8 shows the temporal behavior of the set $\{\bar{z}_3(u(t))\}$ for a representative parameter sample belonging to class 2. After a fast transient phase, the system is bistable, since $u(t)$ is sufficiently large to maintain two stable steady states. However, the second stable steady state vanishes due to a rapidly decreasing $u(t)$. For the trajectory at hand the system becomes monostable already at about $t \approx 102$ min, which is fast compared to its switching time at $t \approx 440$ min. This comes from the fact that, although the system is monostable, $c(\kappa, \theta_i)$ is extremely small about the region of the former second steady state. This causes a very slow dynamic, which can be seen by tracking the normalized state variable $z_3(t)$, as indicated in the Figure. Only at $t \approx 440$ min $z_3(t)$ reaches an area where $|c(\kappa, \theta_i)|$ becomes larger, which results in a subsequent fast convergence to the unique globally stable fixed point at the origin.

Thus, taken together, this analysis suggests that quasi-bistability is caused by traversing a region in the state space in which $\dot{x}$ is extremely small, resulting in a very slow dynamics. The system is only accelerated towards it's single steady state when the state of the system leaves this region. This makes the system behave as a bistable system

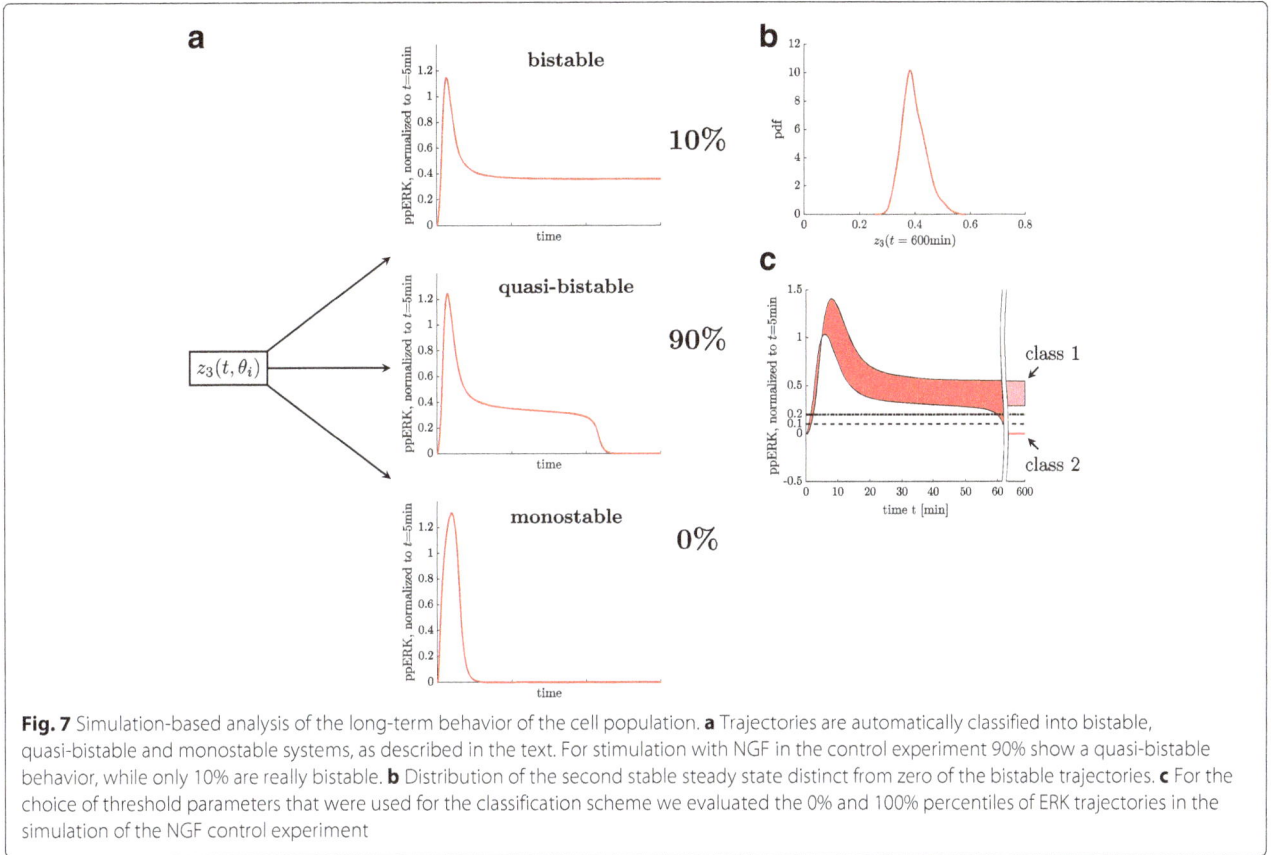

Fig. 7 Simulation-based analysis of the long-term behavior of the cell population. **a** Trajectories are automatically classified into bistable, quasi-bistable and monostable systems, as described in the text. For stimulation with NGF in the control experiment 90% show a quasi-bistable behavior, while only 10% are really bistable. **b** Distribution of the second stable steady state distinct from zero of the bistable trajectories. **c** For the choice of threshold parameters that were used for the classification scheme we evaluated the 0% and 100% percentiles of ERK trajectories in the simulation of the NGF control experiment

for a long time span. This hypothesis was confirmed by a subsequent bifurcation analysis with some representative parameter sets for classes 1 and 2 of the classification scheme, as shown in Fig. 9. Figure 9a illustrates the two effects that act together to delay the response of the system upon a transient stimulus. Figure 9b shows the absolute value of the vector field $\|f(x(t,\theta_i))\|$ of the same trajectory as in Fig. 8, which shows high values a few minutes after stimulation, followed by a long period where $\dot{x}(\theta_i)$ is virtually zero, and a second peak at about $t = 440$ min, where the trajectory is pushed towards the systems unique steady state. A comparison of bifurcation diagrams for representative parameter sets belonging to classes 2 (quasi-bistable) and 1 (bistable) is depicted in Fig. 9c and shows that the difference between these two classes is actually 'smooth' in terms of changes in limit sets.

Discussion

Our bifurcation analysis showed that the transition between the three classes for stability behavior classification (bistable, quasi-bistable and monostable) is actually smooth in terms of locations of bifurcation points and limit sets, and we expect the range of parameters in which quasi-bistability occurs to be rather small.

This expectation was confirmed by a sensitivity analysis for the outcome of the simulation-based classification scheme (Additional file 8), in which we varied all model parameters independently one at a time about the maximum-a-posteriori estimator. The result shows that the appearance of quasi-bistability is highly sensitive to these variations for almost all parameters. Except for the phosphorylation rate of Raf (parameter k_1^+) and the threshold parameter K of the input function, which do not have any influence on the limit sets of the system for $u = 0$, small variations of parameters induce switches to mono- or bistable systems, which is due to the fact that the location of the saddle-node bifurcation and hence the delay time are very sensitive to parameter changes. The fact that most of the samples fall into this seemingly small parameter range shows that the gradient of the posterior distribution must be rather high when varying parameters individually. This is indeed the case, since the fit quality rapidly drops at least for the NGF control experiment when switching from the quasi-bistable to the monostable range, and we observe a similar effect for the switch to the bistable range. Altogether, these results also indicate the existence of strong correlations between parameters in the posterior distribution.

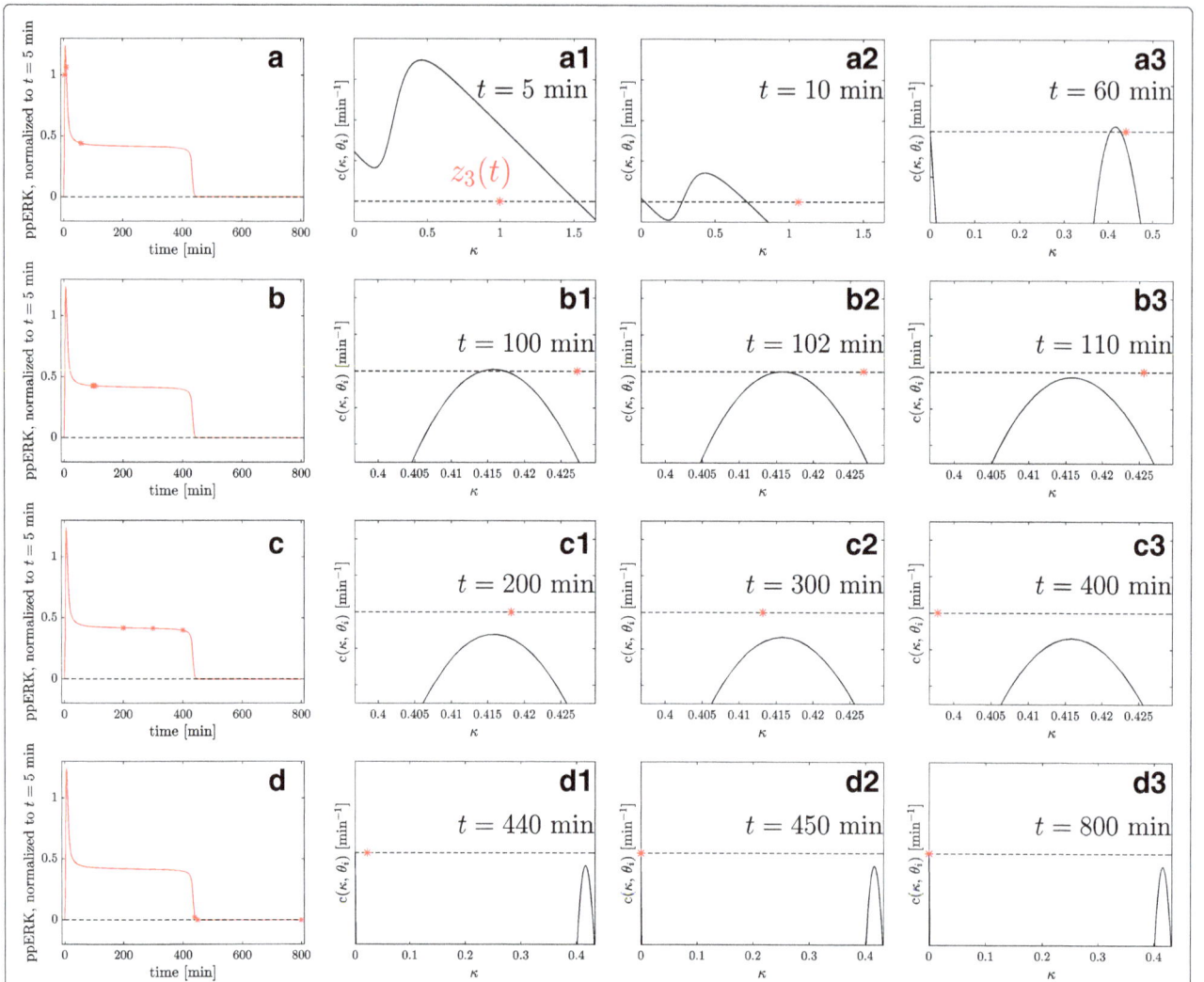

Fig. 8 Quasi-bistability phenomenon. The CBA is used for the investigation of the quasi-bistability phenomenon, in which the system, despite being monostable, shows a very prolonged sustained response. The first column shows the time course of normalized ppERK for a representative parameter sample from class 2 with switching time at $t_{switch} \approx 440$ min. Columns 2,3 and 4 show the circuit-characteristic $c(\kappa, u(t))$, along with the actual normalized state ppERK(t) for 12 different time points. After a fast transient dynamic (**a1**) the circuit-characteristic has three zeros (A2-B1), which disappear at a later time point, here $t = 102$ min (**b2**), via a saddle-node bifurcation. After 60 min the input is almost zero and the vector field and therefore the circuit-characteristic changes only slowly. The system state has almost approached the higher fixed point. **b1**-**c3** are eyeglass views on the dynamics near this second fixed point. These plots show that, even if the fixed point has disappeared, the system trajectory moves very slowly through the state space for a rather long time, since $\dot{x}$ is still small. Only after about 440 min the system has overcome this slow region of the state space, and from here on rapidly moves towards its globally asymptotically stable steady state $\bar{x} = 0$

We note here that the distribution of switching times of quasi-bistable trajectories of our inferred model partly disagrees with observed ERK activities for later time points. While most of the trajectories switch to their steady state between 60 and 120 min in our model, experiments in [37, 38] show that ERK activity is sustained for at least 2–3 h. Since our dataset only contained measurements for up to 60 min, this fact was not taken into account in the model calibration procedure. However, the response duration of the signaling cascade upon stimulation with NGF also seems to show a large variation and to depend in particular on the experimental protocol and distinct clonal PC-12 cell lines, as also stated in [37]. In [39] or [40], for example, MEK (MAP-2 kinase) and ERK activities are almost completely down to the basal level already after 2 h (see [39], Fig. 3 and [40], Fig. 9b, curve without treatment with TPA). Hence it is not completely clear how to describe the activity of the module quantitatively in order to enrich the model with knowledge on the long-term behavior. However,

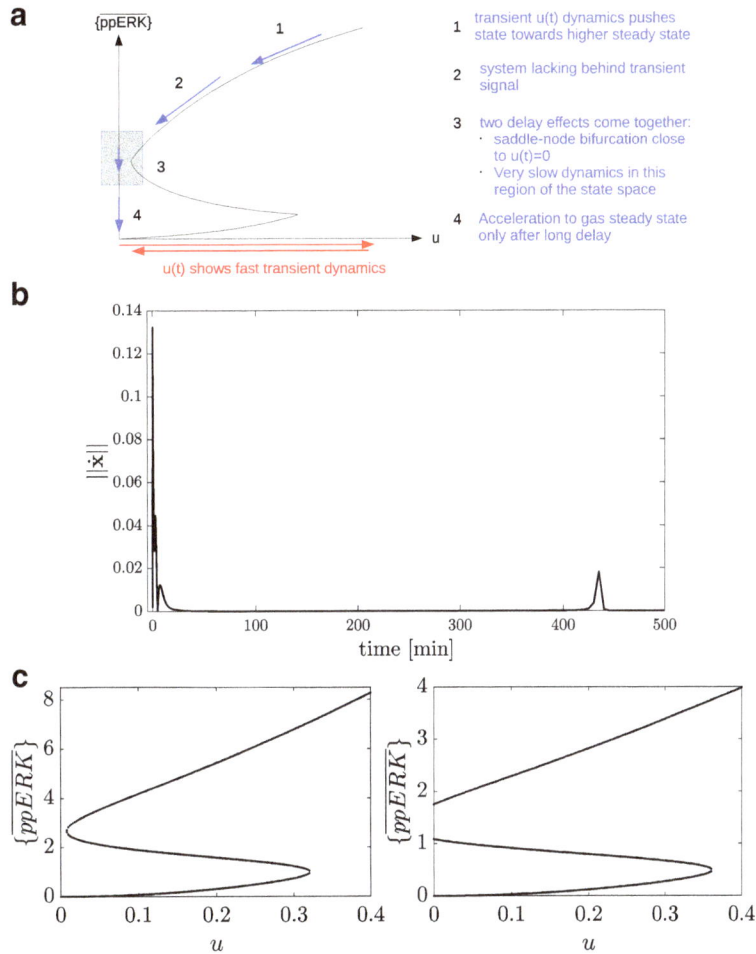

Fig. 9 Combination of two delay mechanisms in quasi-bistable systems. **a** Scheme of a bifurcation diagram for a quasi-bistable system. The system is monostable for $u = 0$ and has a saddle-node bifurcation u^{SNB} close to $u = 0$, where it becomes bistable. A sufficiently strong transient signal $u(t)$ pushes the system state into the basin of attraction of the higher stable steady state (1). As long as the change in $u(t)$ is not slow compared to the dynamics of the system, the system cannot be considered in quasi-steady state, and we observe a transient dynamics (2). When $u(t)$ is almost back to 0, two delay effects lead to quasi-bistable behavior (3). First, the system remains in the upper stable steady state as long as $u(t)$ is still above the saddle-node bifurcation. Second, for $u(t) < u^{SNB}$ the acceleration remains very small in this region of the state space. **b** Absolute value of the vector field along the model trajectory for the same model parameters that have been used in Fig. 8. **c** Two respresentative bifurcation diagrams for a quasi-bistable system belonging to class 2 of the classification scheme, and a bistable system belonging to class 1

we started to investigate the effect of assuming different minimal switching times for ERK activity, which is illustrated in Additional file 9. As expected, filtering for trajectories that still have a substantial remaining activity after two and three hours, respectively, increases the ratio of bistable versus quasi-bistable trajectories, since only quasi-bistable trajectories are filtered out. Thus, we think that our model is generally able also to match the long-term behavior of the cascade.

Furthermore, we have not explicitly taken into account fluctuations in protein content, although we are aware that this is a major source of variability in cell populations. The total amounts of Raf, MEK and ERK do not explicitly

appear any more in the rescaled and normalized model version that we used for our study, hence it is not possible to take absolute fluctuations into account. However, we investigated the effect of varying absolute concentrations by varying the coefficients s_i, $i = 1, 2, 3$ in a narrow range about its nominal values $s_i = 1$ and considered the sensitivity of bistability and quasi-bistability to these parameters. Exemplary results are shown for variations in ERK (s_3) in Additional file 10. Figures for variations in s_1 and s_2 look very similar. Interestingly, the classification of trajectories seems to be very sensitive to these parameters. As can be seen, a moderate reduction in s_i is sufficient to destroy bistability and quasi-bistability almost completely, while bistability is strongly enhanced upon a slight increase in s_i. This is a surprising result, since it is known that stochastic gene expression events can, for example, result in coefficients of variation of about 20-30% in the content of individual proteins [20, 28]. This raises the general question about reliability and robustness of decision processes under such variations. To our knowledge, minimal models for bistability, as used here, are often not robust with respect to such fluctuations and parameter variations, which might trigger further investigations in this direction.

Conclusion

We presented a modeling study that focuses on mechanisms behind sustained responses of signaling pathways upon transient stimulation in PC-12 cells. The model is based on chemical reaction kinetics and was calibrated to a dataset of PC-12 cell lines that were stimulated with EGF and NGF in a control setting and under silencing perturbations. We used a sampling-based Bayesian approach for model calibration, and analyzed model predictions in terms of posterior predictive distributions, which provides complete information about remaining uncertainties. The model was validated by comparing model predictions of new scenarios to experimental data.

Interestingly, the system shows a sustained ERK activity profile upon NGF stimulation, while the response was transient in case of EGF stimulation. This phenomenon has been well-investigated experimentally and theoretically, and it is well believed that the observed sustained response is caused by a bistable system. Here we combine our statistical inference approach with steady state analysis to investigate mechanisms behind this sustained ERK response. Surprisingly, our results indicate that the probability for bistable behavior is far below the observed response, and thus suggest that it is not sufficient to concentrate analysis on steady states only. A simulation-based analysis of the phenomenon revealed the importance of quasi-bistability to shape ERK response. A system is said to be quasi-bistable, if it is monostable but able to maintain a state distinct from its steady state for a long time period. It is known that positive feedback can generally cause quasi-bistability [33], it has however not been shown that this is relevant for decision making in living systems.

For a biological system it might not make a difference at all whether the underlying system is indeed bistable or quasi-bistable, since the system probably acts as an integrator of a response, which starts to trigger further events as soon as a threshold has been reached. However, one has to be careful with the analysis of models for such mechanisms. Our results propose to consider, besides limit sets, also the transient behavior of a system when investigating

processes such as switches, memory effects or decision making.

There is an ongoing debate about relations between ultrasensitivity and/or bistability in responses of single cells on the one hand and the occurrence of bimodality in heterogeneous cell populations on the other hand (see e.g. [9, 36]). It is clear that ultrasensitive and bistable systems can lead to bimodal responses, for example caused by variations in protein contents or stochastic fluctuations. An example that considers the role of mutual inhibition in a gene regulatory network for metastatic transitions and the appearance of stable subpopulations of genetically identical prometastatic cells is described in [41]. On the other hand, for the MAPK pathway it has been shown that bimodality can also emerge from graded single cell responses caused by a broad distribution of ERK pathway activation thresholds [36]. This example reveals that the relation between bistability and bimodality is actually more subtle than a simple one-to-one relation.

We had decided in this study to use a data-driven approach and to adapt model granularity to the data available for model calibration. This results of course in a very simplified model, and the situation in vivo is much more complex. Specific aspects regarding the MAPK signaling pathways are discussed in literature and have also partly been implemented in models. One of the most recent interesting studies investigates the role of feedbacks and their time scales by using pulse experiments on a single cell level (see also [42] for a commentary on this). Kocieniewski et al. [43], for example, focus on the role of the two different MEK isoforms and their contribution in the regulation of the ERK response. According to this study, response duration and amplitude are regulated by the ratio and the total amount of both isoforms, respectively. Moreover, localization of proteins and their regulation via scaffolding proteins, together with nucleoplasmic shuttling, is known to play a major role in the regulation of the pathway [9, 22, 28]. Cross-talk and interactions with other cellular pathways is another important aspect [9, 11], which is difficult to take into account in any modeling approach. However, it is an important and interesting question how single modules such as the MAPK signaling cascade behave embedded in a larger and more complex network. Several studies in recent years hint to the fact that network complexity is intimately linked to functional robustness, meaning that the network structure, and in particular interlinked feedback loops, contribute to a reliable performing of tasks in the presence of perturbations and noise [44–48].

A further critical point in our modeling study is the normalization of model outputs to a particular time point. This normalization was necessary since the dataset used for model calibration only provides relative information. Signals are given in arbitrary units, and the scaling factors are different for each antibody and can also vary across membranes. Thus normalization to a reference experiment is required to make measurements from different experiments comparable and is standard in representing biological data and for modeling [43]. This normalization, however, affects variances of observables, and precludes comparison with experiments where total protein levels matter, such as absolute heights of ppERK peaks under different conditions. Thus, including some information about total protein levels could highly enrich the modeling process in the future.

Finally, recently a new modeling approach, called ODE constraint mixture modeling, was introduced [49]. This approach combines advantages of mechanistic modeling approaches with statistical mixture models to describe heterogeneous cell populations. This framework allows to infer subpopulation structures and dynamics from single cell snapshot data. Since the data used here for model calibration represent only population averages, we did not explicitly take subpopulation structures into account. However, at least the dose response profiles of ERK after stimulation with NGF seem to consist of two or more subpopulations, which was also exploited to mimic the respective dose response curve. Thus, exploiting this framework is another interesting task for future investigations.

Additional files

Additional file 1: A Bayesian framework for ode model calibration.

Additional file 2: Model normalization procedure.

Additional file 3: Formulation of the posterior distribution.

Additional file 4: Details on the MCMC sampling procedure.

Additional file 5: Estimated marginal parameter distributions from the MCMC sample.

Additional file 6: Scatterplot matrix of a subset of the parameters from the MCMC sample.

Additional file 7: Details on the classification scheme with the CBA.

Additional file 8: Sensitivity analysis of the simulation-based classification scheme.

Additional file 9: Simulation-based classification of sample trajectories with varying minimal switching times.

Additional file 10: Classification of trajectories from the MCMC sample with varying concentration of total ERK.

Abbreviations
CBA: Circuit-breaking algorithm; MCMC: Markov chain Monte Carlo; PPD: Posterior predictive distribution

Acknowledgements
We acknowledge support from Dr. Silvia Santos regarding questions concerning the data from [12] used in this study.

Funding
This work was supported by the German Research Foundation (DFG) within the Cluster of Excellence in Simulation Technology (EXC 310/2) at the University of Stuttgart, and from the Federal Ministry of Education and Research (BMBF) through the e:Bio-Innovationswettbewerb Systembiologie project PREDICT (FKZ0316186A). These funding bodies did not play a role in the design of the study and preparation of the manuscript.

Authors' contributions
AJ implemented the sampling procedure for parameter estimation and model predictions, with help from CT. AJ and CT performed the steady state analysis. NR designed the study with input from AJ and CT and wrote the main part of the manuscript. All authors discussed the results and implications and commented on the manuscript at all stages of the project. All authors read and approved the final manuscript.

Competing interests
The authors declare that they have no competing interests.

References
1. Alon U. An introduction to systems biology - design principles of biological circuits. Math and Comput Biol Series. London: Chapman & Hall/CRC; 2006.
2. Lee J, Tiwari A, Shum V, Mills GB, Mancini MA, Igoshin OA, Balázsi G. Unraveling the regulatory connections between two controllers of breast cancer cell fate. Nucleic Acids Res. 2014;42(11):6839–49.
3. Thomas R. On the relation between the logical structure of systems and their ability to generate multiple steady states or sustained oscillations. In: Della-Dora J, Demongeot J, Lacolle B, editors. Numerical methods in the study of critical phenomena. Springer Series in Synergetics, vol. 9. Berlin, Heidelberg: Springer; 1981. p. 180–93.
4. Gouzé J-L. Positive and negative circuits in dynamical systems. J Biol Syst. 1998;6(21):11–15.
5. Freeman M. Feedback control of intracellular signalling in development. Nature. 2000;408:313–19.
6. Alon U. Network motifs: theory and experimental approaches. Nat Rev Genet. 2007;8:450–61.
7. Savageau MA. Comparison of classical and autogenous systems of regulation in inducible operons. Nature. 1974;252(5484):546–49.
8. Grieco L, Calzone L, Bernard-Pierrot I, Radvanyi F, Kahn-Perlés B, Thieffry D. Integrative modelling of the influence of MAPK network on cancer cell fate decision. PLoS Comp Biol. 2013;9(10):e1003286.
9. Kolch W. Coordinating ERK/MAPK signalling through scaffolds and inhibitors. Nat Rev Mol Cell Biol. 2005;6:827–38.
10. Kolch W, Calder M, Gilbert D. When kinases meet mathematics: the systems biology of MAPK signalling. FEBS Lett. 2005;579:1891–95.
11. Orton RJ, Sturm OE, Vyshemirsky V, Calder M, Gilbert DR, Kolch W. Computational modelling of the receptor-tyrosine-kinase-activated MAPK pathway. Biochem J. 2005;392:249–61.
12. Santos SDM, Verveer PJ, Bastiaens PIH. Growth factor-induced MAPK network topology shapes ERK response determining PC-12 cell fate. Nat Cell Biol. 2007;9(3):324–30.
13. Ryu H, Chung M, Dobrzyński M, Fey D, Blum Y, Lee SS, Peter M, Kholodenko BN, Jeon NL, Pertz O. Frequency modulation of ERK activation dynamics rewires cell fate. Mol Syst Biol. 2015;11(838):14.
14. Brightman FA, Fell DA. Differential feedback regulation of the MAPK cascade underlies the quantitative differences in EGF and NGF signalling in PC12 cells. FEBS Lett. 2000;482:169–74.
15. Kholodenko B. Negative feedback and ultrasensitivity can bring about oscillations in the mitogen-activated protein kinase casacde. Eur J Biochem. 2000;267:1583–88.
16. Lavoie H, Therrien M. Regulation of Raf protein kinases in ERK signalling. Nat Rev Mol Cell Biol. 2015;16:281–98.

17. Vaudry D, Stork PJS, Lazarovici P, Eiden LE. Signaling pathways for PC12 cell differentiation: Making the right connections. Science. 2002;296:1648–49.

18. Huang C-YF, Ferrell JE. Ultrasensitivity in the mitogen-activated protein kinase cascade. Proc Natl Acad Sci USA. 1996;93:10078–83.

19. Shin S-Y, Rath O, Choo S-M, Fee F, McFerran B, Kolch W, Cho K-H. Positive- and negative-feedback regulations coordinate the dynamic behavior of the Ras-Raf-MEK-ERK signal transduction pathway. J Cell Sci. 2009;122:425–35.

20. Fritsche-Guenther R, Witzel F, Sieber A, Herr R, Schmidt N, Braun S, Brummer T, Sers C, Blüthgen N. Strong negative feedback from ERK to Raf confers robustness to MAPK signalling. Mol Syst Biol. 2011;7(1):489.

21. Mai Z, Liu H. Random parameter sampling of a generic three-tier MAPK cascade model reveals major factors affecting its versatile dynamics. PLoS ONE. 2013;8(1):e54441.

22. Ahmed S, Grant KG, Edwards LE, Rahman A, Cirit M, Goshe MB, Haugh JM. Data-driven modeling reconciles kinetics of ERK phosphorylation, localization, and activity states. Mol Syst Biol. 2014;10(1):718.

23. Angeli D, Ferrell JE, Sontag ED. Detection of multistability, bifurcations, and hysteresis in a large class of biological positive-feedback systems. Proc Natl Acad Sci USA. 2004;101(7):1822–7.

24. Sasagawa S, Ozaki Y, Fujita K, Kuroda S. Prediction and validation of the distinct dynamics of transient and sustained ERK activation. Nat Cell Biol. 2005;7(4):365–73.

25. Blanchini F, Franco E. Multistability and robustness of the MAPK pathway. In: Proc. of 50th IEEE Conf on Decision and Ctrl and European Control Conf (CDC-ECC). Orlando: IEEE; 2011. p. 2214–19.

26. Franco E, Blanchini F. Structural properties of the MAPK pathway topologies in PC12 cells. J Math Biol. 2013;67:1633–68.

27. Xiong W, Ferrell JE. A positive-feedback-based bistable 'memory module' that governs cell fate decision. Nature. 2003;426:460–65.

28. Qiao L, Nachbar RB, Kevrekidis IG, Shvartsman SY. Bistability and oscillations in the Huang-Ferrell model of MAPK signaling. PLoS Comput Biol. 2007;3(9):1819–26.

29. Legewie S, Schoeberl B, Blüthgen N, Herzel H. Competing docking interactions can bring about bistability in the MAPK cascade. Biophys J. 2007;93:2279–88.

30. Smolen P, Baxter DA, Byrne JH. Bistable MAPK kinase activity: a plausible mechanism contributing to maintenance of late long-term potentiation. Am J Physiol Cell Physiol. 2008;294:C503–15.

31. Blanchini F, Franco E, Giordano G. Determining the structural properties of a class of biological models. In: Proc. of 51st IEEE Conf on Decision and Ctrl (CDC). Maui: IEEE; 2012. p. 5505–10.

32. Radde N. Fixed point characterization of differential equations with complex graph topology. Bioinformatics. 2010;26(22):2874–80.

33. Mitrophanov AY, Groisman EA. Positive feedback in cellular control systems. Bioassays. 2008;30(6):542–55.

34. Kholodenko B, Kiyatkin A, Bruggeman FJ, Sontag E, Westerhoff HV. Untangling the wires: A strategy to trace functional interactions in signaling and gene networks. Proc Natl Acad Sci USA. 2002;99(20):12841–46.

35. Schilling M, Maiwald T, Hengl S, Winter D, Kreutz C, Kolch W, Lehmann W-D, Timmer J, Klingmüller U. Theoretical and experimental analysis links isoform-specific ERK signalling to cell fate decisions. Mol Syst Biol. 2009;5:334.

36. Birtwistle MR, Rauch J, Kiyatkin A, Aksamitiene E, Dobrzynski M, Hoek JB, Kolch W, Ogunnaike BA, Kholodenko BN. Emergence of bimodal

cell population responses from the interplay between analog single-cell signaling and protein expression noise. BMC Syst Biol. 2012;6:109.

37. Nguyen TT, Scimeca J-C, Filloux C, Peraldi P, Carpentier J-L, van Obberghen E. Co-regulation of the mitogen-activated protein kinase, extracellular signal-regulated kinase 1, and the 90-kDa Ribosomal S6 kinase in PC12 cells. J Biol Chem. 1993;268(13):9803–10.

38. Qiu M-S, Green SH. PC12 cell neuronal differentiation is associated with prolonged p21^{ras} activity and consequent prolonged ERK activity. Neuron. 1992;9:705–17.

39. Miyasaka T, Chao MV, Sherline P, Saltiel AR. Nerve growth factor stimulates a protein kinase in PC-12 cells that phosphorylates microtubule-associated protein-2. J Biol Chem. 1990;265:4730–35.

40. Gotoh Y, Nishida E, Yamashita T, Hoshi M, Kawamaki M, Sakai H. Microtubule-associated-protein (MAP) kinase activated by nerve growth factor and epidermal growth factor in PC12 cells. Eur J Biochem. 1990;193:661–9.

41. Lee J, Lee J, Farquhar KS, Yun J, Frankenberger CA, Bevilacqua E, Yeung K, Kim E-J, Balázsi G, Rosner MR. Network of mutually repressive metastasis regulators can promote cell heterogeneity and metastatic transitions. Proc Natl Acad Sci U S A. 2014;111(3):E364–73.

42. Blüthgen N. Signaling output: it's all about timing and feedbacks. Mol Syst Biol. 2015;11(843):2.

43. Kocieniewski P, Lipniacki T. MEK1 and MEK2 differentially control the duration and amplitude of the ERK cascade response. Phys Biol. 2013;10(3):035006.

44. Stelling J, Sauer U, Szallasi Z, Doyle FJ, Doyle J. Robustness of cellular functions. Cell. 2004;118:675–85.

45. Wagner A. Circuit topology and the evolution of robustness in two-gene circadian oscillators. Proc Natl Acad Sci USA. 2005;102(33):11775–80.

46. Barkai N, Shilo B-Z. Variability and robustness in biomolecular systems. Cell. 2007;28:755–60.

47. Cheng P, Yang Y, Liu Y. Interlocked feedback loops contribute to the robustness of the *Neurospora* circadian clock. Proc Natl Acad Sci U S A. 2001;98:7408–13.

48. Clodong S, Dühring U, Kronk L, Wilde A, Axmann I, Herzel H, Kollmann M. Functioning and robustness of a bacterial circadian clock. Mol Syst Biol. 2007;3(90):1–9.

49. Hasenauer J, Hasenauer C, Hucho T, Theis FJ. ODE constraint mixture modelling: A method for unraveling subpopulation structures and dynamics. PLoS Comput Biol. 2014;10(7):e1003686.

3

BCM: toolkit for Bayesian analysis of Computational Models using samplers

Bram Thijssen[1] (ID), Tjeerd M. H. Dijkstra[2,3], Tom Heskes[4] and Lodewyk F. A. Wessels[1,5*]

Abstract

Background: Computational models in biology are characterized by a large degree of uncertainty. This uncertainty can be analyzed with Bayesian statistics, however, the sampling algorithms that are frequently used for calculating Bayesian statistical estimates are computationally demanding, and each algorithm has unique advantages and disadvantages. It is typically unclear, before starting an analysis, which algorithm will perform well on a given computational model.

Results: We present BCM, a toolkit for the Bayesian analysis of Computational Models using samplers. It provides efficient, multithreaded implementations of eleven algorithms for sampling from posterior probability distributions and for calculating marginal likelihoods. BCM includes tools to simplify the process of model specification and scripts for visualizing the results. The flexible architecture allows it to be used on diverse types of biological computational models. In an example inference task using a model of the cell cycle based on ordinary differential equations, BCM is significantly more efficient than existing software packages, allowing more challenging inference problems to be solved.

Conclusions: BCM represents an efficient one-stop-shop for computational modelers wishing to use sampler-based Bayesian statistics.

Keywords: Bayesian statistics, Sampling, Markov chain Monte Carlo, Sequential Monte Carlo, Nested sampling

Background

There is an increasing interest in using Bayesian statistics for the analysis of computational models in biology [1–4]. With Bayesian statistics, the unknown parameters of a computational model are assigned a probability distribution describing their uncertainty. This distribution can be updated from prior information to give the posterior probability distribution, using Bayes' theorem:

$$P(\theta|X, \mathcal{M}) = \frac{P(X|\theta, \mathcal{M})P(\theta|\mathcal{M})}{P(X|\mathcal{M})} \quad (1)$$

where Θ represents the parameters, X the measurement data and $\mathcal{M}$ the computational model. Furthermore, the marginal likelihood, or evidence, can be used to

discriminate between different computational models. It can be calculated by marginalizing the parameters:

$$P(X|\mathcal{M}) = \int P(X|\theta, \mathcal{M})P(\theta|\mathcal{M})d\theta \quad (2)$$

Typically, neither the posterior probability nor the marginal likelihood can be calculated directly, but sampling algorithms can be used to estimate them [5–16]. These sampling algorithms are computationally demanding, especially when the number of parameters is large and when the computational model is expensive to simulate. Typical models in systems biology indeed carry many parameters and are expensive to simulate [17]. Additionally, the posterior probability distributions arising from such models are usually complex, containing multiple modes and ridges that are difficult to traverse [18]. Bayesian analysis of such systems biology models thus requires the use of advanced sampling algorithms. Since these sampling algorithms each have unique characteristics and can be more or less suitable for a particular

* Correspondence: l.wessels@nki.nl
[1]Computational Cancer Biology, The Netherlands Cancer Institute, Plesmanlaan 121, 1066 CX Amsterdam, The Netherlands
[5]Faculty of EEMCS, Delft University of Technology, Mekelweg 4, 2628CD Delft, The Netherlands
Full list of author information is available at the end of the article

task, it would be beneficial to have various algorithm easily available.

BCM, a toolkit for the Bayesian analysis of Computational Models using samplers, provides efficient, multithreaded implementations of eleven algorithms for calculating posterior probabilities and marginal likelihoods.

The BCM toolkit focuses on computational models that involve simulations or extensive calculations. Examples of such computational models are systems of ordinary differential equations describing biochemical reactions; or steady-state signaling networks, where the activity levels may be calculated in diverse ways. These computational models are in contrast to statistical models that can be specified in the BUGS or Stan languages. For such statistical models, excellent software packages already exist [19, 20]. For the computational models that are targeted by BCM, several alternative software packages also exist [5, 21–23]. However, each of these packages implements only a single type of sampling algorithm and most of them focus on one particular type of computational model. In contrast, with BCM the user can choose from eleven sampling algorithms and the plugin architecture allows diverse types of models. Thus, BCM represents a one-stop-shop for Bayesian analysis of systems biology models, where the user has a high chance of finding a suitable algorithm for the analysis of the user-defined model.

Implementation

BCM consists of three components: an inference tool, a model parsing tool and an R script for further analysis and visualization (see Fig. 1a).

The inference tool (*mdlinf*) is the main component of BCM. It uses a specified sampling algorithm to generate samples from the posterior probability distribution and to calculate a marginal likelihood estimate. Error bounds for the marginal likelihood estimate are also provided, which are calculated directly from the samples using a method suitable for the particular algorithm used to calculate the marginal likelihood. As input, the inference tool requires three parts: a configuration file, an XML file specifying the prior, and a dynamic library that evaluates the likelihood function. For constructing the dynamic library that evaluates the likelihood function, BCM provides cross-platform boilerplate code, such that custom model simulation code can be easily adapted for use with BCM. Alternatively, the model parsing tool can be used as described further below.

The inference tool implements three classes of sampling algorithms: Markov chain Monte Carlo (MCMC) [6, 7], sequential Monte Carlo (SMC) [8] and nested sampling [9]. For each class of sampling algorithms, BCM includes several options for proposal distributions, as well as extensions that can increase the sampling efficiency when dealing with complex inference problems, giving a total of eleven different sampling algorithms (Table 1).

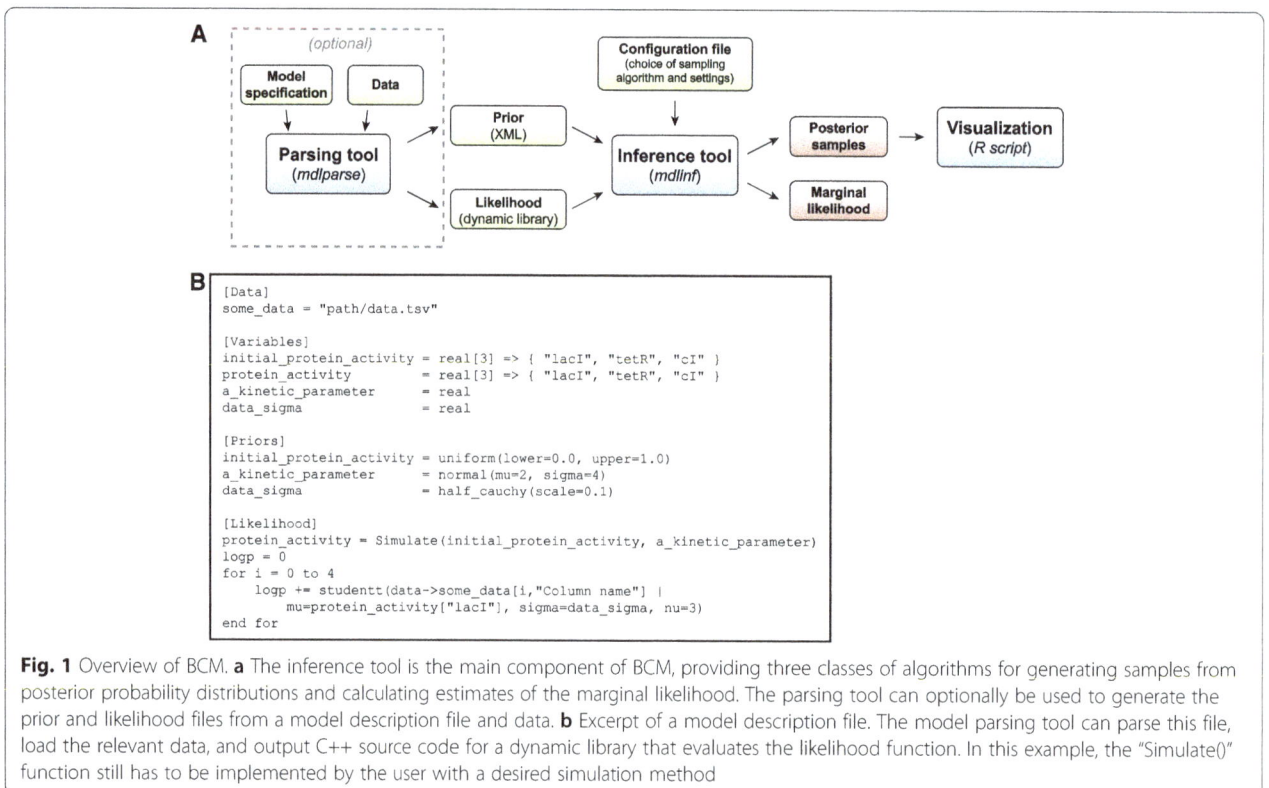

Fig. 1 Overview of BCM. **a** The inference tool is the main component of BCM, providing three classes of algorithms for generating samples from posterior probability distributions and calculating estimates of the marginal likelihood. The parsing tool can optionally be used to generate the prior and likelihood files from a model description file and data. **b** Excerpt of a model description file. The model parsing tool can parse this file, load the relevant data, and output C++ source code for a dynamic library that evaluates the likelihood function. In this example, the "Simulate()" function still has to be implemented by the user with a desired simulation method

Table 1 Sampling algorithms and extensions implemented in BCM

Sampling algorithm	Reference
Markov Chain Monte Carlo	[6, 7]
Parallel tempering	[10]
Adaptive proposals	[11]
Feedback-optimized temperatures	[12]
Thermodynamic integration	[13]
Automated parameter blocking	[14]
Sequential Monte Carlo	[8]
MCMC proposals	[8]
Kernel density estimate proposals	[8]
Automated temperature schedule	[15]
Nested sampling	[9]
MCMC proposals	[9]
Ellipsoid proposals	[16]
MultiNest	[5]

Care has been taken to create efficient, multithreaded implementations of each algorithm. Firstly, the inference tool has been written in C++ and performance bottlenecks have been profiled and optimized. Secondly, each algorithm has been parallelized with a multithreading strategy suitable for that algorithm: for MCMC, multiple chains are distributed across threads, for SMC, particles are distributed in batches across threads, and for nested sampling, a batch of samples is generated at each iteration by all threads which are then re-used in subsequent nested sampling iterations.

The model parsing tool (*mdlparse*) is the second component of BCM. It can be used to generate the prior and likelihood files for the inference tool. The parsing tool reads a model description file that specifies the model, comprising the prior, likelihood and data references, and it outputs C++ source code for a dynamic library that evaluates the prior and likelihood function with the relevant data. This C++ code can then be used as a basis for further modification; or it can be directly compiled into a dynamic library. The input model description file uses a custom format with an easy-to-read syntax. An excerpt of a model description file is shown in Fig. 1b. The use of the model parsing tool is optional and it is meant as an aid in model specification rather than as a comprehensive tool capable of fully specifying all types of models.

Finally, a script is provided to load the output of the inference tool into R for further analysis and for visualization of the results. This script can be used to display kernel density estimates of the posterior probability distribution of the sampled variables, as well as to make plots for visual posterior predictive checking; examples of both of these are shown in Figs. 3 and 4. Basic functionality for convergence diagnostics is included as well, including

autocorrelation functions and trace plots. Functions for conversion of the results to CODA objects [24] and to ggmcmc objects [25], two R packages for MCMC convergence diagnostics and output analysis, are also provided.

Results
Analytically tractable example
To showcase BCM, and to explore how each class of algorithms deals with increasing dimensionality and complex distributions, we first analyzed a problem which is analytically tractable: the Gaussian shells problem described in [5, 26]. While this example is not directly relevant for systems biology, its likelihood function is multimodal and ridge-shaped, resembling the likelihoods often encountered in systems biology models. The likelihood function for this Gaussian shells problem is given by

$$P(\boldsymbol{\theta}) = \sum_{i=1}^{2} \frac{1}{\sqrt{2\pi w^2}} \exp\left(-\frac{(|\boldsymbol{\theta}-\boldsymbol{c}_i|-r)^2}{2w^2}\right) \tag{3}$$

where $r = 2$, $w = 0.1$, and $\boldsymbol{\Theta}$ and $\boldsymbol{c}_i$ are n-dimensional vectors. $\boldsymbol{\Theta}$ is the vector of variables which are to be sampled and $\boldsymbol{c}_i$ are two constant vectors describing the centers of the two peaks and are assigned the values $c_{1,x} = 3.5$, $c_{2,x} = -3.5$ and 0 in the other dimensions. This likelihood function is then composed of two narrow, well-separated, ring-shaped peaks (Fig. 2a), which is a challenging sampling problem.

We tested three sampling algorithms on this problem, one from each class of sampling algorithms: feedback-optimized parallel-tempered Markov chain Monte Carlo (FOPTMC) [12], sequential Monte Carlo (SMC) [8] with the automated temperature schedule selection of [15] but without using Approximate Bayesian Computation, and MultiNest [5].

As shown in Table 2, all three algorithms give the correct estimate for the marginal likelihood within the error bounds. When the number of dimensions is 10 or fewer, MultiNest is extremely efficient: it requires the fewest likelihood evaluations while achieving the tightest error bound. When the number of dimensions is increased beyond 10 however, MultiNest becomes very inefficient. At this point the exponential scaling of the algorithm becomes apparent. In the higher-dimensional setting, the SMC algorithm deals with this problem most efficiently. FOPTMC is least efficient: it requires the largest number of likelihood evaluations and has the largest error bound. FOPTMC can still effectively explore the posterior distribution (as shown in Fig. 2b), however, the temperature schedule of the parallel chains in FOPTMC is optimized for exploration of the posterior rather than for estimation of the marginal likelihood and as a result there is an increasingly large error in the marginal likelihood estimate at higher dimensionality.

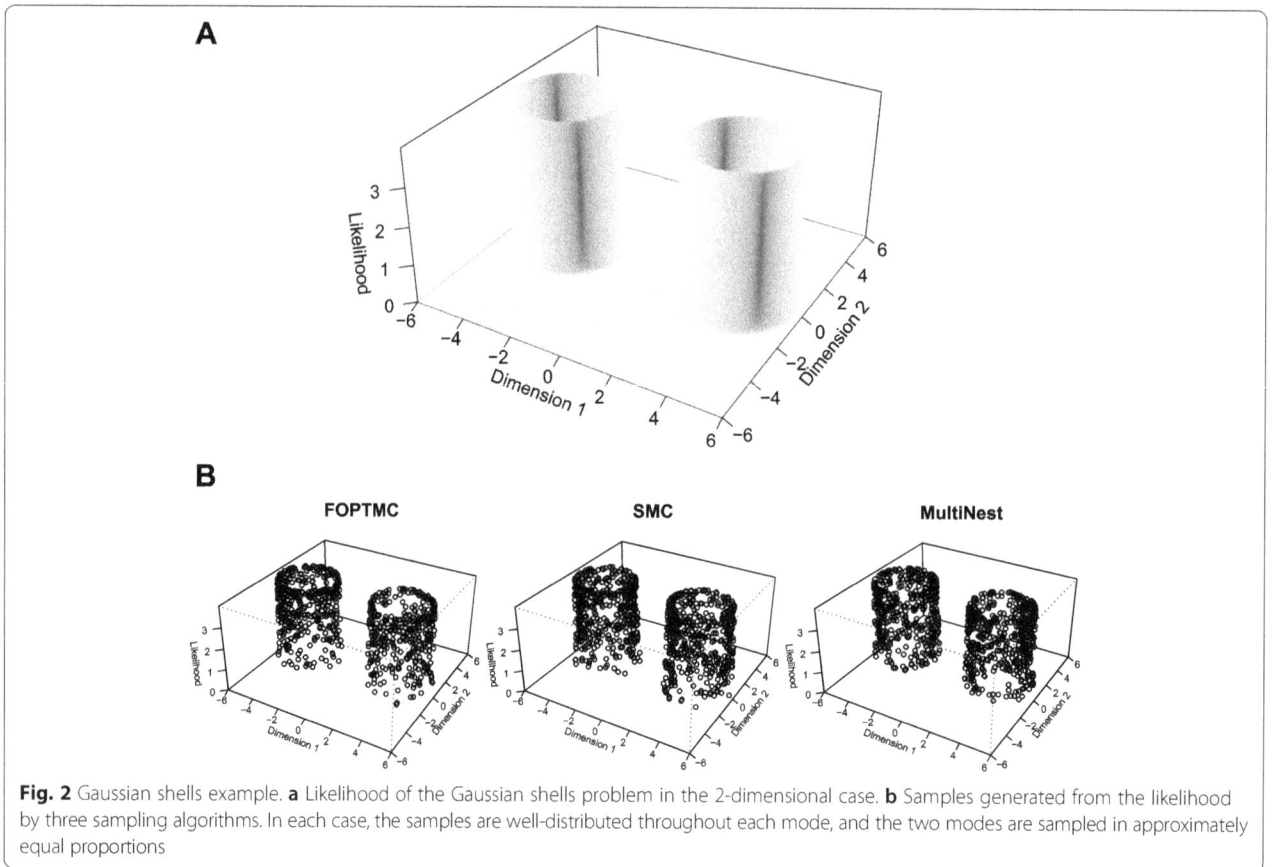

Fig. 2 Gaussian shells example. **a** Likelihood of the Gaussian shells problem in the 2-dimensional case. **b** Samples generated from the likelihood by three sampling algorithms. In each case, the samples are well-distributed throughout each mode, and the two modes are sampled in approximately equal proportions

Kinetic ordinary differential equation model

Having explored the behavior of several sampling algorithms in an analytically tractable example, we now illustrate the use of BCM for analyzing biological computational models. As an example of this, we investigated the inference of the parameters of a model based on a system of ordinary differential equations (ODEs). The 6-variable cell cycle model of Tyson [27] was used, as downloaded from BioModels [17]. A graphical representation of this model is shown in Fig. 3a.

To recreate a typical setting in biology, data was generated from the model at six time points for two observables with three replicates (see Additional file 1). Then BCM was used to infer all 16 parameters of the model (10 kinetic parameters and 6 initial conditions) from these 36 data points. The priors for the kinetic parameters were set to a uniform distribution spanning an order of magnitude on either side of the parameter values that were used to generate the data, and the priors for the initial conditions were set to a uniform distribution between 0 and 1 (see blue curves in Fig. 3c). The likelihood function was set equal to the one that generated the data, that is, a normal distribution with standard deviation 0.05.

Table 2 Performance of three sampling algorithms in calculating the marginal likelihood of an analytically tractable example

Dimensions	Log marginal likelihood				Likelihood evaluations (x1000)		
	Analytical	FOPTMC	SMC	MultiNest	FOPTMC	SMC	MultiNest
2	−1.75	−1.80 ± 0.68	−1.74 ± 0.39	−1.73 ± 0.29	147	79	18
5	−5.67	−5.98 ± 1.65	−5.66 ± 0.47	−5.73 ± 0.38	287	281	28
10	−14.59	−14.92 ± 3.34	−14.64 ± 0.62	−14.13 ± 0.63	969	521	95
30	−60.13	−61.11 ± 9.10	−59.85 ± 0.97	*	6420	1511	*
100	−255.62	−257.7 ± 24.8	−255.8 ± 1.54	*	96,251	4271	*

The following algorithms were used: *FOPTMC* feedback-optimized parallel-tempered Markov Chain Monte Carlo [12], *SMC* automated-temperature sequential Monte Carlo but without ABC approximation [15], and MultiNest [5]. The column 'Analytical' gives the marginal likelihood value calculated analytically. (*) indicates that the computation time exceeded the maximal time of 1 h; the other calculations required at most 5 min

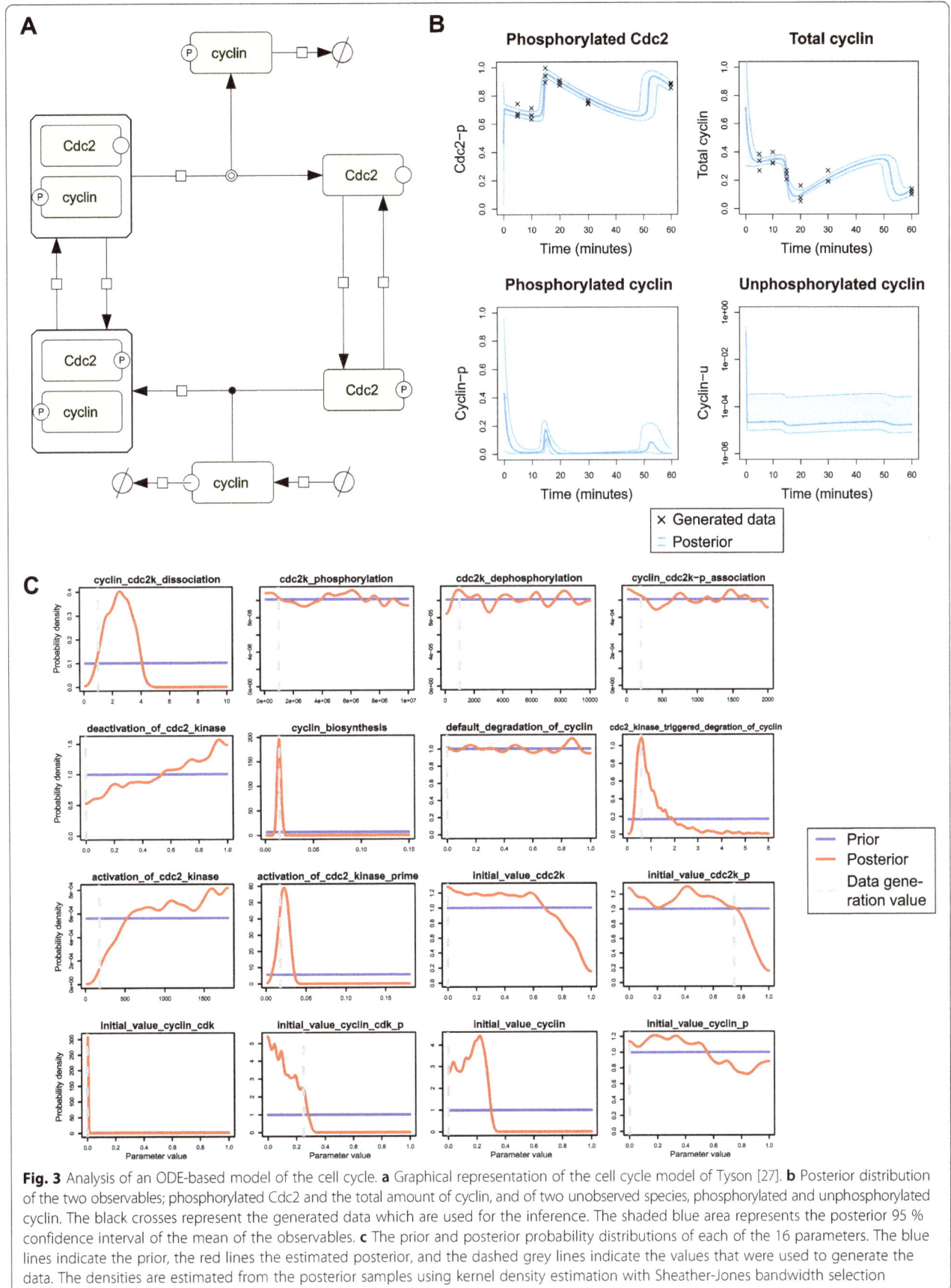

Fig. 3 Analysis of an ODE-based model of the cell cycle. **a** Graphical representation of the cell cycle model of Tyson [27]. **b** Posterior distribution of the two observables; phosphorylated Cdc2 and the total amount of cyclin, and of two unobserved species, phosphorylated and unphosphorylated cyclin. The black crosses represent the generated data which are used for the inference. The shaded blue area represents the posterior 95 % confidence interval of the mean of the observables. **c** The prior and posterior probability distributions of each of the 16 parameters. The blue lines indicate the prior, the red lines the estimated posterior, and the dashed grey lines indicate the values that were used to generate the data. The densities are estimated from the posterior samples using kernel density estimation with Sheather-Jones bandwidth selection

Despite the small size of the model, this inference problem is challenging. Firstly, the ODE system is stiff, and even with the use of an implicit ODE solver it is costly to simulate. Secondly, there are multiple distinct ways in which the model can fit the data, leading to sub-optimal modes in the posterior distribution. Thus, a sampler must be able to escape these local optima, and it must be able to converge to the correct posterior distribution with a limited number of likelihood evaluations due to the computational cost of the simulations.

Four sampling algorithms were tested on this problem: SMC, MultiNest, FOPTMC (now extended with automated parameter blocking [14]), and additionally nested sampling with MCMC proposals (Nested-MCMC) was added as an alternative nested sampling strategy. In this inference task, FOPTMC with automated parameter blocking was most efficient, requiring 14 h to generate 2000 samples from the posterior. SMC required 19 h, while Nested-MCMC required 30 h and MultiNest had to be discontinued as the acceptance rate quickly dropped to essentially zero. The tests were performed using 16 threads on an Intel Xeon E5-2680 processor.

The Bayesian estimates of the parameters and the trajectories of the species can be used to study the uncertainty in the model. Figure 3b shows the posterior distribution of the two observables, as well as of two inferred species for which no observable data was generated, as estimated by FOPTMC. We can see that the data are sufficient to constrain the trajectories of the observed species. For the unobserved species phosphorylated cyclin, the overall trajectory can also be inferred. Nevertheless, for this unobserved species, the second peak is more variable – here the data is insufficient to constrain the precise magnitude of the peak. For the other unobserved species, unphosphorylated cyclin, we see that there is greater uncertainty. The posterior distribution indicates only that the average levels are low, but the precise levels cannot be inferred from these data.

Figure 3c shows the marginal posterior probability distributions of the parameters. It can be seen that for all parameters, the values used to generate the data fall within areas of non-zero probability of the posterior. In most cases the data-generation values also have maximum posterior probability, but interestingly this is not true for all parameters, such as for the activation and deactivation of Cdc2. Furthermore, some parameters are not identifiable, for example the rates of phosphorylation and desphosphorylation of Cdc2 cannot be determined from the data. In general, such lack of identifiability could be for structural reasons, that is, the parameters cannot be inferred in theory given the observed species, due to a redundant parameterization. Alternatively, the parameters may be identifiable in theory, but the data

may provide insufficient information to constrain the parameters in practice.

Overall, the Bayesian estimates provide useful measures of the uncertainty in parameter values, model fit and model predictions.

Comparison with existing software packages

There are several software packages which can perform Bayesian inference of the parameters of ODE-based models: BioBayes [21], ABC-SysBio [22], SYSBIONS [23] and Stan [20]. BioBayes uses parallel-tempered Markov Chain Monte Carlo, ABC-SysBio uses sequential Monte Carlo sampling in combination with Approximate Bayesian Computation, SYSBIONS uses nested sampling, and Stan uses Hamiltonian Monte Carlo and the No-U-Turn sampler (NUTS).

To compare BCM with these software packages, a simplified version of the previous inference problem was used. Instead of inferring all 16 parameters, the initial conditions and 4 of the 10 kinetic parameters were fixed to the values used to generate the data, leaving 6 parameters to be inferred. Figure 4a shows the marginal posterior probability distributions of the simplified problem, as estimated by BCM using FOPTMC (see Additional file 2: Figure S1 for the posteriors estimated by each algorithm/software package). The other software packages were optimized for this problem as much as possible to give a fair comparison (see Additional file 1).

Figure 4b shows the time required to generate 1000 samples from the posterior with each software package and algorithm, using eight threads on an Intel Xeon E5-2680 processor. It is clear that BCM is significantly faster than the other software packages. In particular the MultiNest algorithm in BCM is extremely efficient in this low-dimensional setting, requiring only 75 s. The other algorithms in BCM required between 25 and 50 min, except for ellipsoidal nested sampling which required three hours. From the other software packages, only SYSBIONS and Stan were able to solve this inference problem in a reasonable amount of time. SYSBIONS required five hours using Nested-MCMC, which is approximately six times longer than BCM with the same algorithm. For Stan, using the NUTS algorithm, the sampling with a chain does not always converge as the NUTS algorithm does not have a means to escape sub-optimal modes. This problem was addressed by starting eight separate chains in parallel, in which case most of the chains were sampling the correct, optimal mode. In this case, Stan required approximately six hours to generate the requested samples. BioBayes was able to reach apparent convergence in 4.5 days. For ABC-SysBio, and SYSBIONS using ellipsoidal sampling, the samplers did not reach convergence in 7 days (see Additional file 1).

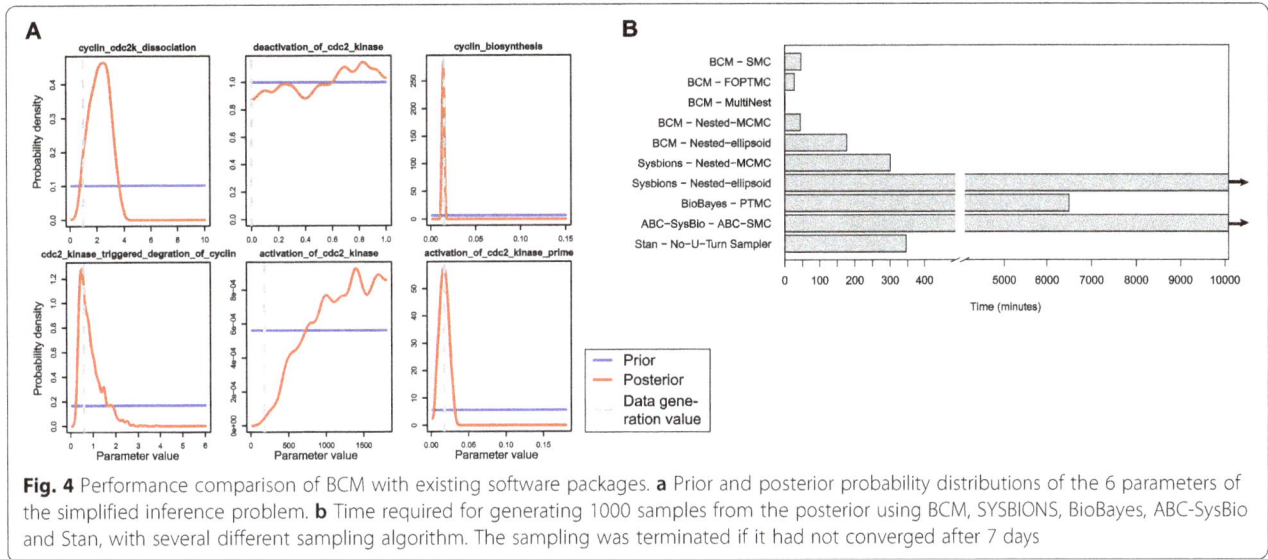

Fig. 4 Performance comparison of BCM with existing software packages. **a** Prior and posterior probability distributions of the 6 parameters of the simplified inference problem. **b** Time required for generating 1000 samples from the posterior using BCM, SYSBIONS, BioBayes, ABC-SysBio and Stan, with several different sampling algorithm. The sampling was terminated if it had not converged after 7 days

Conclusion

The BCM toolkit provides efficient, multithreaded implementations of eleven sampling algorithms for generating posterior samples and calculating marginal likelihoods. Additional tools are included which facilitate the process of specifying models and visualizing the sampling output. This toolkit can be used for analyzing the uncertainty in the parameters and the predictions of computational models using Bayesian statistics.

The examples show that it depends on the problem which sampling algorithm will perform well. In the Gaussian shells example, where the focus was on marginal likelihood estimation, MultiNest performed best in a low-dimensional setting, and in the medium- to high dimensional setting sequential Monte Carlo was most efficient. In the cell cycle example, where the focus was on parameter inference, parallel-tempered Markov chain Monte Carlo was more efficient than sequential Monte Carlo. There are various aspects of the posterior probability distribution which affect the performance of the different algorithms; for example the number of modes, how well the shapes of the modes are approximated by the proposal distributions, and the location and volume of the posterior modes with respect to the prior. These features of the posterior probability distribution will typically not be known for the problem of interest before starting the analysis, and it is then unclear which algorithm might be most suitable. The availability of various algorithms in BCM will therefore be useful in the Bayesian analysis of diverse models.

In the second example, we have shown that BCM can be used to infer the parameters of an ODE-based model of the cell cycle. BCM is significantly more efficient in this task than existing software packages. This increase in efficiency was possible due to the parallelization of

the sampling algorithms in combination with the use of optimized C++ as programming language. Due to the higher efficiency, BCM allows the analysis of larger or more challenging computational models than was previously feasible. In previous cases where Bayesian analysis of complex biological computational models was done, such as in [3, 4, 28], sampling algorithms were newly implemented for each project. The availability of BCM as an efficient, reusable software package can help in streamlining such projects in the future.

Additional files

Additional file 1: Supplementary Information. Description: Supplementary information describing the methodological details and all settings that were used for each inference.

Additional file 2: Figure S1. Description: Overview of the sampling results of each inference of the comparison with existing software packages.

Abbreviations
ABC: Approximate Bayesian Computation; BCM: Toolkit for Bayesian analysis of Computational Models using samplers; FOPTMC: Feedback-optimized parallel-tempered Markov chain Monte Carlo; MCMC: Markov chain Monte Carlo; NUTS: No-U-turn sampler; SMC: Sequential Monte Carlo

Funding
This work was performed within the Cancer Genomics Netherlands Program supported by the Gravitation program of the Netherlands Organization for Scientific Research (NWO).

Authors' contributions
BT wrote the software and performed the analysis. TD, TH and LW supervised the project. BT wrote the manuscript with extensive input from TD, TH and LW. All authors read and approved the final manuscript.

Competing interests
The authors declare that they have no competing interests.

Author details
[1]Computational Cancer Biology, The Netherlands Cancer Institute, Plesmanlaan 121, 1066 CX Amsterdam, The Netherlands. [2]Max Planck Institute for Developmental Biology, Spemannstrasse 35, 72076 Tübingen, Germany. [3]Centre for Integrative Neuroscience, University Clinic Tübingen, Otfried-Müller-Strasse 25, 72076 Tübingen, Germany. [4]Radboud University Nijmegen, Institute for Computing and Information Sciences, Heyendaalseweg 135, 6525 AJ Nijmegen, The Netherlands. [5]Faculty of EEMCS, Delft University of Technology, Mekelweg 4, 2628CD Delft, The Netherlands.

References
1. Wilkinson DJ. Bayesian methods in bioinformatics and computational systems biology. Brief Bioinform. 2007;8:109–16.
2. Vyshemirsky V, Girolami M. Bayesian ranking of biochemical system models. Bioinformatics. 2008;24:833–9.
3. Xu T-R, Vyshemirsky V, Gormand A, von Kriegsheim A, Girolami M, Baillie GS, Ketley D, Dunlop AJ, Milligan G, Houslay MD, Kolch W. Inferring signaling pathway topologies from multiple perturbation measurements of specific biochemical species. Sci Signal. 2010;3:ra20.
4. Eydgahi H, Chen WW, Muhlich JL, Vitkup D, Tsitsiklis JN, Sorger PK. Properties of cell death models calibrated and compared using Bayesian approaches. Mol Syst Biol. 2013;9:644.
5. Feroz F, Hobson MP, Bridges M. MultiNest: an efficient and robust Bayesian inference tool for cosmology and particle physics. Mon Not R Astron Soc. 2009;398:1601–14.
6. Metropolis N, Rosenbluth AW, Rosenbluth MN, Teller AH. Equation of state calculations by fast computing machines. J Chem Phys. 1953;21:1087–92.
7. Hastings WK. Monte Carlo sampling methods using Markov chains and their applications. Biometrika. 1970;57:97–109.
8. Del Moral P, Doucet A, Jasra A. Sequential Monte Carlo samplers. J R Stat Soc Ser B (Stat Methodol). 2006;68:411–36.
9. Skilling J. Nested sampling for general Bayesian computation. Bayesian Anal. 2006;1:833–59.
10. Geyer CJ. Markov chain Monte Carlo maximum likelihood. In: Proceedings of the 23rd symposium interface. 1991. p. 156–63.
11. Haario H, Saksman E, Tamminen J. An adaptive metropolis algorithm. Bernoulli. 2001;7:223–42.
12. Katzgraber HG, Trebst S, Huse DA, Troyer M. Feedback-optimized parallel tempering Monte Carlo. J Stat Mech Theory Exp. 2006. doi:10.1088/1742-5468/2006/03/P03018.
13. Gelman A, Meng X-L. Simulating normalizing constants: from importance sampling to bridge sampling to path sampling. Stat Sci. 1998;13:163–85.
14. Turek D, de Valpine P, Paciorek CJ, Anderson-Bergman C. Automated parameter blocking for efficient Markov chain Monte Carlo sampling. Bayesian Anal. 2016. in press.
15. Del Moral P, Doucet A, Jasra A. An adaptive sequential Monte Carlo method for approximate Bayesian computation. Stat Comput. 2011;22:1009–20.
16. Mukherjee P, Parkinson D, Liddle AR. A nested sampling algorithm for cosmological model selection. Astrophys J. 2006;638:51–4.
17. Chelliah V, Juty N, Ajmera I, Ali R, Dumousseau M, Glont M, Hucka M, Jalowicki G, Keating S, Knight-Schrijver V, Lloret-Villas A, Natarajan KN, Pettit JB, Rodriguez N, Schubert M, Wimalaratne SM, Zhao Y, Hermjakob H, Le Novère N, Laibe C. BioModels: ten-year anniversary. Nucleic Acids Res. 2015;43:D542–8.
18. Girolami M. Bayesian inference for differential equations. Theor Comput Sci. 2008;408:4–16.
19. Lunn D, Spiegelhalter D, Thomas A, Best N. The BUGS project: evolution, critique and future directions. Stat Med. 2009;28:3049–67.
20. Carpenter B, Gelman A, Hoffman M, Lee D, Goodrich B, Betancourt M, Brubaker MA, Guo J, Li P, Riddell A. Stan: a probabilistic programming language. J Stat Softw. 2015. in press.
21. Vyshemirsky V, Girolami M. BioBayes: a software package for Bayesian inference in systems biology. Bioinformatics. 2008;24:1933–4.
22. Liepe J, Barnes C, Cule E, Erguler K, Kirk P, Toni T, Stumpf MPH. ABC-SysBio—approximate bayesian computation in python with GPU support. Bioinformatics. 2010;26:1797–9.
23. Johnson R, Kirk P, Stumpf MPH. SYSBIONS: nested sampling for systems biology. Bioinformatics. 2014;31:604–5.
24. Plummer M, Best N, Cowles K, Vines K. CODA: Convergence Diagnosis and Output Analysis for MCMC. R News. 2006;6:7–11.
25. Fernández-i-Marín X. ggmcmc: analysis of MCMC samples and Bayesian inference. J Stat Softw. 2016;70(9):1–20.
26. Allanach BC, Lester CG. Sampling using a "bank" of clues. Comput Phys Commun. 2008;179:256–66.
27. Tyson JJ. Modeling the cell division cycle: cdc2 and cyclin interactions. Proc Natl Acad Sci U S A. 1991;88:7328–32.
28. Milias-Argeitis A, Oliveira AP, Gerosa L, Falter L, Sauer U, Lygeros J. Elucidation of genetic interactions in the yeast GATA-factor network using Bayesian model selection. PLoS Comput Biol. 2016;12:e1004784.

4

A computational method for the investigation of multistable systems and its application to genetic switches

Miriam Leon[1], Mae L. Woods[1], Alex J. H. Fedorec[1] and Chris P. Barnes[1,2]* (iD)

Abstract

Background: Genetic switches exhibit multistability, form the basis of epigenetic memory, and are found in natural decision making systems, such as cell fate determination in developmental pathways. Synthetic genetic switches can be used for recording the presence of different environmental signals, for changing phenotype using synthetic inputs and as building blocks for higher-level sequential logic circuits. Understanding how multistable switches can be constructed and how they function within larger biological systems is therefore key to synthetic biology.

Results: Here we present a new computational tool, called StabilityFinder, that takes advantage of sequential Monte Carlo methods to identify regions of parameter space capable of producing multistable behaviour, while handling uncertainty in biochemical rate constants and initial conditions. The algorithm works by clustering trajectories in phase space, and iteratively minimizing a distance metric. Here we examine a collection of models of genetic switches, ranging from the deterministic Gardner toggle switch to stochastic models containing different positive feedback connections. We uncover the design principles behind making bistable, tristable and quadristable switches, and find that rate of gene expression is a key parameter. We demonstrate the ability of the framework to examine more complex systems and examine the design principles of a three gene switch. Our framework allows us to relax the assumptions that are often used in genetic switch models and we show that more complex abstractions are still capable of multistable behaviour.

Conclusions: Our results suggest many ways in which genetic switches can be enhanced and offer designs for the construction of novel switches. Our analysis also highlights subtle changes in correlation of experimentally tunable parameters that can lead to bifurcations in deterministic and stochastic systems. Overall we demonstrate that StabilityFinder will be a valuable tool in the future design and construction of novel gene networks.

Keywords: Genetic switches, Sequential Monte Carlo, Design of genetic circuits

Background

Synthetic biology has seen the development of many simple gene circuits such as switches [1–6], oscillators [7–9] and pulse generators [10]. Larger systems have been constructed [11], but the leap from building low-level circuits to assembling them into complex networks is still a major challenge [12, 13]. Efforts to do so are plagued by circuit crosstalk, retroactivity, chassis loading effects,

and cellular noise, which can render synthetic networks non-functional in vivo [14, 15]. Although standardization and better part design can partially lower this barrier [12, 16–19], design processes that enable the informed selection of appropriate parts are crucial [11, 20, 21].

One of the foundational constructs in synthetic biology is the genetic toggle switch. The toggle switch consists of a set of transcription factors that mutually repress each other [1, 22–24]. Genetic switches play a major role in binary cell fate decisions such as stem cell differentiation, as they are capable of exhibiting bistable behaviour, which gives rise to the existence of two distinct phenotypic states. This allows populations of cells to maintain

*Correspondence: christopher.barnes@ucl.ac.uk
[1]Department of Cell and Developmental Biology, University College London, Gower Street, WC1E 6BT, London, UK
[2]Department of Genetics, Evolution and Environment, University College London, Gower Street, WC1E 6BT, London, UK

a heterogeneous response to environmental cues and can increase fitness by bet-hedging [25]. Switches are powerful building blocks; they underlie electronics and logic systems, and have great potential in synthetic biology. The genetic toggle switch has been used for a number of applications including the construction of a synthetic genetic clock [22], the regulation of mammalian gene expression [2, 5], the development of a predictable genetic timer [26], and the formation of biofilms in response to engineered stimuli [27].

The stability of the toggle switch has been investigated extensively in the literature, but the conclusions drawn vary according to model abstraction. Numerous studies have concluded that cooperativity is a necessary condition for bistability to arise [1, 28–31]. However, Lipshtat et al. found that stochastic effects can give rise to bistability even without cooperativity [32]. In another study, Ma et al. found that stochastic fluctuations can stabilize the unstable steady state in the deterministic system, giving rise to tristability [33]. In addition, Biancalani et al. identified multiplicative noise as the source of bistability in the stochastic case [34]. As is clear from the above, there is yet to exist a consensus on the stability a switch is capable of, and the most appropriate method of modelling it. Most of these studies assumed the quasi-steady state approximation (QSSA) [35], which cannot always be assumed to hold in vivo [36].

In terms of system design, extensions of the basic toggle switch motif, including additional positive feedback mechanisms, have been investigated [37, 38], and optimization methods have been used to identify topologies and parameter values for bistable and tristable genetic switches [39–42]. For stochastic switch design, control theoretic approaches [43], and simulation-based frameworks [44], have been developed. However, none of these existing approaches can be be applied to reasonably sized models, under the assumption of deterministic and stochastic dynamics, and identify regions of parameter space for which switching occurs, which we argue is critical in designing systems under considerable uncertainty.

Here we present a computational framework based on sequential Monte Carlo [45] that can determine the parameter region for a given model to produce a given number of (stable) steady states. Uniquely, multistable parameter regions can be identified for both deterministic and stochastic systems, and also complex models with many parameters, thus removing the need for simplifying assumptions. Our framework can be used for comparing the conclusions drawn by various modelling approaches and thus provides a way to investigate appropriate abstractions. This framework is available as a Python package, called StabilityFinder. We investigate genetic toggle switches and uncover the design principles behind making

bistable and tristable switches (all models used in this study are summarised in Table 1.) We find that both production and degradation rates of transcription factors are key parameters for bistability, and outline how the addition of positive autoregulation, combined with particular parameter combinations, can create multistable switching behaviour. Finally we demonstrate the ability of the framework to examine more complex systems and examine the design principles of a three gene switch. These examples demonstrate that StabilityFinder will be a valuable tool in the future design and construction of novel gene networks.

Methods

StabilityFinder is based on a statistical inference method that combines approximate Bayesian computation (ABC) with sequential Monte Carlo [46]. This simulation-based method uses an iterative process to arrive at a distribution of parameter values that can give rise to observed data or a desired system behaviour [44]. ABC methods are used for inferring the posterior distribution in cases where the likelihood is intractable or is too computationally expensive to evaluate. Instead of computing the likelihood, ABC methods simulate the data and then compare the simulated and observed data through a distance function [46]. Given the prior distribution $\pi(\theta)$ we can approximate the posterior distribution, $\pi(\theta|x) \propto f(x|\theta)\pi(\theta)$, where $f(x|\theta)$ is the likelihood of a parameter, θ, given the data, x. There are a number of different variations of the ABC algorithm depending on how the approximate posterior distribution is sampled.

The simplest ABC algorithm is the ABC rejection sampler [47]. In this method, parameters are sampled from the prior and data simulated through the data generating model. For each simulated data set, x^*, a distance from that of the desired behaviour is calculated, $\rho(x^*, y)$, and if greater than a threshold, ϵ, the sample is rejected, otherwise it is accepted.

Algorithm 1 ABC rejection algorithm to generate samples $\{\theta^i; i \leq N\}$ from $\pi(\theta|\rho(x, y) \leq \epsilon)$

1: Initialise $i = 0$
2: Sample a parameter vector θ^* from prior, $\theta \sim \pi(\theta^*)$
3: Simulate a dataset, x^*, from the the model, $x^* \sim f(x|\theta)$
4: Calculate the distance $d = \rho(x^*, y)$.
5: **if** $d \leq \epsilon$ **then**
6:      $\theta^i = \theta^{**}$. $i = i + 1$
7:      **if** $i \leq N$ **then** GoTo step 2
8:      **end if**
9: **end if**

Table 1 Summary of the models used in this study

	Model	Autoregulation (+ve)	Steady states	Reference
Deterministic	Gardner	None	2	[1]
	LU-CS	None	2	[38]
	LU-SP	Single	2	This study
	LU-DP	Double	2, 3, 4	[38]
	Three gene switch	Triple	6	This study
Stochastic	Gardner	None	2	[1]
	MA-CS	None	2, 3	This study
	MA-DP	Double	2, 3	This study

The main disadvantage of this method is that if the prior distribution is very different from the posterior, the acceptance rate is very low [46]. An alternative method is ABC Markov Chain Monte Carlo (MCMC) [48]. The disadvantage of this method is that if it gets stuck in an area of low probability it can be very slow to converge [49]. The method used here is based on sequential Monte Carlo, which avoids both issues faced by the rejection and MCMC methods. It propagates the prior through a series of intermediate distributions in order to arrive at an approximation of the posterior. The tolerance, ϵ, for the distance of the simulated data to the desired data is made smaller at each iteration. When ϵ is sufficiently small, the result will approximate the posterior distribution [46].

To investigate the multistable behaviour of systems, a number of extensions to existing approaches are required. For a given set of parameter values, sample points are taken across initial conditions using latin hypercube sampling [50], and the ensemble system simulated in time until steady state. The distance function in ABC is replaced by a distance on the desired stability of the simulated model. To do this we cluster the steady state coordinates using K-means clustering [51] and use the Gap statistic to determine the number of clusters [52]. At each iteration, the number of steady states is determined by the number of clusters in phase space. A particle is accepted only if the number of clusters present is within an acceptable distance from the threshold ϵ. The algorithm is summarised below.

This algorithm is available as a Python package, called StabilityFinder. The user provides an SBML model file [53, 54] and an input file that contains all the necessary information to run the algorithm, including the desired stability and the final tolerance, ϵ, for the distance from the desired behaviour necessary for the algorithm to terminate. The flow of execution is illustrated in Fig. 1. Since the algorithm is computationally intensive, all deterministic and stochastic simulations are performed using algorithms implemented on Graphics Processing Units

---

Algorithm 2 StabilityFinder algorithm

1: Initialise $t = 0$,
2: $i = 0$
3: **if** t $= 0$ **then**
4:    Sample particle from prior, $\theta^{**} \sim \pi(\theta)$
5: **else**
6:    Sample θ^* from the previous population $\{\theta_{t-1}^i\}$ with weights w_{t-1}.
7:    Perturb the particle, $\theta^{**} \sim K_t(\theta|\theta^*)$ where K_t is the perturbation kernel.
8: **end if**
9: Sample k initial conditions $\{x_0^k\}$ via latin hypercube sampling.
10: Simulate k datasets to steady state, $\{x^{*k}\}$, from the the model, $x^* \sim f(x|\theta, x_0)$
11: Apply clustering in phase space on $\{x^{*k}\}$
12: Calculate the distance $d = \rho(\{x^{*k}\}, y)$.
13: **if** $d \leq \epsilon_t$ **then**
14:    $\theta_t^i = \theta^{**}. i = i + 1$
15:    **if** $i \leq N$ **then** GoTo step 3
16:    **else**
17:        Calculate weight for each accepted θ_t^i
18:        $w_t^{(i)} = \begin{cases} 1, & \text{if } t = 1 \\ \frac{\pi(\theta_t^{(i)})}{\sum_{j=1}^N w_{t-1}^{(j)} K_t(\theta_{t-1}^{(j)}, \theta_t^{(i)})}, & \text{if } t \geq 1. \end{cases}$
19:        Normalise weights
20:        $t = t + 1$.
21:        **if** $t \leq N_t$ **then**
22:            GoTo step 3
23:        **end if**
24:    **end if**
25: **end if**

(GPUs), which are used for mutli-threaded computation [55]. The algorithm returns the final accepted particles and their associated weights, as well as the initial conditions sampled and the steady state values obtained. The

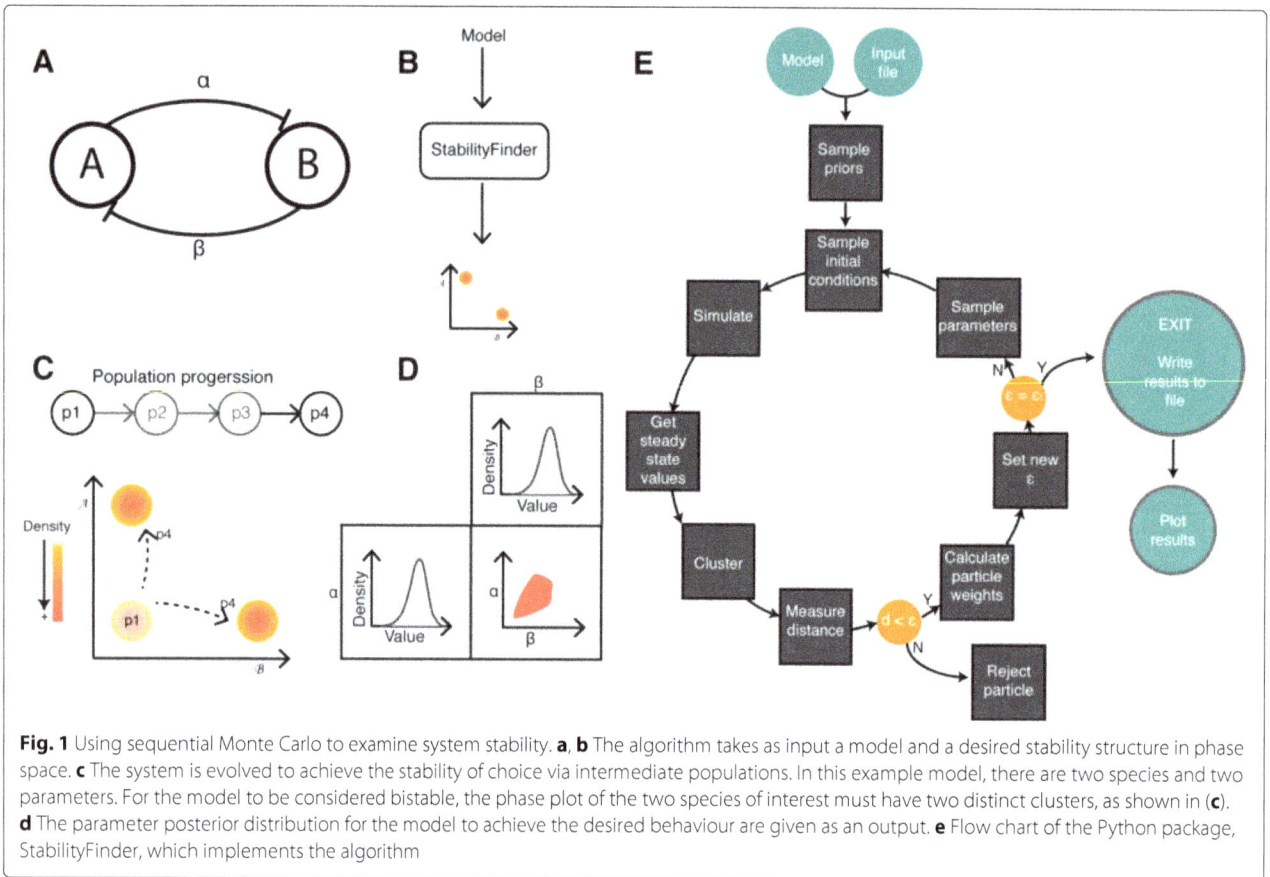

Fig. 1 Using sequential Monte Carlo to examine system stability. **a, b** The algorithm takes as input a model and a desired stability structure in phase space. **c** The system is evolved to achieve the stability of choice via intermediate populations. In this example model, there are two species and two parameters. For the model to be considered bistable, the phase plot of the two species of interest must have two distinct clusters, as shown in (**c**). **d** The parameter posterior distribution for the model to achieve the desired behaviour are given as an output. **e** Flow chart of the Python package, StabilityFinder, which implements the algorithm

final accepted particles can be used to study the characteristics of the posterior distribution. The sampled initial conditions and the resulting steady state values can be used to study the basins of attraction of the system.

Results and discussion

The Gardner switch under deterministic and stochastic dynamics

The first synthetic genetic toggle switch was constructed in *E. coli* by Gardner et al. and consisted of two mutually repressing transcription factors [1]. The model used to design and interpret the system is shown in Fig. 2a, and in the deterministic case is defined by the following ODEs

$$\frac{du}{dt} = \frac{\alpha_1}{1 + v^\beta} - u$$
$$\frac{dv}{dt} = \frac{\alpha_2}{1 + u^\gamma} - v,$$

where u is the concentration of repressor 1, v the concentration of repressor 2, α_1 and α_2 denote the effective rates of synthesis of repressors 1 and 2 respectively, and β and γ are the cooperativity of repression of promoter 1 and of repressor 2 respectively. Gardner et al. studied the deterministic case and concluded that there are two conditions for bistability for this model; that α_1 and α_2

are balanced and that β, γ >1 [1]. In order to test StabilityFinder we used it to find the posterior distribution for which this model exhibits bistable behaviour. We therefore set the desired behaviour to two (stable) steady states, and using a wide range of values as priors as shown in the Additional file 1, we used StabilityFinder to find the parameter values necessary for bistability to occur. The posterior distribution calculated by StabilityFinder for the Gardner deterministic case is shown in Fig. 2c. These results agree with the results reported by Gardner et al. [1]. For this switch to be bistable α_1 and α_2 must be balanced while β and γ must both be >1, as can be seen in the marginal distributions of β and γ in Fig. 2c.

We next applied StabilityFinder to the case of the Gardner switch under stochastic dynamics using the same priors as the deterministic case, and again searched the parameter space for bistable behaviour. The posterior is shown in Fig. 2d. We can see that the conditions on the parameters required for bistability in the deterministic case generally still stand in the stochastic case. There appears to be slightly looser requirements on the parameters of the stochastic model (wider marginal distributions), which is expected due to the nature of clustering deterministic steady states versus stochastic steady states. The Gap statistic is used in the case of the stochastic case,

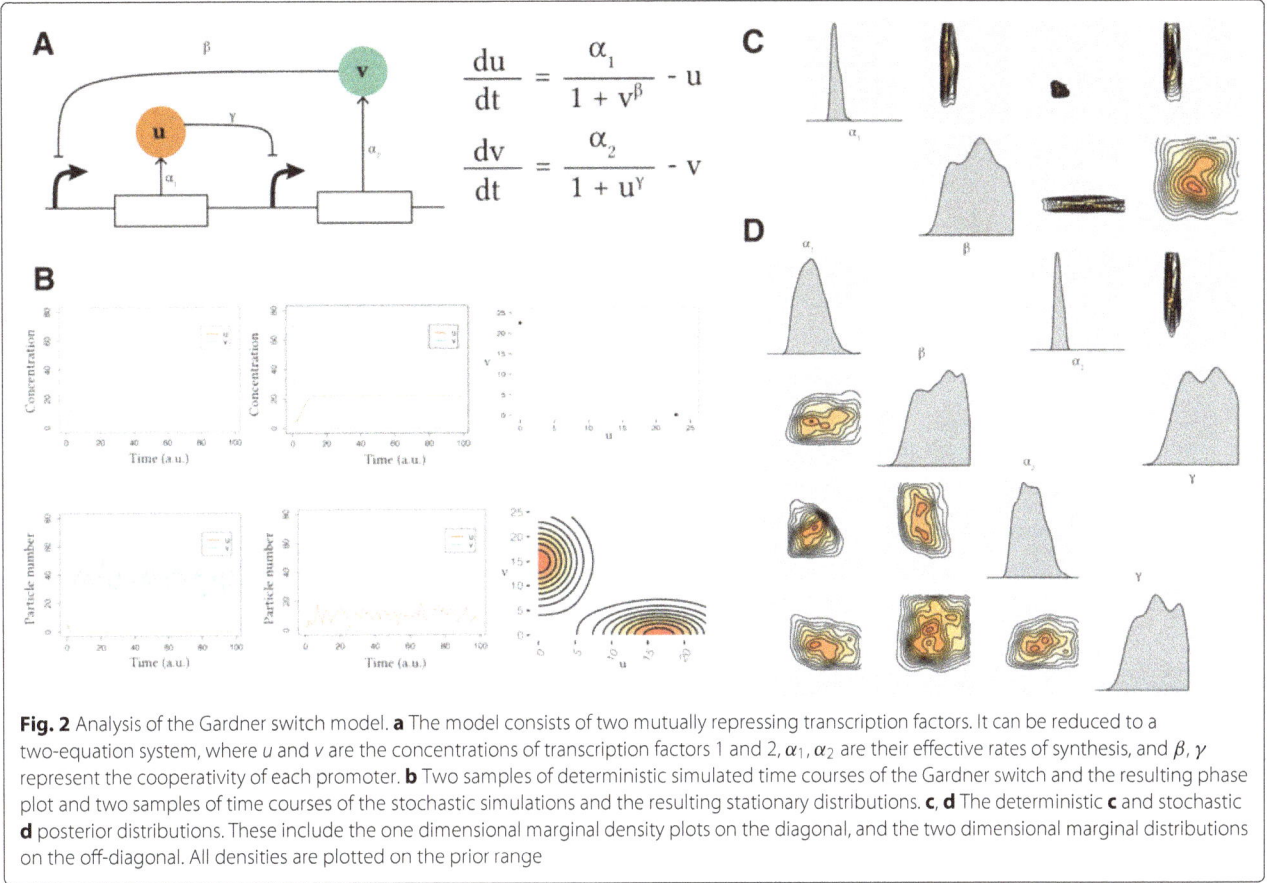

Fig. 2 Analysis of the Gardner switch model. **a** The model consists of two mutually repressing transcription factors. It can be reduced to a two-equation system, where u and v are the concentrations of transcription factors 1 and 2, α_1, α_2 are their effective rates of synthesis, and β, γ represent the cooperativity of each promoter. **b** Two samples of deterministic simulated time courses of the Gardner switch and the resulting phase plot and two samples of time courses of the stochastic simulations and the resulting stationary distributions. **c, d** The deterministic **c** and stochastic **d** posterior distributions. These include the one dimensional marginal density plots on the diagonal, and the two dimensional marginal distributions on the off-diagonal. All densities are plotted on the prior range

as it is capable of dealing with noisier data whereas a simpler and faster algorithm is used for clustering the deterministic solutions (see Additional file 1). These results demonstrate that StabilityFinder can be used to find the parameter values that produce a desired stability and allow us to confidently apply the methodology to more complex models.

Repressor degradation rates are key for achieving bistablity
We next analyzed an extension of the Gardner switch model previously studied by Lu et al. [38]. They considered two types of switches, the classic switch consisting of two mutually repressing transcription factors (model LU-CS), as well as a switch with double positive autoregulation (model LU-DP). The LU-CS switch was found to be bistable given the set of parameters used, while the LU-DP switch was found to be tristable [38]. The classical model used in their study is given by the following system of ODEs

$$\frac{dx}{dt} = g_x H^S_{xy}(y) - k_x x$$
$$\frac{dy}{dy} = g_y H^S_{yx}(x) - k_y y,$$

where

$$H^S_I(x) = H^-_I(x) + \lambda_I H^+_I(x)$$
$$H^-_I(x) = 1 / \left[1 + (x/x_I)^{n_I} \right]$$
$$H^+_I(x) = 1 - H^-_I(x),$$

and g_I represents the production rate, k_I the degradation rate, n_I the Hill coefficient, x_I the Hill threshold concentration and λ_I the fold change of the transcription rates, and $I \in \{xy, yx, xx, yy\}$ (see Additional file 1 for the details of all models used).

For the parameter values used in the Lu study, the classical switch exhibits three steady states (Fig. 3), two of which are stable and one is unstable. Using StabilityFinder with priors centred around the parameter values used in the original paper (see Additional file 1), we can identify the most important parameters for achieving bistability. The posterior distribution of these models are shown in Fig. 3a. We find that the parameters representing the rates of degradation of the transcription factors in the system (k_x, k_y) must both be large in relation to the prior range, and approximately equal, for bistability to occur. Protein degradation rates have been shown to be important for many system behaviours including oscillations [7, 56, 57]. We also find that the steady states of the LU-CS model

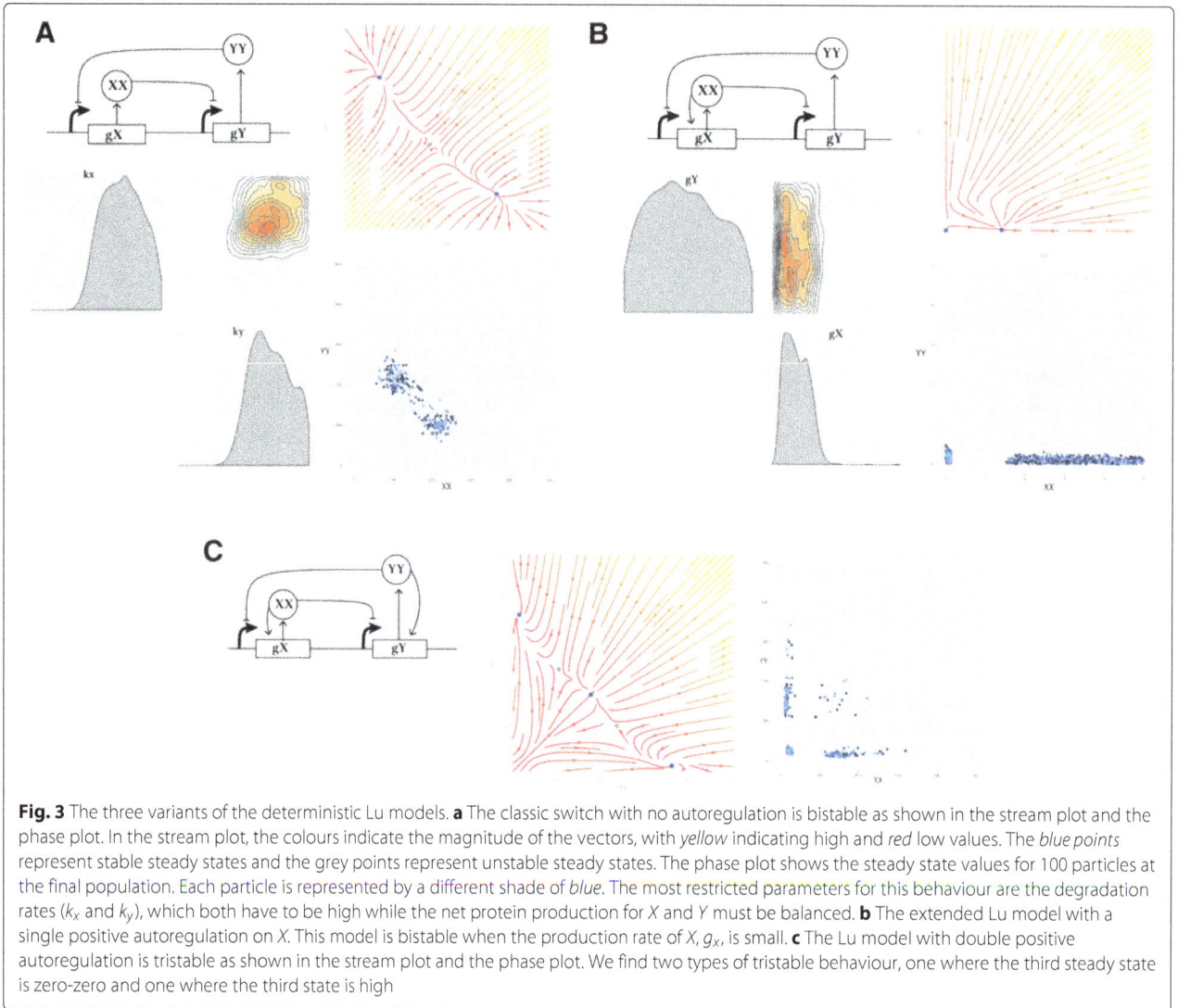

Fig. 3 The three variants of the deterministic Lu models. **a** The classic switch with no autoregulation is bistable as shown in the stream plot and the phase plot. In the stream plot, the colours indicate the magnitude of the vectors, with *yellow* indicating high and *red* low values. The *blue points* represent stable steady states and the grey points represent unstable steady states. The phase plot shows the steady state values for 100 particles at the final population. Each particle is represented by a different shade of *blue*. The most restricted parameters for this behaviour are the degradation rates (k_x and k_y), which both have to be high while the net protein production for X and Y must be balanced. **b** The extended Lu model with a single positive autoregulation on X. This model is bistable when the production rate of X, g_x, is small. **c** The Lu model with double positive autoregulation is tristable as shown in the stream plot and the phase plot. We find two types of tristable behaviour, one where the third steady state is zero-zero and one where the third state is high

are symmetric: the values for the dominant and repressed species are equivalent in both steady states.

The addition of symmetric and asymmetric positive autoregulation

It is known that the addition of positive autoregulation to the classical toggle switch can induce tristability [37, 38, 58]. Here we investigate the interplay of positive autoregulation on the values of the other parameters in the model. We extended the analysis presented in Lu et al. by including the switch with single positive autoregulation (model LU-SP), where a single positive autoregulatory feedback is present on one of the genes. This system topology has also been constructed previously [23, 59]. The advantage of using StabilityFinder over traditional bifurcation analysis is that the full parameter space is explored rather than solving the system for a single set of parameters. This allows us to deduce model properties that could

not otherwise be identified. Robustness to parameter fluctuations can be explored, as well as parameter correlations and restrictions on the values they can take while still producing the desired behaviour.

The LU-DP model is given by

$$\frac{dx}{dt} = g_x H_{xy}^S(y) H_{xx}^S(x) - k_x x$$

$$\frac{dy}{dt} = g_y H_{yx}^S(x) H_{yy}^S(y) - k_y y,$$

whereas the LU-SP switch is modelled using the following ODE system

$$\frac{dx}{dt} = g_x H_{xy}^S(y) H_{xx}^S(x) - k_x x$$

$$\frac{dy}{dt} = g_y H_{yx}^S(x) - k_y y.$$

We find that the switch with single positive autoregulation is capable of bistable behaviour as seen in Fig. 3b, but this is only possible when the strength of the promoter under positive autoregulation, g_x, is small. There appear to be no such constraints on the strength of the original, unmodified, promoter, g_y. We also find that unlike the LU-CS and LU-DP models, the steady states of the bistable LU-SP are not symmetric. The levels of Y are around zero and always lower than the levels of X. The levels of both are lower than those found in the LU-CS and LU-DP models.

Upon examination of the LU-DP model, we also find that tristability in the switch is relatively robust, as this phenotype is found across a large range of parameter values, with no parameters strongly constrained (see Additional file 1) but the two parameters for gene expression, g_x and g_y tend to be small compared to the priors. Two types of tristable behaviour are identified, one where the third steady state is at (0,0) and and one where the third steady state has non-zero values. This result agrees with previous work where it was found that a switch can exhibit two kinds of tristability, one in which the third steady state is high (III_H) and one in which it is low (III_L) [37].

Design principles for a switch capable of two, three and four steady states

The LU-DP switch is capable of both bistable and tristable behaviour as well as four coexisting states under deterministic dynamics (quadristability) [37]. It is of great interest to understand the conditions under which these three behaviours occur. We carried out a bifurcation analysis of the DP switch using the PyDSTool [60] in order to get an indication of the stabilities this model is capable of, and at which parameter ranges these are found (Fig. 4b). This shows that by varying the parameter for gene expression (g_x) while all other parameters remain constant we can obtain all three behaviours. We find that if $100 \leq g_x \leq 120$ the system exhibits four steady states, if $9 \leq g_x \leq 10$ the system is tristable and if $10 \leq g_x \leq 100$ the system is bistable (see Additional file 1).

Using StabilityFinder we obtained posterior distributions for the bistable, tristable and quadristable phenotypes (Fig. 4). Upon examination of the these distributions, we observe that a subset of the parameter values are different for the three behaviours, although the differences are surprisingly subtle. We find differences in the univariate distribution of the parameters for gene expression, g_x, as highlighted in Fig. 4c, box 1. This parameter must be small for four steady states to occur but there are no such restraints for a bistable or a tristable switch. Furthermore, parameter x_{xx} (the dissociation constant for autoregulation) must be small for tristable and quadristable behaviour to be achieved, but there are no such restraints for a bistable switch, as can be seen in Fig. 4c, box 2.

Additionally, we find a difference in the bivariate distributions in the posterior. Most notably, we find that parameters x_{xx} and g_x are tightly constrained in the tristable and quadristable cases, where both parameters are required to be small, but less so in the bistable case (Fig. 4c, box 3). Another notable difference is between parameters x_{xx} and n_{xx} shown in Fig. 4c, box 5, where they are constrained in the tristable and quadristable cases but not the bistable case. Interestingly, we also find parameter correlations conserved between the three behaviours, as seen in Fig. 4c, box 4, where parameters l_{xx} and g_x, (positive autoregulation and gene expression) are negatively correlated in both cases. This highlights the importance of treating unknown parameters as distributions rather than fixed values when studying complex system behaviour. These ensemble based methods are capable of uncovering not only the ranges and values required for certain behaviour, but also the correlations between parameters, which would be missed by optimisation methods.

Bistability and tristability using more realistic abstractions

In order to study the switch system in the most realistic way, we avoid using the quasi-steady state approximation (QSSA) that is often used in modelling the toggle switch. Under the assumption of mass action kinetics, the two-equation system becomes a system of 14 equations and 9 parameters in the classical switch case. These are shown in the Additional file 1. In the cellular environment stochastic effects can be non-negligible and should therefore be taken into account when trying to elucidate the behaviours that a system is capable of. Therefore, we model these switches using stochastic dynamics.

We find that under stochastic dynamics, both the simple switch, CS-MA, and positive autoregulation switch, DP-MA, are capable of bistable and tristable behaviour (Fig. 5). The fact that tristability can occur in the classical model is consistent with the effect of small molecule numbers; if gene expression remains low, it provides the opportunity for small number effects to be observed, and the third unstable steady state to stabilise [33]. To verify the robustness of the tristability found in the stochastic case, we re-sampled the posterior distributions, simulated to steady state and confirmed tristable behaviour. As can be seen in Fig. 5, differences in the parameter values are observed between the bistable and tristable switches, in both CS-MA and DP-MA models. The design principles for both the CS-MA model and the DP-MA model are summarised in Table 2.

Achieving higher multistability

To further demonstrate the flexibility of our framework we investigated a system capable of higher stabilities. Multistability is found in some differentiating pathways, such as the myeloid differentiation pathway [61, 62]. We allow

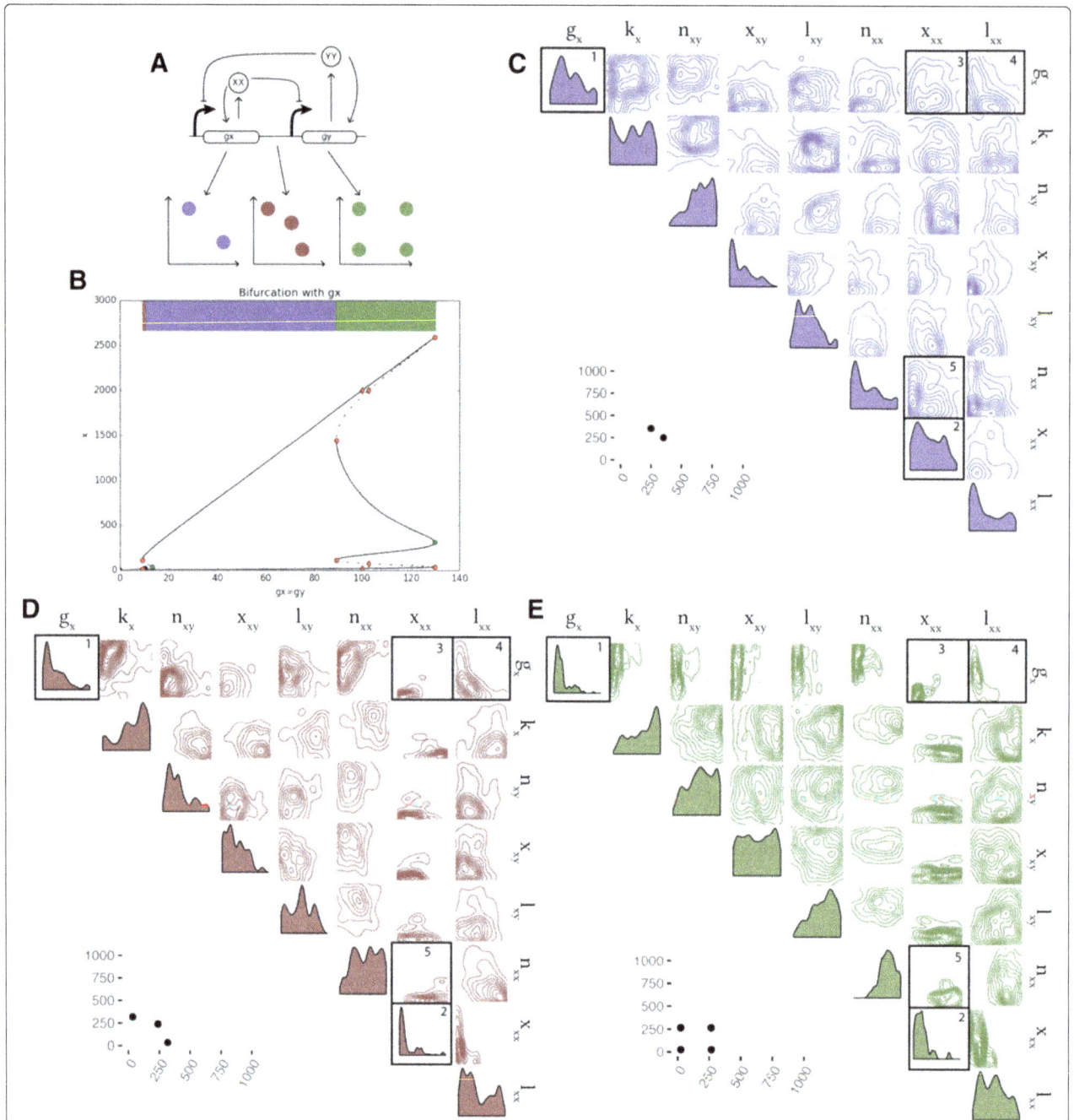

Fig. 4 Design principles of multistable switches. **a** Using the Lu model with added positive autoregulation we uncover the design principles dictating if a switch will be bistable, tristable, or quadristable. **b** An initial bifurcation analysis of the LU-DP switch uncovers the stabilities it is capable when varying the parameter for gene expression (while keeping all other parameters fixed). **c–e** By considering the bivariate distributions of the parameters we can uncover the differences in the parameters of a bistable switch compared to a tristable switch, compared to a quadristable switch. The posterior distribution of the bistable switch is shown in purple, the tristable switch in *red* and the quadristable switch in *green*, all plotted on the prior ranges. The bivariate distributions for which a difference is observed between the stabilities are in black boxes. An example of the phenotype (phase plot) from each switch is shown next to the corresponding posterior distribution. Parameter legend key: g_x production rates; k_x degradation rates; n_{xy} Hill coefficients; x_{xy} Hill threshold concentrations; l_{xy} transcription rate fold change; n_{xx} autoregulation Hill coefficients; x_{xx} autoregulation Hill threshold concentrations; l_{xx} autoregulation transcription rate fold change

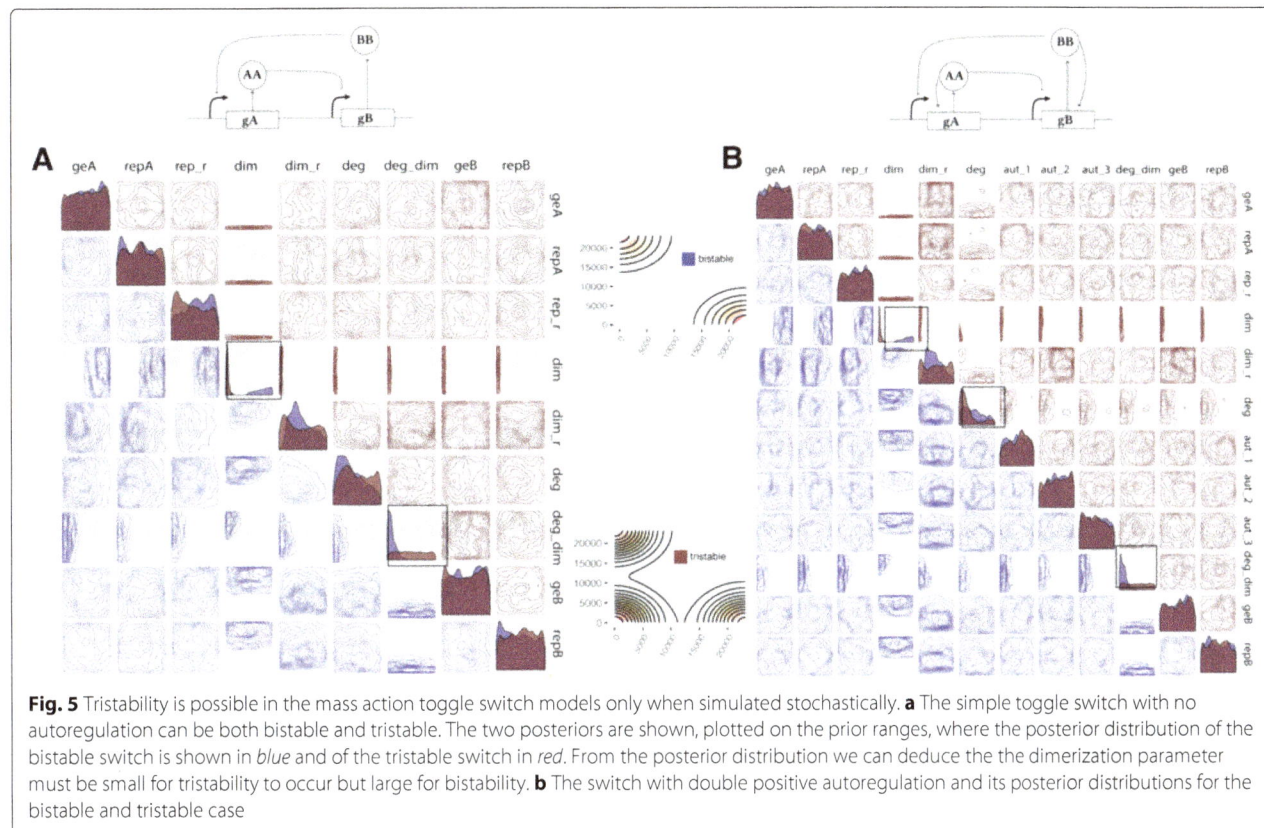

Fig. 5 Tristability is possible in the mass action toggle switch models only when simulated stochastically. **a** The simple toggle switch with no autoregulation can be both bistable and tristable. The two posteriors are shown, plotted on the prior ranges, where the posterior distribution of the bistable switch is shown in *blue* and of the tristable switch in *red*. From the posterior distribution we can deduce the the dimerization parameter must be small for tristability to occur but large for bistability. **b** The switch with double positive autoregulation and its posterior distributions for the bistable and tristable case

for these more complex dynamics by extending the LU-DP model by adding another gene, making it a three gene switch. This new system is depicted in Fig. 6a. In StabilityFinder we look for six steady states, the output being in nodes X and Y and use the priors shown in the Additional file 1. We successfully find that the system is capable of six steady states, as shown in Fig. 6c. Consistent with the LU-DP switch capable of 2, 3 and 4 steady states, we find that the steady states are symmetric (Fig. 6c). Each of the six steady states exists in symmetry with another one, in tightly constrained regions. We find that the most constrained parameters for this behaviour are again the degradation rate of the proteins, k_x. If they are too large or too small the system will not exhibit hexastability. Additionally we find that the Hill coefficients for the repressors, n_{xy}, are constrained to be smaller than 1.5

as seen in Fig. 6d. This example demonstrates that StabilityFinder can be used to elucidate the dynamics of more complex network architectures, which will be key to the successful design and construction of novel gene networks as synthetic biology advances.

Conclusions

We have developed an algorithm that can identify the parameter regions necessary for a model to achieve a given number of stable steady states. The novelty in our framework over existing methodology is that complex models can be analyzed assuming both deterministic and stochastic dynamics. We have shown that the algorithm can be used to infer the parameter ranges that give rise to specific behaviour in various models. We uncovered the design principles that make a bistable, a tristable

Table 2 Design principles of bistable and tristable switches

	CS-MA		DP-MA	
	Bistable	*Tristable*	*Bistable*	*Tristable*
Dimerisation	High	Low	High	Low
Protein degradation	-	-	-	Low
Dimer degradation	Low	-	Low	-

Differences are observed between the parameter values of the bistable and tristable CS-MA and DP-MA switches

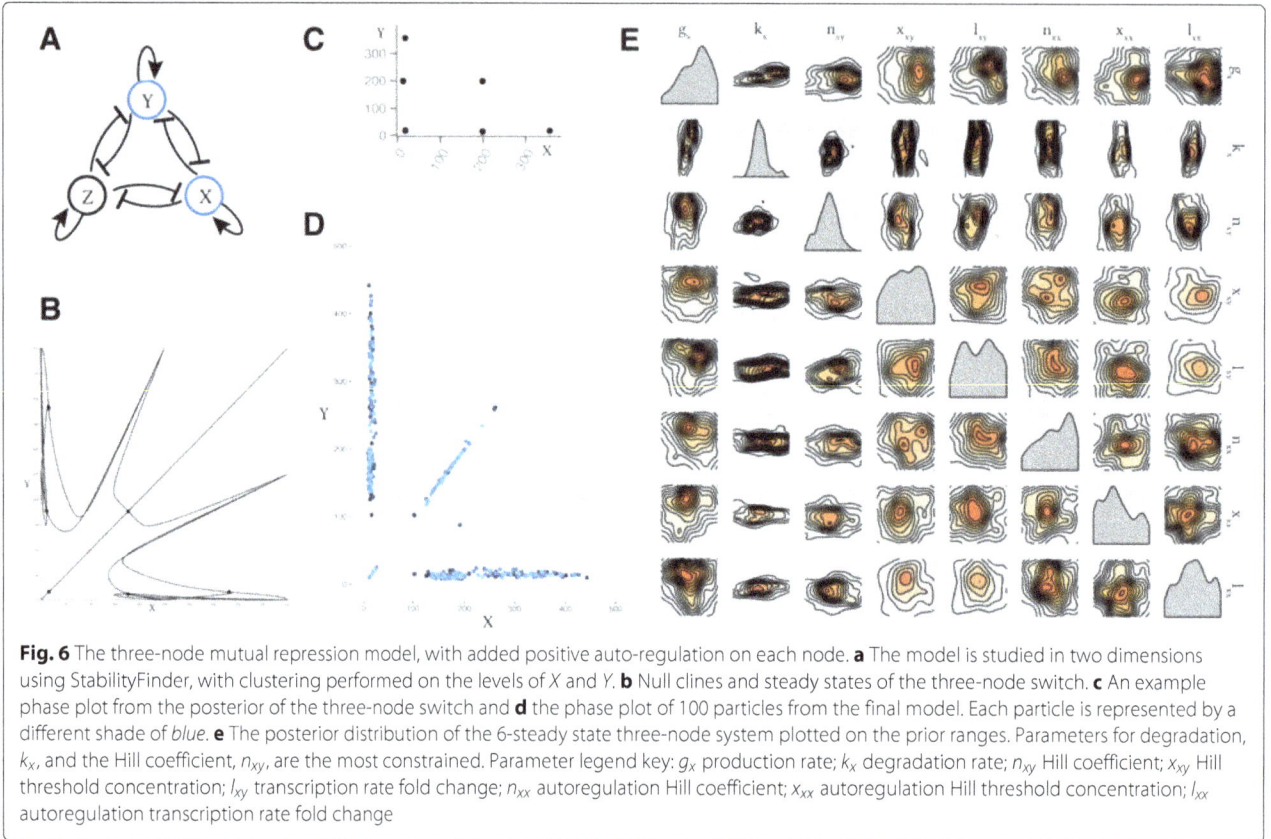

Fig. 6 The three-node mutual repression model, with added positive auto-regulation on each node. **a** The model is studied in two dimensions using StabilityFinder, with clustering performed on the levels of X and Y. **b** Null clines and steady states of the three-node switch. **c** An example phase plot from the posterior of the three-node switch and **d** the phase plot of 100 particles from the final model. Each particle is represented by a different shade of *blue*. **e** The posterior distribution of the 6-steady state three-node system plotted on the prior ranges. Parameters for degradation, k_x, and the Hill coefficient, n_{xy}, are the most constrained. Parameter legend key: g_x production rate; k_x degradation rate; n_{xy} Hill coefficient; x_{xy} Hill threshold concentration; l_{xy} transcription rate fold change; n_{xx} autoregulation Hill coefficient; x_{xx} autoregulation Hill threshold concentration; l_{xx} autoregulation transcription rate fold change

and a quadristable switch. We also found that a three-node switch is capable of hexastability. Importantly, we removed assumptions made to simplify the switch models and showed that they are still capable of bistable and tristable behaviour.

Although we only examined models containing combined transcription and translation, our approach could be applied to any models of switching behaviour, including more detailed kinetic models and more complex multistable switches that exist in natural biological systems, such as developmental pathways. We also limited our framework to the objective behaviour of a given number of stable steady states. However, this approach is extremely flexible, and could be extended to find systems with a given switching rate, or systems robust to a particular set of perturbations, both of which could be of great importance for building more complex genetic circuits.

One limitation of our approach is that we cannot rule out a specific behaviour; it is always possible that some part of parameter space remains unexplored, or because the search space must be predefined, interesting regions are not included in the search. In the Bayesian sense, this predefined space is the prior distribution for the parameters that give rise to the stability under investigation. In principle, once our knowledge of these biochemical rate constants grows, we can incorporate these data into the

prior regions for exploration. Another limitation is that of scalability. Our framework can currently be applied to small and medium size gene networks since the computational time is exponential in size, whereas optimization methods are more scalable [39–41]. This is a manifestation of a general tradeoff between finding an optimal value and exploring a parameter space. However, we argue that for current and relevant problems in synthetic biology, this computational burden is acceptable.

Approaches based on parameter space exploration are indispensable tools for providing understanding of general system properties and guiding more detailed experimental and theoretical studies. They will also be key for the design and construction of synthetic gene networks. By selecting standardized parts accordingly, such as promoters, RBS sequences and other untranslated regions [18, 63–65], in vivo systems can be matched to parameter regions with a high probability of function.

More generally our results highlight that changing the level of abstraction, in addition to the modification of the feedback structure and parameter values, can significantly alter the qualitative behaviour of a system model. These results advocate the need for a programme of experimental work, combined with systems modelling, to understand the rules of thumb for abstraction in model based design of synthetic biological systems.

Acknowledgements
All the authors would like to acknowledge that the work presented here made use of the Emerald High Performance Computing facility made available by the Centre for Innovation. The Centre is formed by the universities of Oxford, Southampton, Bristol, and University College London in partnership with the STFC Rutherford-Appleton Laboratory.

Funding
CPB and MLW acknowledge funding from the Wellcome Trust through a Research Career Development Fellowship (097319/Z/11/Z). ML and AJHF acknowledge funding through the UCL Impact Award scheme and the UCL CoMPLEX doctoral training centre respectively.

Authors' contributions
The project was conceived by CPB. ML developed the StabilityFinder code, developed the switch models, ran the Monte Carlo analyses and interpreted the results. MLW, AJHF and CPB performed additional analysis of the results. ML and CPB wrote the manuscript. All authors read and approved the final manuscript.

Competing interests
The authors declare that they have no competing interests.

References

1. Gardner TS, Cantor CR, Collins JJ. Construction of a genetic toggle switch in Escherichia coli. Nature. 2000;403(6767):339–42.
2. Kramer BP, Viretta AU, Daoud-El-Baba M, Aubel D, Weber W, Fussenegger M. An engineered epigenetic transgene switch in mammalian cells. Nat Biotechnol. 2004;22(7):867–70.
3. Isaacs FJ, Hasty J, Cantor CR, Collins JJ. Proc Nat Acad Sci USA. 2003;100(13):7714–9.
4. Ham TS, Lee SK, Keasling JD, Arkin AP. Design and Construction of a Double Inversion Recombination Switch for Heritable Sequential Genetic Memory. PLoS ONE. 2008;3(7):2815.
5. Deans TL, Cantor CR, Collins JJ. A Tunable Genetic Switch Based on RNAi and Repressor Proteins for Regulating Gene Expression in Mammalian Cells. Cell. 2007;130(2):363–72.
6. Friedland AE, Lu TK, Wang X, Shi D, Church G, Collins JJ. Synthetic gene networks that count. Science. 2009;324(5931):1199–202.
7. Stricker J, Cookson S, Bennett MR, Mather WH, Tsimring LS, Hasty J. A fast, robust and tunable synthetic gene oscillator. Nature. 2008;456(7221):516–9.
8. Fung E, Wong WW, Suen JK, Bulter T, Lee SG, Liao JC. A synthetic gene–metabolic oscillator. Nature. 2005;435(7038):118–22.
9. Tigges M, Marquez-Lago TT, Stelling J, Fussenegger M. A tunable synthetic mammalian oscillator. Nature. 2009;457(7227):309–12.
10. Basu S, Mehreja R, Thiberge S, Chen MT, Weiss R. Proc Nat Acad Sci USA. 2004;101(17):6355–360.
11. Nielsen AA, Der BS, Shin J, Vaidyanathan P, Paralanov V, Strychalski EA, Ross D, Densmore D, Voigt CA. Genetic circuit design automation. Science. 2016;352(6281):aac7341. doi:10.1126/science.aac7341.
12. Lu TK, Khalil AS, Collins JJ. Next-generation synthetic gene networks. Nat Biotechnol. 2009;27(12):1139–50.
13. Cardinale S, Arkin AP. Contextualizing context for synthetic biology Ũ identifying causes of failure of synthetic biological systems. Biotechnol J.2012. doi:10.1002/biot.201200085.
14. Del Vecchio D. Modularity, context-dependence, and insulation in engineered biological circuits. Trends Biotechnol. 2015;33(2):111–9.
15. Ceroni F, Algar R, Stan GB, Ellis T. Quantifying cellular capacity identifies gene expression designs with reduced burden. Nat Methods. 2015;12(5):415–8. doi:10.1038/nmeth.3339.
16. Shetty RP, Endy D, Knight TF. Engineering BioBrick vectors from BioBrick parts. J Biol Eng. 2008;2:5–5.
17. Galdzicki M, Rodriguez C, Chandran D, Sauro HM, Gennari JH. Standard biological parts knowledgebase. PLoS ONE. 2011;6(2):17005.
18. Mutalik VK, Guimaraes JC, Cambray G, Lam C, Christoffersen MJJ, Mai Q-AA, Tran AB, Paull M, Keasling JD, Arkin AP, Endy D. Precise and reliable gene expression via standard transcription and translation initiation elements. Nat Methods. 2013;10(4):354–60. doi:10.1038/nmeth.2404.
19. Nielsen AA, Segall-Shapiro TH, Voigt CA. Advances in genetic circuit design: novel biochemistries, deep part mining, and precision gene expression. Curr Opinion Chem Biol. 2013;17(6):878–92. doi:10.1016/j.cbpa.2013.10.003.
20. Beal J, Weiss R, Densmore D, Adler A, Appleton E, Babb J, Bhatia S, Davidsohn N, Haddock T, Loyall J, Schantz R, Vasilev V, Yaman F. An end-to-end workflow for engineering of biological networks from high-level specifications. ACS Synthetic Biol. 2012;1(8):317–31. doi:10.1021/sb300030d.
21. Yaman F, Bhatia S, Adler A, Densmore D, Beal J. Automated selection of synthetic biology parts for genetic regulatory networks. ACS Synthetic Biol. 2012;1(8):332–44.
22. Atkinson MR, Savageau MA, Myers JT, Ninfa AJ. Development of genetic circuitry exhibiting toggle switch or oscillatory behavior in Escherichia coli. Cell. 2003;113(5):597–607.
23. Lou C, Liu X, Ni M, Huang Y, Huang Q, Huang L, Jiang L, Lu D, Wang M, Liu C, Chen D, Chen C, Chen X, Yang L, Ma H, Chen J, Ouyang Q. Synthesizing a novel genetic sequential logic circuit: a push-on push-off switch. Mol Syst Biol. 2010;6:. doi:10.1038/msb.2010.2.
24. Litcofsky KD, Afeyan RB, Krom RJ, Khalil AS, Collins JJ. Iterative plug-and-play methodology for constructing and modifying synthetic gene networks. Nat Methods. 2012;9(11):1077–80.
25. Veening JW, Smits WK, Kuipers OP. Bistability, epigenetics, and bet-hedging in bacteria. Microbiology. 2008;62:193–210.
26. Ellis T, Wang X, Collins JJ. Diversity-based, model-guided construction of synthetic gene networks with predicted functions. Nat Biotechnol. 2009;27(5):465–71.
27. Kobayashi H, Kaern M, Araki M, Chung K, Gardner TS, Cantor CR, Collins JJ. Programmable cells: interfacing natural and engineered gene networks. Proc Nat Acad Sci USA. 2004;101(22):8414–419.
28. Cherry JL, Adler FR. How to make a biological switch,. J Theor Biol. 2000;203(2):117–33.
29. Warren PB, ten Wolde PR. Enhancement of the Stability of Genetic Switches by Overlapping Upstream Regulatory Domains. Phys Rev Lett. 2004;92(12):128101.
30. Walczak AM, Onuchic JN, Wolynes PG. Absolute rate theories of epigenetic stability. Proc Nat Acad Sci USA. 2005;102(52):18926–31.
31. Warren PB, ten Wolde PR. Chemical models of genetic toggle switches. J Phys Chem B. 2005;109(14):6812–23.
32. Lipshtat A, Loinger A, Balaban NQ, Biham O. Genetic toggle switch without cooperative binding. Phys Rev Lett. 2006;96(18):188101.
33. Ma R, Wang J, Hou Z, Liu H. Small-number effects: a third stable state in a genetic bistable toggle switch. Phys Rev Lett. 2012;109(24):248107.
34. Biancalani T, Assaf M. Genetic Toggle Switch in the Absence of Cooperative Binding: Exact Results. Phys Rev Lett. 2015;115:208101.
35. Loinger A, Lipshtat A, Balaban NQ, Biham O. Stochastic simulations of genetic switch systems. Phys Rev E Stat Nonlin Soft Matter Phys. 2007;75(2 Pt 1):021904.
36. Pedersen MG, Bersani AM, Bersani E. Quasi steady-state approximations in complex intracellular signal transduction networks – a word of caution. J Math Chem. 2007;43(4):1318–44.
37. Guantes R, Poyatos JF. Multistable decision switches for flexible control of epigenetic differentiation. PLoS Comput Biol. 2008;4(11):1000235.
38. Lu M, Onuchic J, Ben-Jacob E. Construction of an Effective Landscape for Multistate Genetic Switches. Phys Rev Lett. 2014;113(7):078102.
39. Dasika MS, Maranas CD. Optcircuit: an optimization based method for computational design of genetic circuits. BMC Syst Biol. 2008;2(1):1.
40. Otero-Muras I, Banga JR. Multicriteria global optimization for biocircuit design. BMC Syst Biol. 2014;8:113.

41. Otero-Muras I, Banga JR. Exploring design principles of gene regulatory networks via pareto optimality. IFAC-PapersOnLine. 2016;49(7):809–14. doi:10.1016/j.ifacol.2016.07.289. 11th {IFAC} Symposium on Dynamics and Control of Process SystemsIncluding Biosystems DYCOPS-CAB 2016Trondheim, Norway, 6–8 June 2016.

42. Rodrigo G, Carrera J, Jaramillo A. Computational design of synthetic regulatory networks from a genetic library to characterize the designability of dynamical behaviors. Nucleic Acids Res. 2011;39(20): 138–8.

43. Baetica AA, Yuan Y, Gonçalves JM, Murray RM. A stochastic framework for the design of transient and steady state behavior of biochemical reaction networks. In: 54th IEEE Conference on Decision and Control, CDC 2015, Osaka, Japan, December 15–18, 2015; 2015. p. 3199–205. doi:10.1109/CDC.2015.7402699. http://dx.doi.org/10.1109/CDC.2015.7402699.

44. Barnes CP, Silk D, Sheng X, Stumpf MPH. Proc Nat Acad Sci USA. 2011;108(37):15190–15195.

45. Del Moral P, Doucet A, Jasra A. Sequential monte carlo samplers. J R Stat Soc Series B (Statistical Methodology). 2006;68(3):411–36.

46. Toni T, Welch D, Strelkowa N, Ipsen A, Stumpf MPH. Approximate Bayesian computation scheme for parameter inference and model selection in dynamical systems. J R Soc Interface / R Soc. 2009;6(31): 187–202.

47. Pritchard JK, Seielstad MT, Perez-Lezaun A, Feldman MW. Population growth of human Y chromosomes: A study of Y chromosome microsatellites. Mol Biol Evol. 1999;16(12):1791–8.

48. Marjoram P, Molitor J, Plagnol V, Tavare S. Markov chain Monte Carlo without likelihoods. Proc Nat Acad Sci USA. 2003;100(26):15324–8.

49. Sisson SA, Fan Y, Tanaka MM. Sequential Monte Carlo without likelihoods. Proc Nat Acad Sci USA. 2007;104(6):1760–1765.

50. McKay MD, Beckman RJ, Conover WJ. A comparison of three methods for selecting values of input variables in the analysis of output from a computer code. Technometrics. 2000;42(1):55–61.

51. Lloyd SP. Least squares quantization in PCM. IEEE Trans Inf Theory. 1982;28(2):129–137.

52. Tibshirani R, Walther G, Hastie T. Estimating the number of clusters in a data set via the gap statistic. J R Stat Soc B. 2001;63:411–23.

53. Hucka M, Finney A, Sauro HM, Bolouri H, Doyle JC, Kitano H, Arkin AP, Bornstein BJ, Bray D, Cornish-Bowden A, Cuellar AA, Dronov S, Gilles ED, Ginkel M, Gor V, Goryanin II, Hedley WJ, Hodgman TC, Hofmeyr JH, Hunter PJ, Juty NS, Kasberger JL, Kremling A, Kummer U, Le Novère N, Loew LM, Lucio D, Mendes P, Minch E, Mjolsness ED, Nakayama Y, Nelson MR, Nielsen PF, Sakurada T, Schaff JC, Shapiro BE, Shimizu TS, Spence HD, Stelling J, Takahashi K, Tomita M, Wagner J, Wang J. SBML Forum. Bioinformatics. 2003;19(4):524–31.

54. Bornstein BJ, Keating SM, Jouraku A, Hucka M. LibSBML: An API Library for SBML. Bioinformatics. 2008;24(6):880–1. doi:10.1093/bioinformatics/btn051.

55. Kirk DB, Hwu W-mW. Programming Massively Parallel Processors. A Hands-on Approach. Burlington: Morgan Kaufmann; 2010.

56. Wong WW, Tsai TY, Liao JC. Single-cell zeroth-order protein degradation enhances the robustness of synthetic oscillator. Mol Syst Biol. 2007; 3:130.

57. Woods ML, Leon M, Perez-Carrasco R, Barnes CP. A Statistical Approach Reveals Designs for the Most Robust Stochastic Gene Oscillators. ACS Synthetic Biol. 2016;5(6):459–70.

58. Lu M, Jolly MK, Gomoto R, Huang B, Onuchic J, Ben-Jacob E. Tristability in cancer-associated microRNA-TF chimera toggle switch. J Phys Chem B. 2013;117(42):13164–74.

59. Huang D, Holtz WJ, Maharbiz MM. A genetic bistable switch utilizing nonlinear protein degradation. J Biol Eng. 2012;6(1):1–13. doi:10.1186/1754-1611-6-9.

60. Clewley R. Hybrid models and biological model reduction with PyDSTool. PLoS Comput Biol. 2012;8(8):1002628.

61. Ghaffarizadeh A, Flann NS, Podgorski GJ. Multistable switches and their role in cellular differentiation networks. BMC Bioinformatics. 2014;15 Suppl 7:7.

62. Cinquin O, Demongeot J. High-dimensional switches and the modelling of cellular differentiation. J Theor Biol. 2005;233(3):391–411.

63. Canton B, Labno A, Endy D. Refinement and standardization of synthetic biological parts and devices. Nat Biotechnol. 2008;26(7):787–93.

64. Kelly JR, Rubin AJ, Davis JH, Ajo-Franklin CM, Cumbers J, Czar MJ, de Mora K, Glieberman AL, Monie DD, Endy D. Measuring the activity of BioBrick promoters using an in vivo reference standard. J Biol Eng. 2009;3(1):4.

65. Salis HM, Mirsky EA, Voigt CA. Automated design of synthetic ribosome binding sites to control protein expression. Nat Biotechnol. 2009;27(10): 946–50.

5

Predicting protein-binding regions in RNA using nucleotide profiles and compositions

Daesik Choi[†], Byungkyu Park[†], Hanju Chae, Wook Lee and Kyungsook Han[*]

Abstract

Background: Motivated by the increased amount of data on protein-RNA interactions and the availability of complete genome sequences of several organisms, many computational methods have been proposed to predict binding sites in protein-RNA interactions. However, most computational methods are limited to finding RNA-binding sites in proteins instead of protein-binding sites in RNAs. Predicting protein-binding sites in RNA is more challenging than predicting RNA-binding sites in proteins. Recent computational methods for finding protein-binding sites in RNAs have several drawbacks for practical use.

Results: We developed a new support vector machine (SVM) model for predicting protein-binding regions in mRNA sequences. The model uses sequence profiles constructed from log-odds scores of mono- and di-nucleotides and nucleotide compositions. The model was evaluated by standard 10-fold cross validation, leave-one-protein-out (LOPO) cross validation and independent testing. Since actual mRNA sequences have more non-binding regions than protein-binding regions, we tested the model on several datasets with different ratios of protein-binding regions to non-binding regions. The best performance of the model was obtained in a balanced dataset of positive and negative instances. 10-fold cross validation with a balanced dataset achieved a sensitivity of 91.6%, a specificity of 92.4%, an accuracy of 92.0%, a positive predictive value (PPV) of 91.7%, a negative predictive value (NPV) of 92.3% and a Matthews correlation coefficient (MCC) of 0.840. LOPO cross validation showed a lower performance than the 10-fold cross validation, but the performance remains high (87.6% accuracy and 0.752 MCC). In testing the model on independent datasets, it achieved an accuracy of 82.2% and an MCC of 0.656. Testing of our model and other state-of-the-art methods on a same dataset showed that our model is better than the others.

Conclusions: Sequence profiles of log-odds scores of mono- and di-nucleotides were much more powerful features than nucleotide compositions in finding protein-binding regions in RNA sequences. But, a slight performance gain was obtained when using the sequence profiles along with nucleotide compositions. These are preliminary results of ongoing research, but demonstrate the potential of our approach as a powerful predictor of protein-binding regions in RNA. The program and supporting data are available at http://bclab.inha.ac.kr/RBPbinding.

Keywords: Protein-binding region, RNA-protein interaction, Prediction method

*Correspondence: khan@inha.ac.kr
[†]Equal Contributors
Department of Computer Science and Engineering, Inha University, 22212 Incheon, South Korea

Background

Interactions between protein and RNA molecules are essential to various cellular processes, such as post transcriptional gene regulation, translation, and alternative splicing [1]. Many studies have been conducted to identify RNA-binding proteins (RBPs) or binding sites in protein and RNA molecules. In particular, recent advances in high-throughput experimental technologies, including next-generation sequencing technologies and cross-linking and immunoprecipitation (CLIP), have accelerated the discovery of RBPs and their target RNAs. Despite the increased number of known RBPs and their target RNAs, the mechanism of protein-RNA interactions is not fully uncovered and a large number of RBPs and their target RNAs remain to be uncovered. For example, for the $\sim 20,500$ protein-coding genes in humans, only 1,542 RBPs (7.5%) and their target RNAs have been identified so far [2].

As a complement to experimental methods, several computational methods have been proposed, which are largely motivated by the increased amount of data on protein-RNA interactions and the availability of complete genome sequences of several organisms. Computational methods in general are much less time-consuming and costly than experimental methods.

Most existing computational methods are primarily limited to finding RNA-binding sites in proteins instead of protein-binding sites in RNAs. For instance, BindN+ [3], an upgraded version of BindN [4], uses a support vector machine (SVM) to predict the RNA- or DNA-binding residues from biochemical features and evolutionary information of protein sequences. RNABindRPlus [5] also predicts RNA-binding residues in a protein sequence by combining predictions from an optimized SVM and those from a sequence homology method. aaRNA [6] predicts RNA binding residues in protein using sequence- and structure-based features.

Compared to the task of predicting RNA-binding sites in proteins, predicting protein-binding sites in RNA is more challenging for several reasons [7]. Until very recently, there were few computational methods that can predict protein-binding sites in RNA. catRAPID estimates the binding propensity of RNA and protein molecules by combining secondary structure, hydrogen bonding and van der Waals contributions [8]. It often predicts an entire RNA sequence as a binding site even for an RNA sequence of 50 or more nucleotides. DeepBind [9] is known to out-perform state-of-the-art experimental and computational methods. It uses deep convolutional neural networks, trained on a huge amount of data from high-throughput experiments. For the problem of predicting RBP-binding sites in RNA sequences, DeepBind was trained on data from RNAcompete, CLIP-seq and RIP-seq [10]. It contains ~ 200 distinct models, each for different RBPs, so

the user should try all of them in the absence of prior information on RBP. As output, it only provides a predictive binding score without protein-binding sites in the input RNA sequence. A new prediction model called PRIdictor [11, 12] predicts binding sites in RNA and protein sequences at the nucleotide and residue level. Wong et al. [13] developed a method that predicts interacting nucleotides and residues between DNA and proteins.

In this paper, we propose a new method for predicting protein-binding regions in mRNA, which are associated with post-transcriptional regulation of gene expression. The method uses sequence profiles constructed from log-odds scores of mono- and di-nucleotides and sequence compositions of mono-, di- and tri-nucleotides. As shown in the paper, the proposed method showed a high performance in testing on a large number of human RNA sequences and was substantially better than other methods. The rest of the paper presents the details of our approach and its experimental results.

Methods
Datasets

We obtained protein-binding sites in RNAs from CLIPdb [14], which provides curated published CLIP-seq data sets for four species (human, mouse, worm, and yeast). To obtain a sufficient amount of reliable data, we restricted the data to those binding regions of 25 nucleotides in '+' strands of human mRNAs, which were identified by PAR-CLIP technology [15] and have the binding affinity score > 0.9 in PARalyzer [16]. Human mRNAs were selected against others because the largest amount of RBP binding sites is known in human mRNAs. Different RBPs are known to have different binding preferences within an mRNA. We examined the type of RBP binding regions in the extracted human mRNAs by mapping the Ensembl transcripts to the GRCh37 assembly. Coding sequence (CDS) regions of mRNA are the most frequent binding regions of RBPs, followed by 3' UTR (Additional file 1).

The reason for selecting 25 nucleotides as the size of a binding region is because protein-binding regions identified by PAR-CLIP are typically between 21 and 35 nucleotides in length, and binding regions of 25 nucleotides resulted in the larger amount of data from CLIPdb than other choices for the size (see Additional file 2 for the distribution of the length of RBP-binding regions). After extracting a total of 5,145 RBP-binding regions for 14 RBPs, we assembled RNA sequences using the reference human genome GRCh37/hg19. These RNA sequences were used as positive data in our study (Additional file 3). RBP sequences were obtained from NCBI GEO (http://www.ncbi.nlm.nih.gov/geo/).

For negative data, we selected 51,450 (10-fold of the positive data) non-binding regions of 25 nucleotides in the same reference human genome GRCh37/hg19.

The human genome contains more non-binding regions than protein-binding regions, so we constructed several datasets with different ratios of binding to non-binding regions (called 1:1, 1:2, 1:4, 1:6, 1:8 and 1:10 datasets hereafter).

In order to remove redundancy in the datasets, we first executed CD-HIT-EST [17] on each of the six datasets (1:1, 1:2, 1:4, 1:6, 1:8 and 1:10 datasets) and removed those with a sequence similarity of 80% or higher. After removing similar sequences, 4372 sequences out of the 5,145 RBP-binding sequences were left. The remaining 4372 RBP-binding sequences were partitioned into two datasets: training dataset (70% of the remaining RBP-binding sequences) and test dataset (30%). Thus, there are no similar RNA sequences between training and test datasets and within training or test datasets. Table 1 shows the number of sequences in the training and test datasets with different ratios of positive to negative instances. Since the redundancy removal was enforced separately in the 1:1, 1:2, 1:4, 1:6, 1:8 and 1:10 datasets, the ratio of positive to negative instances may not be exactly $1 : n$ ($n = 1, 2, 4, 6, 8, 10$) (see Additional files 4 and 5).

Nucleotide profiles and compositions

We constructed position weight matrices (PWMs) of two types: (1) mono-nucleotide position weight matrix (mPWM) and (2) di-nucleotide position weight matrix (dPWM). $mPWM(i, j)$ represents the log-odds score of the i-th nucleotide ($i = 1, 2, 3, 4$) in the j-th position ($j = 1, 2, \ldots$, sequence length n), which is defined by Eq. 1. Likewise, $dPWM(di, j)$ represents the log-odds score of the di-th di-nucleotide ($di = 1, 2, \ldots, 16$) in the j-th position ($j = 1, 2, \ldots, n - 1$), defined by Eq. 2.

$$mPWM(i, j) = \ln \left(\frac{frequency^+(i, j)}{frequency^-(i, j)} \right) \quad (1)$$

$$dPWM(di, j) = \ln \left(\frac{frequency^+(di, j)}{frequency^-(di, j)} \right) \quad (2)$$

The PWM of mono-nucleotides, also known as position specific score matrix (PSSM) or sequence profile, is frequently used with slightly different definitions [3, 18]. We computed PWM^+ and PWM^- from a training dataset of protein-binding sequences and non-binding sequences, respectively (see Fig. 1). Each element of PWM^+ and PWM^- represents the frequency of i-th nucleotide (i is any one of A, C, G and U) in the j-th position of RNA of n nucleotides. We combined PWM^+ and PWM^- of a training dataset into mPWM by Eq. 1, which represents the log-odds score of the i-th nucleotide in the j-th position.

The PWM of di-nucleotides (dPWM) is less commonly used than PWM of mono-nucleotides, but can elucidate higher order structures of protein-binding sequences. We built dPWM in a similar way to mPWM. We first constructed $dPWM^+$ and $dPWM^-$ from a training dataset of protein-binding sequences and non-binding sequences, respectively. Each element of $dPWM^+$ and $dPWM^-$ represents the frequency of the di-th di-nucleotide (di is any one of AA, AC, $\ldots$, UU) in the j-th position ($j = 1, 2, \ldots, n - 1$) of RNA of n nucleotides. $dPWM^+$ and $dPWM^-$ of a training dataset were combined into dPWM, which represents log-odds score the di-th di-nucleotide in the j-th position. The same mPWM and dPWM generated from a training dataset were used in both training and testing the prediction model.

In addition to the position weight matrices of two types, we computed nucleotide compositions of three types: mono-nucleotide composition (mC), di-nucleotide composition (dC) and tri-nucleotide composition (tC). Thus, a single RNA sequence of n nucleotides is represented in a feature vector with $2n + 83$ elements (n elements for mPWM, $n - 1$ elements for dPWM, and 84 elements for nucleotide compositions). For a sequence of 25 nucleotides, a single feature vector contains 133 elements (see Fig. 2 for the structure of a feature vector).

Protein features

To represent a protein sequence, 20 amino acids are first clustered into 7 groups {A, G, V}, {C}, {M, S, T, Y}, {F, I, L, P}, {H, N, Q, W}, {K, R} and {D, E} based on their dipoles and volumes [19]. Every amino acid in each protein sequence is transformed into an index representing

Table 1 Number of RNA sequences in training and test datasets

P:N	1:1	1:2	1:4	1:6	1:8	1:10
Training						
Dataset	3,372:3,679	3,372:7,200	3,372:13,611	3,372:19,065	3,372:22,826	3,372:26,212
Subtotal	7,051	10,572	16,983	22,473	26,198	29,584
Test						
Dataset	1,000:1,000	1,000:2,000	1,000:3,998	1,000:5,998	1,000:7,998	1,000:9,998
Subtotal	2,000	3,000	4,998	6,998	8,998	10,998
Total	9,051	13,572	21,981	29,435	35,196	40,582

Since similar sequences were removed separately in each 1:n dataset, the number of negative data (N) is not an exact multiple of the number of positive data (P)

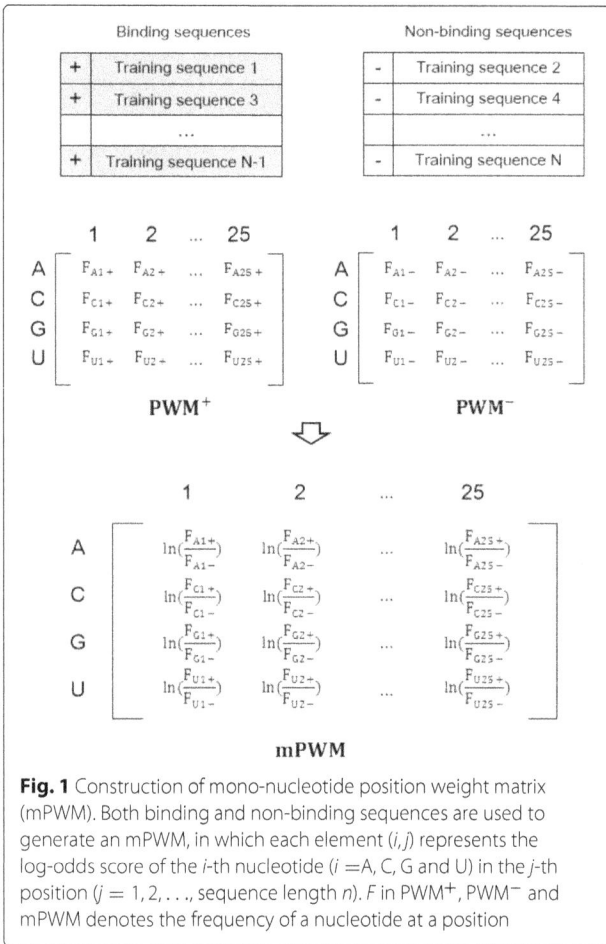

Fig. 1 Construction of mono-nucleotide position weight matrix (mPWM). Both binding and non-binding sequences are used to generate an mPWM, in which each element (i, j) represents the log-odds score of the i-th nucleotide (i =A, C, G and U) in the j-th position ($j = 1, 2, \ldots$, sequence length n). F in PWM$^+$, PWM$^-$ and mPWM denotes the frequency of a nucleotide at a position

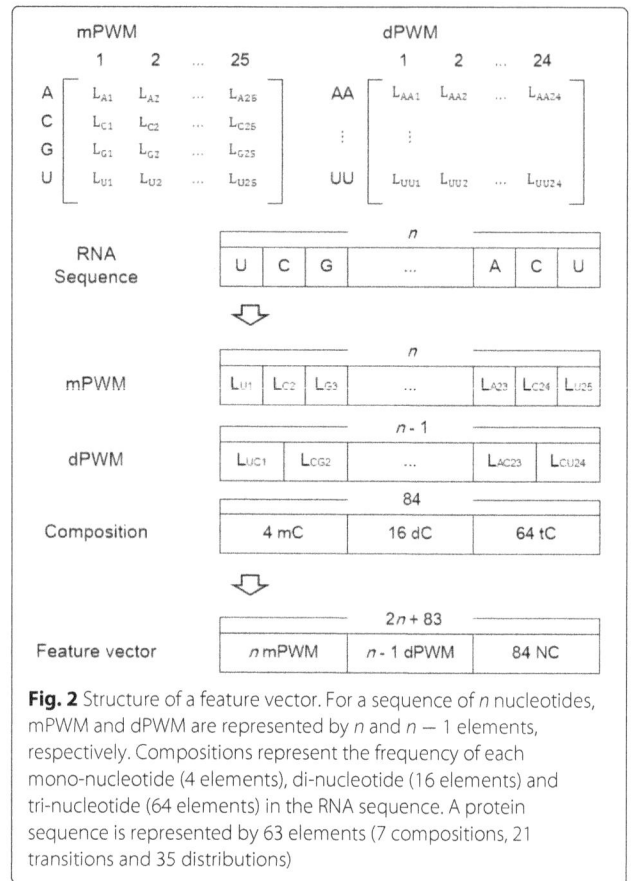

Fig. 2 Structure of a feature vector. For a sequence of n nucleotides, mPWM and dPWM are represented by n and $n - 1$ elements, respectively. Compositions represent the frequency of each mono-nucleotide (4 elements), di-nucleotide (16 elements) and tri-nucleotide (64 elements) in the RNA sequence. A protein sequence is represented by 63 elements (7 compositions, 21 transitions and 35 distributions)

an amino acid group. For each protein sequence, the composition, transition, and distribution of amino acid groups are represented in a feature vector [19]. The composition is the normalized frequency of each group in the protein sequence. The transition is the normalized frequency of transition between each group in the protein sequence. The distribution is the normalized position of the first, 25, 50, 75 and 100%-th amino acid of each group in the protein sequence. A protein sequence is represented by a feature vector with 63 elements (7 compositions, 21 transitions, and 35 distributions). Thus, a model that predicts RBP binding sites using both RNA and proteins features require 63 more elements in a feature vector than that using RNA features only.

Prediction model

We built a support vector machine (SVM) model using a library for support vector machine (LIBSVM) [20]. As a kernel the radial basis function (RBF) was selected instead of the linear kernel because the number of instances (> 100,000 RNA sequences) in our dataset is much larger than the number of features (≈ 200). Besides, it is known

that there is no need to consider linear SVM if complete model selection has been conducted using the Gaussian kernel [21].

The SVM model with the RBF kernel has two parameters, cost (C) and γ. We determined the best parameter values (C = 32 and $\gamma = 0.0078125$) by running the grid search tool of LIBSVM on the training dataset. Unless specified otherwise, all the results shown in this paper were obtained with C = 32 and $\gamma = 0.0078125$.

For comparative purposes, we also built another model using WEKA random forest (http://www.cs.waikato.ac.nz/ml/weka/). As discussed later in the Result section, the SVM model was chosen as the final model for the web server after it was compared with the random forest model. The results of the random forest model shown in this paper were obtained with 60 trees and 25 features, which resulted in the best performance.

Evaluation of the model

The performance of the SVM and random forest models was evaluated using six measures: sensitivity, specificity, accuracy, positive predictive value (PPV), negative predictive value (NPV), and Matthews correlation coefficient (MCC), which are defined as follows.

$$Sensitivity = \frac{TP}{TP + FN} \qquad (3)$$

$$Specificity = \frac{TN}{TN + FP} \qquad (4)$$

$$Accuracy = \frac{TP + TN}{TP + FP + TN + FN} \qquad (5)$$

$$PPV = \frac{TP}{TP + FP} \qquad (6)$$

$$NPV = \frac{TN}{TN + FN} \qquad (7)$$

$$MCC = \frac{(TP \times TN) - (FP \times FN)}{\sqrt{(TP + FP)(TP + FN)(TN + FP)(TN + FN)}} \qquad (8)$$

True positives (TP), true negatives (TN), false positives (FP), and false negative (FN) represent correctly predicted binding regions, correctly predicted non-binding regions, non-binding regions that are incorrectly predicted as binding, and binding regions that are incorrectly predicted as non-binding, respectively.

As described above, our prediction model uses PWM of two types and nucleotide compositions as RNA features. To examine the contribution of the features to the prediction performance, we tried different combinations of features in 10-fold cross validation.

We evaluated the model in several different ways. First, we performed two types of cross validation: (1) standard 10-fold cross validation with six different training datasets (1:1, 1:2, 1:4, 1:6, 1:8 and 1:10 training datasets) and (2) leave-one-protein-out (LOPO) cross validation [22] with the 1:1 training dataset. The reason for performing LOPO cross validation is because typical k-fold cross validation tends to over-estimate predictive performances for paired inputs such as protein-protein interactions (PPIs) or protein-RNA interactions. Recently Park and Marcottee [23] and Hamp and Rost [24] have demonstrated that both standard and refined cross validations lead to inflated accuracy of PPI prediction methods. In LOPO cross validation with respect to RBPs, all RNA sequences (both RBP-binding and non-binding sequences) for one RBP are taken out for testing and remaining RNA sequences are used for training.

In addition to cross validations of two types, we also tested the SVM model on independent datasets, which were not used in training the model. We also compared our SVM model with DeepBind [9] and catRAPID [8] using another test dataset. Out of the 14 RBPs used in our study, DeepBind provides 7 distinct models, one for each of 7 RBPs (FUS, FXR1, FXR2, IGF2BP2, LIN28A, QKI,

TARDBP). For a fair comparison, we extracted new 700 RBP-binding regions of 25 nucleotides from CLIPdb (100 RBP-binding regions for each of the 7 RBPs). To remove redundancy between the 700 RNA sequences and the training dataset, we executed CD-HIT-EST-2D on them with a cut-off value of 0.8. (see Table 2 for the number of remaining RNA sequences after running CD-HIT-EST-2D).

Since catRAPID requires an RNA sequence of at least 50 nucleotides, we extended the RBP-binding regions by including 13 nucleotides on each side of the binding regions in their original genome sequences. Redundancy between the extended RNA sequences and the training dataset was removed by running CD-HIT-EST-2D on them with a cut-off value of 0.9 because instead of 0.8 since the cut-off value of 0.8 removed too many RNA sequences (see Table 3 for the number of remaining RNA sequences after running CD-HIT-EST-2D). As negative data for the 700 RNA sequences, we extracted additional 100 non-binding regions of 25 and 51 nucleotides in the reference human genome GRCh37/hg19.

Results and discussion
Evaluation of feature contribution
Table 4 compares different combinations of features in 10-fold cross validation of our SVM model with the 1:1 training dataset. Among the single features, mPWM and dPWM were much better than nucleotide compositions. With mPWM or dPWM alone, the SVM model achieved an accuracy above 89% and an MCC above 0.79. This result indicates that mPWM and dPWM are very powerful features in predicting protein-binding regions in RNA sequences. Compared to using single features alone, using two different features resulted in performance improvement in sensitivity, accuracy, NPV and MCC. Nucleotide compositions alone achieved a much lower performance than sequence profiles of log-odds scores of mono-nucleotides and those of di-nucleotides, but performance gain was obtained with combination of nucleotide compositions and sequence profiles (sensitivity of 91.61%, specificity of 92.39%, accuracy of 92.02%, PPV of 91.69%, NPV of 92.31% and MCC of 0.840).

Cross validations
Table 5 shows the results of the standard 10-fold cross validations of the SVM model with the RBF kernel and random forest model with the 1:1, 1:2, 1:4, 1:6, 1:8 and 1:10 training datasets. The best performance of the SVM model observed in the balanced dataset with 1:1 ratio of positive to negative instances (sensitivity of 91.61%, specificity of 92.39%, accuracy of 92.02%, PPV of 91.69%, NPV of 92.31% and MCC of 0.840). As expected, running the SVM model on unbalanced datasets resulted in lower performances on average than running it on the balanced

Table 2 Results of testing our model and DeepBind on RNA sequences of 25 nucleotides. catRAPID could not be tested on RNA sequences of 25 nucleotides since the minimum length of an RNA sequence required by catRAPID is 50 nucleotides

RBP	#RBP-binding RNA regions	Sensitivity	Specificity	Accuracy	PPV	NPV	MCC
Our model							
FUS	64	93.75%	94.00%	93.90%	90.91%	95.92%	0.873
FXR1	67	97.01%	94.00%	95.21%	91.55%	97.92%	0.902
FXR2	80	66.25%	94.00%	81.67%	89.83%	77.69%	0.638
IGF2BP2	79	74.68%	94.00%	85.47%	90.77%	82.46%	0.709
LIN28A	82	85.37%	94.00%	90.11%	92.11%	88.68%	0.801
QKI	77	84.42%	94.00%	89.83%	91.55%	88.68%	0.793
TARDBP	94	12.77%	94.00%	54.64%	66.67%	53.41%	0.117
Weighted average		**70.72%**	**94.00%**	**83.83%**	**90.14%**	**80.54%**	**0.676**
DeepBind							
FUS	64	32.81%	42.00%	38.41%	26.58%	49.41%	−0.246
FXR1	67	11.94%	44.00%	31.14%	12.50%	42.72%	−0.444
FXR2	80	15.00%	55.00%	37.22%	21.05%	44.72%	−0.320
IGF2BP2	79	41.77%	51.00%	46.93%	40.24%	42.58%	−0.072
LIN28A	82	12.20%	52.00%	34.07%	17.24%	41.94%	−0.382
QKI	77	83.12%	75.00%	78.53%	71.91%	85.23%	0.576
TARDBP	94	52.13%	92.00%	72.68%	85.96%	67.15%	0.484
Weighted average		**36.28%**	**58.71%**	**48.91%**	**40.53%**	**54.29%**	**−0.051**

The specificity of our method is the same for all RBPs because it used a same set of negative data for all RBPs with a single model, whereas DeepBind has distinct models for each RBP

dataset with 1:1 ratio of positive to negative instances. In particular, PPV and MCC were significantly decreased as the ratio of negative instances was increased. But, NPV was rather increased slightly.

As the dataset contains more negative instances, sensitivity, PPV and MCC of the random forest model were decreased. In particular, it showed a substantial decrease in sensitivity. Since there are much more non-binding sites than binding sites in actual RNA sequences, we determined that finding all possible binding sites at the expense of low PPV is better than missing the binding sites. Thus, we selected the SVM model as the final model for the web server.

As stated earlier, the SVM model with the RBF kernel is known to be better than the SVM with linear kernel when the number of instances is much larger than the number of features. For comparative purposes, we built an SVM model with linear kernel and performed 10-fold cross validation of the model (Additional file 6). The SVM model with linear kernel showed a slightly lower performance than the SVM model with the RBF kernel.

Our SVM model uses the protein sequence as an additional information when it is available. Additional file 7 shows the results of 10-fold cross validation of the SVM model when it is given a protein sequence in addition to

an RNA sequence. The best performance was observed in the balanced dataset with 1:1 ratio of positive to negative instances (sensitivity of 93.18%, specificity of 92.01%, accuracy of 92.57%, PPV of 91.44%, NPV of 93.64% and MCC of 0.851).

Results of LOPO cross validation with respect to RBPs in the 1:1 training dataset are shown in Table 6. Since different RBPs have very different numbers of known RBP-binding regions, we examined a weighted average of performance measures instead of a simple average of them. The weighted average was computed from the total values of TP, FP, TN and FN of all runs. In LOPO cross validation, the model showed a sensitivity of 85.54%, a specificity of 89.53%, an accuracy of 87.60%, a PPV of 88.42%, an NPV of 86.89% and an MCC of 0.752. This result indicates that LOPO cross validation of our SVM model obtained a lower performance than 10-fold cross validation, but its average performance is reasonably high.

Independent tests

For rigorous evaluation of our model, we tested it on independent datasets (30% of the entire data), which were not used in training the model. As in the 10-fold cross validation, we tested it on six test datasets with different ratios of positive to negative instances (called 1:1, 1:2, 1:4, 1:6,

Table 3 Results of testing our model, DeepBind and catRAPID on RNA sequences of 51 nucleotides

RBP	#RBP-binding RNA regions	Sensitivity	Specificity	Accuracy	PPV	NPV	MCC
Our model							
FUS	100	79.00%	70.00%	74.50%	72.48%	76.92%	0.492
FXR1	97	88.66%	70.00%	79.19%	74.14%	86.42%	0.596
FXR2	93	69.89%	70.00%	69.95%	68.42%	71.43%	0.399
IGF2BP2	94	55.32%	70.00%	62.89%	63.41%	62.50%	0.256
LIN28A	96	58.33%	70.00%	64.29%	65.12%	63.64%	0.285
QKI	100	78.00%	70.00%	74.00%	72.22%	76.09%	0.482
TARDBP	100	22.00%	70.00%	46.00%	42.31%	47.30%	−0.091
Weighted average		**64.41%**	**70.00%**	**67.25%**	**67.59%**	**66.94%**	**0.345**
DeepBind							
FUS	100	32.00%	33.00%	32.50%	32.32%	32.67%	−0.350
FXR1	97	32.99%	42.00%	37.56%	35.56%	39.25%	−0.251
FXR2	93	43.01%	73.00%	58.55%	59.70%	57.94%	0.168
IGF2BP2	94	48.94%	59.00%	54.12%	52.87%	55.14%	0.080
LIN28A	96	36.46%	53.00%	44.90%	42.68%	46.49%	-0.107
QKI	100	82.00%	81.00%	81.50%	81.19%	81.82%	0.630
TARDBP	100	50.00%	86.00%	68.00%	78.12%	63.24%	0.386
Weighted average		**46.62%**	**61.00%**	**53.91%**	**53.73%**	**54.05%**	**0.077**
catRAPID		DP value					
FUS	10	16.40%	–	–	–	–	–
FXR1	10	17.60%	–	–	–	–	–
FXR2	10	22.30%	–	–	–	–	–
IGF2BP2	10	16.70%	–	–	–	–	–
LIN28A	10	19.10%	–	–	–	–	–
QKI	10	15.50%	–	–	–	–	–
TARDBP	10	18.10%	–	–	–	–	–
Weighted average		**18.22%**	–	–	–	–	–

Sensitivity is shown for our model and DeepBind, and discriminative power (DP) value is shown for catRAPID. The specificity of our method is the same for all RBPs because it used a same set of negative data for all RBPs with a single model, whereas DeepBind has distinct models for each RBP. Due to the speed of the catRAPID server, catRAPID was tested on 10 RBP-binding sequences of 51 nucleotides for each RBP, whereas both our model and DeepBind were tested on all the RBP-binding sequences. Detailed results are available in Additional file 12

1:8, and 1:10 test datasets hereafter). As shown in Table 7, the specificity, PPV and MCC were decreased as the ratio of negative instances was increased.

In particular, PPV and MCC were significantly decreased as the dataset contains more negative instances. This trend was also observed in 10-fold cross validation. However, other performance measures (sensitivity, accuracy, and NPV) were rather increased, and specificity was decreased slightly.

Figure 3 shows the ROC curves of 10-fold cross validation and independent testing of the SVM models. In 10-fold cross validation, the SVM model with the RBF kernel yielded a slightly larger area under the ROC curve (AUC = 0.9732) than the SVM model with linear kernel (AUC = 0.9714). Likewise, in independent testing the SVM model with RBF kernel showed a slightly larger AUC (0.8912) than the SVM with linear kernel (0.8878).

Since the prediction model was trained with RBP-binding RNA sequences of 25 nucleotides, we examined whether it is applicable to RNAs of different sizes. For RNAs of k nucleotides ($k < 25$), we extracted a total of 12,576 RBP-binding RNAs from CLIPdb. When testing the model on each RNA sequence with < 25 nucleotides, we selected a position in the RNA sequence which

Table 4 Comparison of different combinations of features in 10-fold cross validation

	Sensitivity	Specificity	Accuracy	PPV	NPV	MCC
mPWM	89.09%	90.60%	89.87%	89.67%	90.06%	0.797
dPWM	90.48%	92.06%	91.31%	91.27%	91.34%	0.826
compositions	71.44%	88.23%	80.20%	84.76%	77.12%	0.608
mPWM + dPWM	91.46%	91.98%	91.73%	91.27%	92.16%	0.834
mPWM + compositions	91.31%	91.55%	91.43%	90.83%	92.00%	0.828
dPWM + compositions	91.07%	92.53%	91.83%	91.78%	91.88%	0.836
mPWM + dPWM + compositions	**91.61%**	**92.39%**	**92.02%**	**91.69%**	**92.31%**	**0.840**

Using all 3 features showed the best performance. mPWM: mono-nucleotide position weight matrix, dPWM: di-nucleotide position weight matrix, compositions: frequency of mono-nucleotides, di-nucleotides, and tri-nucleotides in the RNA sequence

results in the maximum sum of log-odds scores from an ungapped alignment of the sequence with mPWM. Based on the selected position, we encoded both mPWM and dPWM features and filled zeros for matrix elements that have no corresponding nucleotides in the RNA sequence to make the size of the feature vector comparable to those for 25-mer RNAs. Nucleotide compositions of short RNA sequences were encoded in the same way as RNA sequences of 25 nucleotides. The prediction performance with short RNA sequences was lower than that with 25-mer RNAs, but its accuracy is as high as 74.4% (Additional file 8). We also tested the prediction model on RNA sequences with > 25 nucleotides, and details are discussed in the next section. Additional file 9 shows the change in accuracy of the model for RNA sequences with lengths between 21 and 40 nucleotides.

Without changing the original mPWM and dPWM, we tested our model for new RBPs that were not considered in constructing datasets. It showed a low performance for some RBPs but obtained a high performance for some RBPs (Additional file 10). The best performance was observed for HNRNPD (sensitivity of 94.29%, specificity of 94.37%, accuracy of 94.33%, PPV of 92.52%, NPV of 95.71% and MCC of 0.884).

A negative dataset in our study was constructed by random selection. For comparative purposes, we constructed different negative datasets by extracting a subsequence in the upstream region of each RBP binding region. We tried several different distances ranging from 1 to 1001 nucleotides between the negative instance and the positive instance (i.e., RBP binding region) in a same RNA sequence. The performance of our model with a new negative dataset was as high as that with the previous

Table 5 Results of 10-fold cross validations of SVM and random forest on 6 datasets with different P:N ratios of positive to negative instances

P:N	Sensitivity	Specificity	Accuracy	PPV	NPV	MCC
SVM						
1:1	**91.61%**	92.39%	92.02%	91.69%	92.31%	0.840
1:2	**91.37%**	92.17%	91.91%	84.53%	95.80%	0.819
1:4	**91.13%**	92.33%	92.09%	74.64%	97.68%	0.777
1:6	**91.22%**	91.95%	91.84%	66.71%	98.34%	0.736
1:8	**91.22%**	91.92%	91.83%	62.52%	98.61%	0.713
1:10	**91.19%**	91.54%	91.50%	58.11%	98.78%	0.686
Random forest						
1:1	**91.13%**	92.06%	91.62%	91.32%	91.89%	0.832
1:2	**85.44%**	95.21%	92.09%	89.31%	93.32%	0.816
1:4	**80.40%**	97.18%	93.85%	87.59%	95.24%	0.802
1:6	**77.88%**	97.77%	94.78%	86.01%	96.16%	0.788
1:8	**76.01%**	98.01%	95.18%	84.95%	96.51%	0.777
1:10	**75.24%**	98.14%	95.53%	83.90%	96.86%	0.770

PPV positive prediction value, *NPV* negative prediction value, *MCC* Matthews correlation coefficient

Table 6 Results of LOPO cross validation of our method with respect to 14 RBPs

	TP	TN	FP	FN	Sensitivity	Specificity	Accuracy	PPV	NPV	MCC
AGO1	37	50	3	18	67.27%	94.34%	80.56%	92.50%	73.53%	0.638
AGO2	39	49	2	18	68.42%	96.08%	81.48%	95.12%	73.13%	0.664
EWSR1	200	198	14	14	93.46%	93.40%	93.43%	93.46%	93.40%	0.869
FUS	468	534	46	19	96.10%	92.07%	93.91%	91.05%	96.56%	0.879
FXR1	3	7	0	1	75.00%	100.00%	90.91%	100.00%	87.50%	0.810
FXR2	25	33	1	11	69.44%	97.06%	82.86%	96.15%	75.00%	0.688
IGF2BP2	57	55	7	15	79.17%	88.71%	83.58%	89.06%	78.57%	0.678
LIN28A	221	263	25	57	79.50%	91.32%	85.51%	89.84%	82.19%	0.714
LIN28B	2214	2343	329	227	90.70%	87.69%	89.13%	87.06%	91.17%	0.783
QKI	3	5	0	1	75.00%	100.00%	88.89%	100.00%	83.33%	0.791
TAF15	11	16	1	2	84.62%	94.12%	90.00%	91.67%	88.89%	0.796
TARDBP	39	159	14	149	20.74%	91.91%	54.85%	73.58%	51.62%	0.179
YTHDF2	35	39	5	6	85.37%	88.64%	87.06%	87.50%	86.67%	0.741
ZC3H7B	388	438	43	94	80.50%	91.06%	85.77%	90.02%	82.33%	0.720
Total	3,740	4,189	490	632						
Weighted average					**85.54%**	**89.53%**	**87.60%**	**88.42%**	**86.89%**	**0.752**

The weighted average was computed from the total values of TP, TN, FP and FN of all runs. TP: true positive, *TN* true negative, *FP* false positive, *FN* false negative, *PPV* positive prediction value, *NPV* negative prediction value, *MCC* Matthews correlation coefficient

dataset in which negative instances were sampled randomly. The specificity has been increased slightly with the new negative dataset. Details are available in Additional file 11.

Comparison with other methods

For the comparison with DeepBind and catRAPID, we prepared two new datasets of RBP-binding RNA sequences. The first test dataset consists of RNA sequences of 25 nucleotides extracted from CLIPdb. In the first dataset, similar sequences with any in the training dataset were removed by running CD-HIT-EST with a cut-off value of 0.8. The second test dataset was constructed by adding 13 nucleotides in the original genome sequence at both ends of the 25-mer RNAs in the first dataset. The reason that we could not use RBP-binding RNA sequences of 51 nucleotides in CLIPdb is because

DeepBind does not provide a prediction model for RBP-binding RNA sequences of 51 nucleotides (DeepBind provides distinct models for each RBP). For negative data of the test datasets, we selected 100 non-binding regions of 25 and 51 nucleotides in the reference human genome GRCh37/hg19.

When testing the model on each RNA sequence with > 25 nucleotides, we found a 25-mer subsequence of the RNA sequence which results in the maximum sum of log-odds scores from an alignment of the 25-mer subsequence with mPWM. In a feature vector, we encoded both mPWM and dPWM features of the selected 25-mer subsequence along with nucleotide compositions of the entire RNA sequence.

Table 2 shows the results of testing our model and DeepBind on RBP-binding sequences for 7 RBPs. In predicting RBP-binding regions of 25 nucleotides, our model

Table 7 Results of independent testing of our method on 6 datasets with different P:N ratios of positive to negative instances

P:N	Sensitivity	Specificity	Accuracy	PPV	NPV	MCC
1:1	**72.50%**	**91.90%**	**82.20%**	**89.95%**	**76.97%**	**0.656**
1:2	72.40%	91.80%	85.33%	**81.53%**	86.93%	**0.663**
1:4	74.10%	91.10%	87.70%	**67.55%**	83.36%	**0.630**
1:6	77.00%	90.26%	88.37%	**56.87%**	95.92%	**0.596**
1:8	77.80%	89.68%	88.36%	**48.53%**	97.00%	**0.554**
1:10	79.10%	89.70%	88.73%	**43.44%**	97.72%	**0.532**

PPV positive prediction value, *NPV* negative prediction value, *MCC* Matthews correlation coefficient

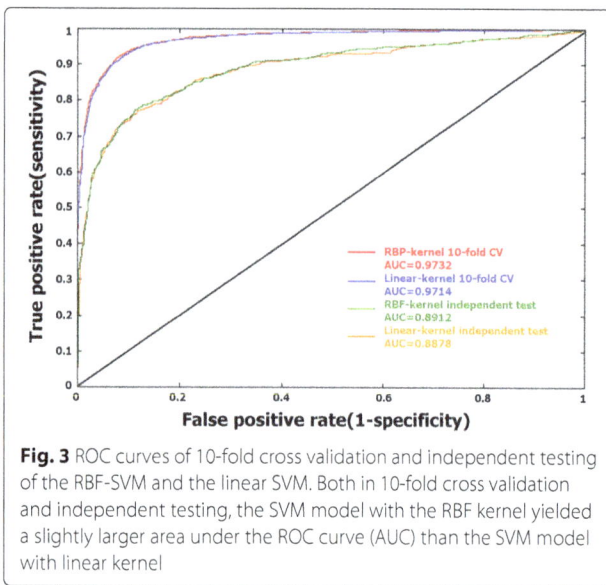

Fig. 3 ROC curves of 10-fold cross validation and independent testing of the RBF-SVM and the linear SVM. Both in 10-fold cross validation and independent testing, the SVM model with the RBF kernel yielded a slightly larger area under the ROC curve (AUC) than the SVM model with linear kernel

achieved an average sensitivity of 70.72%, specificity of 94.00%, accuracy of 83.83%, PPV of 90.14%, NPV of 80.54% and MCC of 0.676. DeepBind showed very low scores for most RBP-binding sequences, but the scores of DeepBind are known to be on an arbitrary scale [9]. Thus, for a fair comparison, we computed Z-scores of DeepBind scores. If an RNA sequence tested by Deep-Bind had a Z-score > 0, it was considered as RBP-binding; otherwise, it was considered as non-binding. DeepBind showed an average sensitivity of 36.28%, specificity of 58.71%, accuracy of 48.91%, PPV of 40.53%, NPV of 54.29% and MCC of -0.051, which is much lower than ours.

In testing on RBP-binding regions of 51 nucleotides, our model showed a much better performance than DeepBind (Table 3). Our model obtained an average sensitivity of 64.41%, specificity of 70.00%, accuracy of 67.25%, PPV of 67.59%, NPV of 66.94% and MCC of 0.345, whereas DeepBind showed an average sensitiv-ity of 46.42%, specificity of 61.00%, accuracy of 53.91%, PPV of 53.73%, NPV of 54.05% and MCC of 0.077. The catRAPID server was too slow to test all RBP-binding sequences shown in Table 3, so it was tested on 10 RBP-binding sequences for each RBP. catRAPID showed low discriminative power (DP) values in most test cases. Since DP of catRAPID represents the interaction propen-sity of a protein—RNA pair with respect to the training sets [8], the result of testing catRAPID on RBP-binding sequences indicates a low confidence level of the pre-diction. Details of the RBP-binding sequences used for comparison of three methods and raw data obtained from execution of the three methods are available in Additional file 12.

Conclusion

In this paper we proposed a new computational method to predict protein-binding regions in mRNA sequences using sequence profiles constructed from log-odds scores of mono- and di-nucleotides and nucleotide composi-tions. The method has been implemented in SVM models and evaluated in several ways, including standard 10-fold cross validation on six datasets with different ratios of positive to negative instances, LOPO cross validation, and independent testing with six datasets of different ratios of positive to negative instances. We also compared our method with DeepBind and catRAPID using another test dataset.

Results of cross validation and independent testing of the method on actual RBP-binding regions in human mRNAs showed that sequence profiles of log-odds scores of mono- and di-nucleotides are much more powerful features than nucleotide compositions in finding protein-binding regions in RNA sequences. Nucleotide com-positions alone achieved a much lower performance than sequence profiles of log-odds scores of mono-nucleotides and those of di-nucleotides, but performance gain was obtained with combination of nucleotide com-positions and sequence profiles. The best performance was observed in a balanced dataset of positive and neg-ative instances. 10-fold cross validation with a balanced dataset achieved a sensitivity of 91.6%, a specificity of 92.4%, an accuracy of 92.0%, a PPV of 91.7%, an NPV of 92.3% and an MCC of 0.84. 10-fold cross valida-tion of RNA and protein sequence feature vector model with a balanced dataset achieved a sensitivity of 93.2%, a specificity of 92.0%, an accuracy of 92.6%, a PPV of 91.4%, an NPV of 93.6% and an MCC of 0.85. LOPO cross validation showed a lower performance than the 10-fold cross validation, but the performance remains high (sensitivity of 85.5%, specificity of 89.5%, accuracy of 87.6%, PPV of 88.4%, NPV of 86.9% and MCC of 0.752). In testing the model on independent datasets, it achieved a sensitivity of 72.5%, a specificity of 91.9%, an accuracy of 82.2%, a PPV of 89.9%, an NPV of 77.0% and an MCC of 0.66. Testing of our model and two other methods showed that our model is better than the others.

The results shown in this paper are preliminary, but demonstrate the potential of our method to predict RBP-binding regions in mRNA. Given that the average length of human mRNAs is about 2 kb and that different RBPs have different binding preferences within an mRNA, it is not straightforward to find RBP binding regions in mRNAs. A computational method like ours will help biol-ogists save time and effort in designing and performing their in vivo or in vitro experiments to detect protein-RNA binding sites by narrowing down candidate binding regions on target RNAs.

Additional files

Additional file 1: Type of RBP binding regions. Type of RBP binding regions in human mRNAs.

Additional file 2: Histogram of the length of RBP-binding regions in CLIPdb. Distribution of the length of RNA sequences binding with 14 RBPs. nt: length in nucleotides of the RBP-binding regions.

Additional file 3: 5,145 RBP-binding regions. 5,145 RBP-binding regions in human mRNA sequences obtained from CLIPdb. For each binding region, RBP name, chromosome name, the starting position of the binding region in the chromosome, the ending position of the binding region in the chromosome, binding affinity score, and strand information are specified.

Additional file 4: 6 training datasets with different ratios of positive to negative instances. 6 training datasets with different ratios of positive to negative instances (called 1:1, 1:2, 1:4, 1:6, 1:8 and 1:10 training datasets).

Additional file 5: 6 test datasets with different ratios of positive to negative instances. 6 test datasets with different ratios of positive to negative instances (called 1:1, 1:2, 1:4, 1:6, 1:8 and 1:10 test datasets).

Additional file 6: Results of 10-fold cross validation of the SVM model with linear kernel with 6 train datasets. The performance of the SVM model with linear kernel with different ratios of positive to negative instances.

Additional file 7: Results of 10-fold cross validation of the SVM model using both RNA and protein features. The performance of the SVM model that uses protein features as well as RNA features in 6 different datasets.

Additional file 8: Results of testing our model on RNA sequences shorter than 25 nucleotides. The performance of the SVM model with RNA sequences shorter than 25 nucleotides.

Additional file 9: Results of testing our model on RNA sequences with length between 21 and 40 nucleotides.

Additional file 10: Results of testing our model for new RBPs. Results of testing our model on predicting RBP binding regions in RNA for new RBPs.

Additional file 11: Results of testing our model on RNA sequences with different negative datasets. The performance of our model with different negative datasets whose instances were selected in the upstream region of each RBP binding region.

Additional file 12: Results of testing DeepBind and catRAPID on RNA sequences of 25 and 51 nucleotides. RBP-binding sequences used for comparison of DeepBind and catRAPID prediction methods and raw data obtained from execution of the three methods.

Abbreviations
CDS: Coding sequence; CLIP: Cross-linking and immunoprecipitation; dC: Di-nucleotide composition; dPWM: Di-nucleotide position weight matrix; FN: False negative; FP: False positive; LIBSVM: Library for support vector machine; LOPO: Leave-one-protein-out; mC: Mono-nucleotide composition; MCC: Matthews correlation coefficient; mPWM: Mono-nucleotide position weight matrix; NPV: Negitive predictive value; PPI: Protrin-protein interaction; PPV: Positive predictive value; PWM: Position weight matrix; RBF: Radial basis function; RBP: RNA-binding protein; SVM: Support vector machine; tC: Tri-nucleotide composition; TN: True negative; TP: True positive

Acknowledgments
Not applicable.

Declarations
This article has been published as part of *BMC Systems Biology* Volume 11 Supplement 2, 2017. Selected articles from the 15th Asia Pacific Bioinformatics Conference (APBC 2017): systems biology. The full contents of the supplement are available online

https://bmcsystbiol.biomedcentral.com/articles/supplements/volume-11-supplement-2.

Funding
This work was supported by INHA UNIVERSITY Research Grant. The publication costs of this article were funded by Inha University.

Authors' contributions
DC designed and implemented the prediction model and prepared the initial manuscript. BP implemented a web server, analyzed the data on binding regions, and compared the prediction model with other methods. HC constructed data sets and prepared the initial manuscript. WL assisted the work and examined the results. KH supervised the work and wrote the manuscript. All authors read and approved the final manuscript.

Authors' information
Department of Computer Science and Engineering, Inha University, 22212, Incheon, South Korea.

Competing interests
The authors declare that they have no competing interests.

References
1. König J, Zarnack K, Luscombe NM, Ule J. Protein-RNA interactions: new genomic technologies and perspectives. Nat Rev Genet. 2012;13:77–83.
2. Gerstberger S, Hafner M, Tuschl T. A census of human RNA-binding proteins. Nat Rev Genet. 2014;15(0):829–845.
3. Wang L, Huang C, Yang MQ, Yang JY. BindN+ for accurate prediction of DNA and RNA-binding residues from protein sequence features. BMC Syst Biol. 2010;4(Suppl 1):S3.
4. Wang L, Brown SJ. BindN: a web-based tool for efficient prediction of DNA and RNA binding sites in amino acid sequences. Nucleic Acids Res. 2006;34:243–8.
5. Walia RR, Xue LC, Wilkins K, El-Manzalawy Y, Dobbs D, Honavar V. RNABindRPlus: A predictor that combines machine learning and sequence homology-based methods to improve the reliability of predicted RNA-binding residues in proteins. PLOS ONE. 2014;9(5):e97725.
6. Li S, Yamashita K, Amada KM, Standley DM. Quantifying sequence and structural features of protein—RNA interactions. Nucleic Acids Res. 2014;42:10086–98.
7. Choi S, Han K. Predicting protein-binding RNA nucleotides using the feature-based removal of data redundancy and the interaction propensity of nucleotide triplets. Comput Biol Med. 2013;43(11):1687–97.
8. Bellucci M, Agostini F, Masin M, Tartaglia GG. Predicting protein associations with long noncoding RNAs. Nat Methods. 2011;8(6):444–6.
9. Alipanahi B, Delong A, Weirauch MT, Frey BJ. Predicting the sequence specificities of DNA- and RNA-binding proteins by deep learning. Nat Biotechnol. 2015;33:831–8.
10. Ray D, Kazan H, Cook KB, Weirauch MT, Najafabadi HS, Li X, Gueroussov S, Albu M, Zheng H, Yang A, Na H, Irimia M, Matzat LH, Dale RK, Smith SA, Yarosh CA, Kelly SM, Nabet B, Mecenas D, Li W, Laishram RS, Qiao M, Lipshitz HD, Piano F, Corbett AH, Carstens RP, Frey BJ, Anderson RA, Lynch KW, Penalva LOF, et al. A compendium of RNA-binding motifs for decoding gene regulation. Nature. 2013;499:172–7.
11. Tuvshinjargal N, Lee W, Park B, Han K. R N A Predicting protein-binding nucleotides with consideration of binding partners. Comput Methods Prog Biomed. 2015;120(1):3–15.
12. Tuvshinjargal N, Lee W, Park B, Han K. PRIdictor: Protein-RNA Interaction predictor. BioSystems. 2016;139:17–22.

13. Wong KC, Li Y, Peng C, Moses AM, Zhang Z. Computational learning on specificity-determining residue-nucleotide interactions. Nucleic Acids Res. 2015;43(21):10180–9.
14. Yang Y-CT, Di C, Hu B, Zhou M, Liu Y, Song N, Li Y, Umetsu J, Lu ZJ. CLIPdb: A CLIP-seq database for protein-RNA interactions. BMC Genomics. 2015;16:51.
15. Hafner M, Landthaler M, Burger L, Khorshid M, Hausser J, Berninger P, Rothballer A, Ascano M, Jungkamp A-C, Munschauer M, Ulrich A, Wardle GS, Dewell S, Zavolan M, Tuschl T. PAR-CliP - a method to identify transcriptome-wide the binding sites of RNA binding proteins. J Visualized Exp. 2010;(41):2034.
16. Corcoran DL, Georgiev S, Mukherjee N, Gottwein E, Skalsky RL, Keene JD, Ohler U. PARalyzer: Definition of RNA binding sites from PAR-CLIP short-read sequence data. Genome Biol. 2011;12(8):R79.
17. Huang Y, Niu B, Gao Y, Fu L, Li W. Cd-hit suite: A web server for clustering and comparing biological sequences. Bioinformatics. 2010;26(5):680–2.
18. Ahmad S, Sarai A. PSSM-based prediction of DNA binding sites in proteins. BMC Bioinforma. 2005;6(33):6.
19. Zhu-Honh Y, Keith CCC, Pengwei H. Predicting protein-protein interactions from primary protein sequences using a novel multi-scale local feature representation scheme and the random forest. PLoS ONE. 2015;10(5):e0125811.
20. Chang C-C, Lin C-J. LIBSVM: A library for support vector machines. ACM Trans Intell Syst Technol. 2011;2(3):27.
21. Keerthi SS, Lin C-J. Asymptotic behaviors of support vector machines with Gaussian kernel. MIT Press. 2003;15(7):1667–89.
22. Abbasi WA, Minhas FUAA. Issues in performance evaluation for host-pathogen protein interaction prediction. J Bioinforma Comput Biol. 2016;14(3):1650011.
23. Park Y, Marcotte EM. A flaw in the typical evaluation scheme for pair-input computational predictions. Nat Methods. 2012;9(12):1134–6.
24. Hamp T, Rost B. More challenges for machine-learning protein interactions. Bioinformatics. 2015;31(10):1521–5.

6

Identifying model error in metabolic flux analysis – a generalized least squares approach

Stanislav Sokolenko, Marco Quattrociocchi and Marc G. Aucoin[*] ⓘD

Abstract

Background: The estimation of intracellular flux through traditional metabolic flux analysis (MFA) using an overdetermined system of equations is a well established practice in metabolic engineering. Despite the continued evolution of the methodology since its introduction, there has been little focus on validation and identification of poor model fit outside of identifying "gross measurement error". The growing complexity of metabolic models, which are increasingly generated from genome-level data, has necessitated robust validation that can directly assess model fit.

Results: In this work, MFA calculation is framed as a generalized least squares (GLS) problem, highlighting the applicability of the common t-test for model validation. To differentiate between measurement and model error, we simulate ideal flux profiles directly from the model, perturb them with estimated measurement error, and compare their validation to real data. Application of this strategy to an established Chinese Hamster Ovary (CHO) cell model shows how fluxes validated by traditional means may be largely non-significant due to a lack of model fit. With further simulation, we explore how t-test significance relates to calculation error and show that fluxes found to be non-significant have 2-4 fold larger error (if measurement uncertainty is in the 5–10 % range).

Conclusions: The proposed validation method goes beyond traditional detection of "gross measurement error" to identify lack of fit between model and data. Although the focus of this work is on t-test validation and traditional MFA, the presented framework is readily applicable to other regression analysis methods and MFA formulations.

Keywords: Metabolic flux analysis (MFA), Generalized least squares (GLS), Measurement uncertainty, t-test

Background

As the metabolic phenotype of the cell, the flow of material through intracellular reactions (or metabolic flux) represents the sum total of all underlying cellular processes. The accurate determination of metabolic flux is becoming increasingly important for assessing the impact of metabolic engineering or feeding strategies on cellular metabolism [1]. In lieu of in vivo observation, the inference of intracellular fluxes is commonly accomplished through metabolic flux analysis (MFA). At its most basic, MFA refers to the process of modeling intracellular flux via a stoichiometric balance of metabolic reaction and transport rates (assuming a "pseudo steady-state" in the form of negligible molecule accumulation) [2]. The original applications of the technique centered on using simple element balances as a means to correct unreliable measurements [3]. However, the increasing availability of data from multi-omic technologies has led to the development of metabolic flux models that extend far beyond these foundations.

The basis of MFA is the stoichiometry matrix. In the typical arrangement, rows represent balances on molecular species, with each column encoding the stoichiometry of a reaction (see [2] for details). As cellular reaction networks generally have more reactions than species, the resulting stoichiometry matrix is typically underdetermined. The estimation of a single flux profile requires that the number of unknown reaction rates be equal to or less than the number of molecular species, and this has traditionally been accomplished by observing as many extracellular transport rates as possible. However, the growing availability of genomic data has opened the door to developing models that may contain thousands of reactions, complicating the calculation of a unique flux profile.

*Correspondence: maucoin@uwaterloo.ca
Department of Chemical Engineering, University of Waterloo, 200 University Avenue West, N2L 3G1 Waterloo ON, Canada

A considerable amount of metabolic information can be gathered without calculating a unique flux profile through constraint-based reconstruction and analysis (COBRA) methods. The combination of mass balance constraints from stoichiometric relations as well as other factors such as enzyme capacity and reaction thermodynamics can be used to generate a feasible solution space for cellular metabolism. If a unique flux profile is required, one can be estimated by assuming an objective function such as cell growth maximization. However, it is also possible to study the solution space directly (for a detailed review, see [4]). The popularity of COBRA methods has resulted in the development of a large number of software packages that have considerably simplified analysis (see [5]). However, the complexity of genome-scale models remains an ongoing challenge.

Despite the recent advances, the process of translating genomic information to cellular reactions is still under development. Even the well-studied genomes of *Escherichia coli* and *Saccharomyces cerevisiae* had approximately 20 % of their open reading frames (ORFs) uncharacterized as recently as 2010 [6] and the development of reaction networks requires a significant amount of curation [6–8]. Furthermore, the relation between the presence of a gene sequence and enzymatic activity is not always obvious [7]. A combined transcriptomic-metabolomic modeling study of *E. coli* has revealed the existence of redundant gene expression where no flux was observed [9]. Meanwhile, a study of lysine-producing *Corynebacterium glutamicum* metabolism suggested that while the expression of some genes appears tightly coupled to metabolic fluxes, others can remain practically constant despite considerable changes in metabolic flux [10]. The popular Chinese Hamster Ovary (CHO) cell line has an added problem of high genetic variability that may question the generality of a given model [11, 12]. Taken together, these issues add a considerable amount of uncertainty to modeling efforts, especially for less studied expression systems.

The addition of isotopically labelled substrate and the analysis of resulting metabolites through ^{13}C-MFA can be a powerful means to gain better understanding of a metabolic system. But despite the ready availability of algorithms and software packages to assist with everything from identifying optimal labelling strategies to final analysis (as reviewed in [13, 14]), ^{13}C-MFA is not always practical. Isotopic labelling is expensive, especially for large volume bioreactor cultivation, and can not be used to monitor ongoing production processes. Moreover, studying transient labelling patterns requires accurate intracellular metabolite quantification, which is not always straightforward [15], and increased computational resources [13].

As such, one approach to dealing with genome-scale model uncertainty and complexity has been to simplify the models to a level where they can be solved directly from measured extracellular transport rates [16, 17], continuing the use of traditional overdetermined MFA. The simplification can be aided by software such as CellNetAnalyzer that can deal with both underdetermined COBRA models and overdetermined MFA formulations [18, 19]. Recent developments have also led to an automation of the model simplification process [20]. Despite increasing model size, overdetermined MFA has continued to see use over the last 10 years [16, 21–27], especially for less commonly used cell lines that lack well curated genomic and transcriptomic data. However, the reduction of genome-levels models in this fashion is an inversion of the original MFA foundations. In contrast to the use of a simple, reductive model for the reconciliation of questionable data, it is the accuracy of the model that is becoming increasingly variable – making it necessary to rigorously assess the validity of model simplification.

A number of strategies are currently available for model validation. The stoichiometric matrix can be probed directly by checking its condition number [28] or by determining the sensitivity of calculated fluxes to measurement error [29]. The incorporation of measurement flux uncertainty allows the use of gross measurement error detection [30], which identifies whether deviations between observed and fit data are normally distributed through a χ^2-test. While useful for identifying singular errors of large magnitude, this statistic does not asses the overall quality of fit – errors may be unreasonably large while remaining normally distributed. Despite the increasing consideration of confidence intervals around calculated fluxes in recent studies [31, 32], the question of whether a set of data fits a given metabolic model has thus far remained open.

In this work, we propose the use of a standard *t*-test as a natural extension of the least-squares calculation that underpins traditional MFA calculation. Applying MFA to a Chinese Hamster Ovary (CHO) cell culture, *t*-tests were used to determine whether each calculated flux could be deemed sufficiently distinct from zero. Once nonsignificant fluxes were identified, we explored whether the uncertainty in calculated fluxes could be explained by measurement uncertainty alone, or if a lack of model fit could be to blame. To do this, the solution space of the stoichiometric model was constrained by observed flux ranges and hypothetical flux profiles were generated directly from the model. The profiles were perturbed by measurement error and collected to establish a baseline of calculated flux significance given perfect model fit.

Methods

Theoretical principles[1]

The material balance on molecular species that forms the basis of MFA is typically expressed as

$$Sv = 0 \qquad (1)$$

where S is the stoichiometric matrix and v is the vector of fluxes that correspond to reactions defined by columns of S. This formulation proceeds from a pseudo steady-state assumption that changes in metabolite pools (as a result of cell division or other processes) are much smaller than metabolite production and consumption fluxes and can therefore be ignored. The Sv matrix can be be separated into $S_c v_c + S_o v_o$, where c stands for calculated flux and o for observed flux.

$$S_c v_c + S_o v_o = 0 \qquad (2)$$

$$-S_o v_o = S_c v_c \qquad (3)$$

Since v_o is a vector of observed data, $S_o v_o$ can be calculated directly. The dimension of S_c depends on how many fluxes can be observed, i.e., the length of v_o. S_c must have no more columns than rows to calculate a unique flux profile, although the observation of more fluxes (and the accompanying reduction in the number of S_c columns) is useful for error estimation[2]. Pooling cyclic or parallel pathways may be required in the initial formulation of S to ensure the required form of S_c is obtained.

Assuming that an overdetermined form of S_c can be formulated (with sufficient information to calculate v_c), Eq. (3) is equivalent to linear regression and can be solved in a similar fashion.

Linear regression   MFA

$$y = X\beta + \varepsilon \quad (4) \qquad -S_o v_o = S_c v_c + \varepsilon \quad (5)$$

$$\hat{\beta} = \left(X^T X\right)^{-1} X^T y \quad (6) \qquad \hat{v}_c = -\left(S_c^T S_c\right)^{-1} S_c^T S_o v_o \quad (7)$$

With this formulation, ε represents the deviation between observed and calculated fluxes that may be the result of either measurement error or lack of model fit. Equation (7) assumes ε is independently and identically distributed, which is unlikely to be the case. The variance-covariance matrix $\text{Cov}(\varepsilon)$ can be expressed as a scalar σ^2 multiplied by a matrix of relative covariance terms V, i.e., $\text{Cov}(\varepsilon) = \sigma^2 V$. If observed fluxes do not covary and have equal variance, then $V = I$, where I is the identity matrix.

Otherwise, Eq. (5) needs to be rescaled by the matrix square root of V. Taking $V = PP$, the scaled form of Eq. (5) is:

$$-P^{-1} S_o v_o = P^{-1} S_c v_c + P^{-1} \varepsilon \qquad (8)$$

where $P^{-1} \varepsilon$ now satisfies the assumptions of linear regression. Formally, this is equivalent to generalized least squares (GLS) regression; however, incorporating P^{-1} directly into each term allows the use of all ordinary least squares techniques. Letting $P^{-1} S_o = S_o'$, $P^{-1} S_c = S_c'$, and $P^{-1} \varepsilon = \varepsilon'$:

$$\hat{v}_c = -\left(S_c'^T S_c'\right)^{-1} S_c'^T S_o' v_o \qquad (9)$$

The calculation of P^{-1} requires the estimation of $\text{Cov}(\varepsilon)$ from the variance of observed fluxes. Calculating the covariance-variance matrix of both sides of Eq. (5):

$$\text{Cov}\left(-S_o v_o\right) = \text{Cov}\left(S_c v_c + \varepsilon\right) \qquad (10)$$

$$\text{Cov}(\varepsilon) = S_o \text{Cov}(v_o) S_o^T \qquad (11)$$

Since $\text{Cov}(\varepsilon) = \sigma^2 V$ for any value of σ, σ is set to 1 so that $V = \text{Cov}(\varepsilon)$. In practice, $\text{Cov}(v_o)$ need only capture the relative magnitudes of observed flux variances as $\hat{\sigma}$ is estimated during regression. Balances around molecular species that do not include an observed flux v_o will have a row of zeros in $\text{Cov}(\varepsilon)$, which prevents the calculation of a matrix inverse (required to get P^{-1}). Although this mathematically equates to a variance of zero for those balances, a better interpretation is that there is an unknown variance around the "observation" of no net flux. The simplest solution is to add a small non-zero value to each diagonal entry of $\text{Cov}(\varepsilon)$, representing the confidence of the calculated fluxes being fully balanced. If there is more uncertainty around some balances than others, this information could be encoded in the magnitude of the added variance. P can then be calculated via a matrix square root of estimated $\text{Cov}(\varepsilon)$. Since a variance (covariance) matrix is positive semi-definite, P is known to be unique.

Whereas calculated fluxes $\hat{v}_c$ are commonly estimated using a very similar "weighted" least squares approach, the use of validation methods that are part of the regression framework have yet to be explored. The common χ^2 test can still be used to detect gross measurement errors in estimated residuals ($\hat{\varepsilon}$); however, the validation of a regression model also requires the use of t-tests to ensure the significance of calculated fluxes. Confidence and prediction intervals are also highly relevant to MFA. Estimated fluxes require a confidence interval to report the uncertainty of calculation, while a prediction interval around a predicted balance can be used to judge the validity of that balance being closed. The calculation of a t-statistic follows from normal regression:

Linear regression $\qquad$ MFA

$$t_{\hat{\beta}_i} = \frac{\hat{\beta}_i}{\text{se}(\hat{\beta}_i)} \qquad (12) \qquad t_{\hat{v}_{c,i}} = \frac{\hat{v}_{c,i}}{\text{se}(\hat{v}_{c,i})} \qquad (13)$$

Thus:

$$t_{\hat{v}_{c,i}} = \frac{\left(-\left(S_c'^T S_c'\right)^{-1} S_c'^T S_o' v_o\right)_i}{\hat{\sigma}\sqrt{\left(S_c'^T S_c'\right)^{-1}_{i,i}}} \qquad (14)$$

The estimated standard deviation of ε (or $\hat{\sigma}$) is calculated as follows:

$$\hat{\sigma}^2 = \frac{\sum\left(\hat{\varepsilon}_i'\right)^2}{n_b - n_c - 1} \qquad (15)$$

where:

$$\hat{\varepsilon}' = -S_o' v_o + S_c'\left(S_c'^T S_c'\right)^{-1} S_c'^T S_o' v_o \qquad (16)$$

and n_b is the number of balances (rows of S_c') while n_c is the number of fluxes to be calculated (columns of S_c'). If the model is correct and $\text{Cov}(\varepsilon)$ was correctly estimated, $\hat{\sigma}^2$ should be approximately equal to 1. Once the t-value is calculated, a flux can be judged statistically significant if $|t_{\hat{v}_{c,i}}| \geq t_{\alpha/2, n_b - n_c - 1}$ where α is the significance level.

The identification of non-significant flux may be interpreted in two ways. The measurement error around observed fluxes may be too high to allow robust flux calculation. In that case, non-significant fluxes should be treated as having a flux of zero and excluded from the model or further analysis. Alternatively, non-significance may be the result of excess variability from a lack of fit between the model and observed data, requiring model correction. To distinguish between these cases, it is necessary to separate model error from measurement uncertainty. One way to accomplish this is to reduce measurement uncertainty through added replication; however, the required effort can make this approach practically infeasible. Another solution is to simulate a set of feasible fluxes directly from the stoichiometric model (and therefore free of model error) for comparison to the observed data.

The simulation of feasible fluxes can be simplified by eliminating flux equality constraints expressed by the stoichiometry matrix. Essentially, only $n_c - n_b$ fluxes have to specified in order to generate all the other values. More formally, the relationships between the fluxes can be succinctly summarized through the nullspace (or kernel) of S, which describes all flux balance conservations in the model. This makes it possible to calculate all fluxes from a smaller set of variables referred to as the basis. Unlike fluxes, which must satisfy constraints imposed by

$Sv = 0$, the basis can take any arbitrary value to generate fluxes that satisfy all required constraints. Expressed mathematically,

$$\text{Null}(S) = K \qquad (17)$$
$$Kb = v \qquad (18)$$

where b is a basis vector of any value with the same number of rows as columns of K. While all values of b satisfy $Sv = 0$, it is still necessary to constrain fluxes to a set of realistic values representative of a cell cultivation. The space of all feasible fluxes v can be constrained by defining upper and lower bounds on each observed flux:

$$v = Kb$$
$$\text{subject to } K_i b \leq v_i + a \cdot \text{sd}(v_i) \qquad (19)$$
$$K_i b \geq v_i - a \cdot \text{sd}(v_i)$$

where v_i is an observed flux, K_i is the corresponding row of K, and a is a scaling constant that can be set to $t_{\alpha/2, df}$ to specify a confidence interval around v_i. As the basis solution space is only constrained by inequalities, it is readily amenable to stochastic sampling. All values of v that satisfy Eq. (19) represent feasible fluxes that would perfectly satisfy the stoichiometric model while remaining within measurement uncertainty of real observations. If the resulting space is infeasible, then the observed data does not fit the specified model. Otherwise, a random sample of feasible fluxes can be taken for comparison to observed results. If the addition of measurement error to simulated fluxes results in less uncertainty than from observed results, then model error is to blame.

Cell culture

CHO-BRI cells were grown in a 3 L bioreactor (Applikon Biotechnology Inc., Foster City, CA) in serum-free BioGro-CHO media (BioGro Technologies Inc., Winnipeg, Canada) with an in-house amino acid supplement (manuscript submitted). The culture was seeded at $0.3 \cdot 10^6$ cells/ml with a working volume of 2 L. Temperature, pH, dissolved oxygen, and agitation speed were held at 37 °C, 7.4, 50 %, and 120 RPM respectively. Samples were taken three times a day for offline analysis. Cell density was determined using a Coulter Counter Z2 (Beckman Coulter, Miami, FL) calibrated to results from trypan blue exclusion analysis. Aliquots were centrifuged, with the supernatant collected and stored at -80 °C until Nuclear Magnetic Resonance (NMR) analysis. Dry cell mass was calculated by vacuum filtering 15 mL of cell culture through a type A/D glass filter (Pall Corporation, Port Washington, NY) and weighing the filter after drying it for 24 hours at 50 °C.

Metabolite quantification

NMR spectra acquisition, metabolite quantification, and internal standard correction are described in [33]. In brief, samples were scanned on a Bruker Avance 600 MHz spectrometer using the first increment of a 1D-NOESY pulse sequence with metabolite quantification carried out using Chenomx NMR Suite 8.1 (Chenomx Inc., Edmonton, Canada). GlutaMAX™ was added manually to the software library using the Chenomx NMR Suite's 'compound builder' tool. All compounds were profiled in triplicate. Ammonia measurements were taken using an Orion Star™Plus ISE Meter (Thermo Fisher Scientific, Waltham, MA).

MFA model

A CHO cell MFA model was taken from [34]. New transport fluxes were added for acetate, formate, pyruvate, citrate, malate, pyroglutamate, and GlutaMAX™ (the fluxes of which could all be observed via NMR). The transport of GlutaMAX™ was grouped together with the conversion of the dipeptide into glutamine and alanine. The transport of cystine was grouped together with the reduction of cystine into cysteine. A new reaction was added for the conversion of glutamate into pyroglutamate [35] (via a number of possible enzymatic and non-enzymatic reactions). New reactions were also added for acetyl-CoA hydrolase and formate-tetrahydrofolate ligase to explain acetate and formate production. Along with a Systems Biology Markup Language (SBML) representation of the model, a full list of reactions and an outline of metabolite flow are provided as Additional files 1, 2 and 3. As in the original formulation, a number of unbalanced species were removed from the model before analysis, including O_2, CO_2, ATP, NADH, NADPH, and FADH (NADH and NADPH were later reintroduced in a modified form of the model).

Flux estimation

Metabolite and cell concentration timecourse data was fit by a regression spline with 4 cubic basis functions (provided by the gam function [36] in the R programming language [37]). Measurement error was estimated by calculating the variance of observation deviation from the fit. 1000 predicted concentration timecourses were simulated for each trend by adding normally distributed error corresponding to the sum of regression and measurement variance. A new regression split fit was calculated for each of the simulated timecourses. Metabolite transport fluxes were calculated by dividing the derivative of the metabolite concentration fit by cell concentration ($v_o = \frac{1}{X}\frac{dC_o}{dt}$). The mean and variance of the simulated fluxes at each time-point were used for all MFA analysis. Biomass fluxes were calculated as in [34], with the exception that dry cell mass measured to be 0.24 mg/10^6 cells. A single mid-exponential time-point of 66 hours was chosen for MFA analysis to fulfill pseudo steady-state conditions.

Implementation

All MFA calculations, validation, and sampling were carried out using the omfapy Python package, developed in-house. The package as well as analysis code is available on github (https://github.com/ssokolen/omfapy). Basic functionality was based on theoretical principles presented in [2]. Sampling of a feasible flux space was implemented using the random direction algorithm [38] as well as the mirror algorithm presented in [39]. Although slower, the mirror algorithm was able to generate more even coverage of the sampling space.

Results
Identification of model error

Observed uptake fluxes and their corresponding coefficients of variation 66 hours post inoculation are shown in Table 1, with overall metabolite concentration profiles and cell density in Fig. 1. As usual for CHO cells, the metabolic profile was dominated by large fluxes of glucose and lactate. Considerable fluxes of alanine, GlutaMAX^{TM}, ammonia, and glutamine were also observed. The median coefficient of variation was found to be 9.3 %. Although this was similar to previously reported estimates for concentration quantification via NMR [40], incorporating the uncertainty of derivative calculation resulted in a somewhat larger probability of high variance values. As in [40], the singularly high variability of glutamate flux was primarily due to its low concentration and heavy spectral convolution.

The incorporation of the observed fluxes into the MFA model showed no issues using typical metrics. The condition number of the reduced stoichiometry matrix was considerably below 1000 and the χ^2 p-value was 0.93, indicating little evidence of gross measurement error. However, t-test analysis on the calculated fluxes using the GLS framework revealed that only 15 of 47 fluxes were statistically significant (at the standard 5 % significance level). The statistically significant fluxes were primarily those that related to glycolysis – offering only a shallow look at cellular metabolism. All of the TCA and many of the amino acid degradation fluxes were deemed non-significant. To determine whether measurement variability or model error was to blame, 100 flux profiles were sampled from the stoichiometric matrix bounded by 99 % confidence intervals on the measured fluxes (fluxes generated directly from the model in this way will be referred to as "balanced"). Ninety-nine percent intervals were chosen to include practically all possible flux values. The sampled fluxes had good coverage of the constraint space, suggesting that the model was flexible enough to fit fluxes similar to those observed. Each balanced flux profile was then perturbed 100 times using normally distributed noise generated from observed flux standard deviations. The result was 10 000 sets of fluxes

Table 1 Observed uptake fluxes and coefficients of variation (standard deviation of flux divided by flux) 66 hours post inoculation

	Flux $\left(\frac{nmol}{10^6 cells \cdot h}\right)$	CV (%)
Acetate	-1.03	5.08
Alanine	-33.95	3.32
Ammonia	-17.65	23.10
Arginine	2.52	16.33
Asparagine	2.21	7.64
Aspartate	2.14	7.09
Carbohydrates	-2.13	12.25
Citrate	-1.56	7.14
Cystine	0.33	19.04
DNA	-0.31	13.15
Formate	-7.52	2.06
Glucose	161.87	2.89
Glutamate	-0.17	213.18
Glutamax	17.98	10.69
Glutamine	7.35	12.48
Glycine	-2.25	8.79
Histidine	1.02	14.92
Isoleucine	1.52	8.13
Lactate	-283.53	3.19
Leucine	2.66	9.47
Lipids	-1.36	14.86
Lysine	1.80	8.05
Malate	-0.40	13.78
Methionine	0.89	6.44
Phenylalanine	1.19	7.04
Proline	1.94	9.17
Protein	-32.69	13.11
Pyroglutamate	-3.86	3.86
Pyruvate	-2.62	5.74
RNA	-0.89	13.77
Serine	2.64	12.36
Succinate	-0.15	15.52
Threonine	1.70	11.45
Tryptophan	0.34	17.40
Tyrosine	1.11	6.57
Valine	2.24	5.26

subject to observed measurement error but no model error.

Figure 2 compares the percentage of simulated (balanced) fluxes found to be non-significant to the results from observed data. The simulation revealed that approximately half of the calculated fluxes (and all TCA fluxes) are entirely non-significant even when there is no model error (Fig. 2b). Many of the other fluxes were only significant for 50 % of the simulations or fewer. The lack of significance showed that the model was incapable of providing high confidence results for the collected data. Along with the overall low significance, evidence of model error could also be observed. Focusing on approximately 20 of the lowest magnitude fluxes, all were deemed to be non-significant based on the observed data. Comparing the simulated data, the same fluxes were rejected as non-significant 50–95 % of the time. Taken together, the probability of all the low magnitude fluxes being observed as non-significant is extremely low, giving strong indication of poor fit beyond the effect of measurement error alone, i.e., as a result of model error. Although model correction is outside the scope of this work, the proposed methodology was successful in identifying a considerable degree of uncertainty overlooked by commonly used validation methods.

Effect of measurement noise

An extended simulation was carried out to determine whether the lack of statistical significance was due to measurement variability. The flux constraints were extended beyond 66 hours post inoculation to consider the broader applicability of the model. 99 % confidence intervals were generated for all fluxes 18-80 hours post inoculation with the minimum and maximum values for each flux used to bound the flux solution space. 100 balanced flux profiles were generated with 100 sets of measurement error drawn from a normal distribution using 5, 10, 15, and 20 % coefficients of variation for each flux. The 45 calculated fluxes spanned more than 3 logarithms of values from approximately 0.1 $\frac{nmol}{10^6 cells \cdot h}$ to 400 $\frac{nmol}{10^6 cells \cdot h}$ (Fig. 3a). Fluxes had variable magnitudes across the simulations, so all analysis was performed as a function of flux rank, where a rank of 1 indicates the smallest magnitude flux in a given flux profile.

All the simulated flux profiles were subject to a χ^2 test, with only 5 % of the simulations rejected (equal to the false positive rate). The remainder of the fluxes are shown in Fig. 3. As the simulated fluxes included both observed and calculated values, a percent error could be calculated for each calculated flux. Despite passing the χ^2 test, most fluxes were characterized by median errors of 10–20 % (Fig. 3b), increasing with measurement variability. It should be noted that the median is a relatively conservative statistic. By definition, half of the calculated fluxes featured much greater errors than the reported values. The pronounced jump in error for flux ranks of 36 to 44 was traced to the TCA fluxes, which had high error despite large flux magnitudes. Similar to median error, the percentage of fluxes identified as non-significant increased with measurement variability (Fig. 3c). However, even

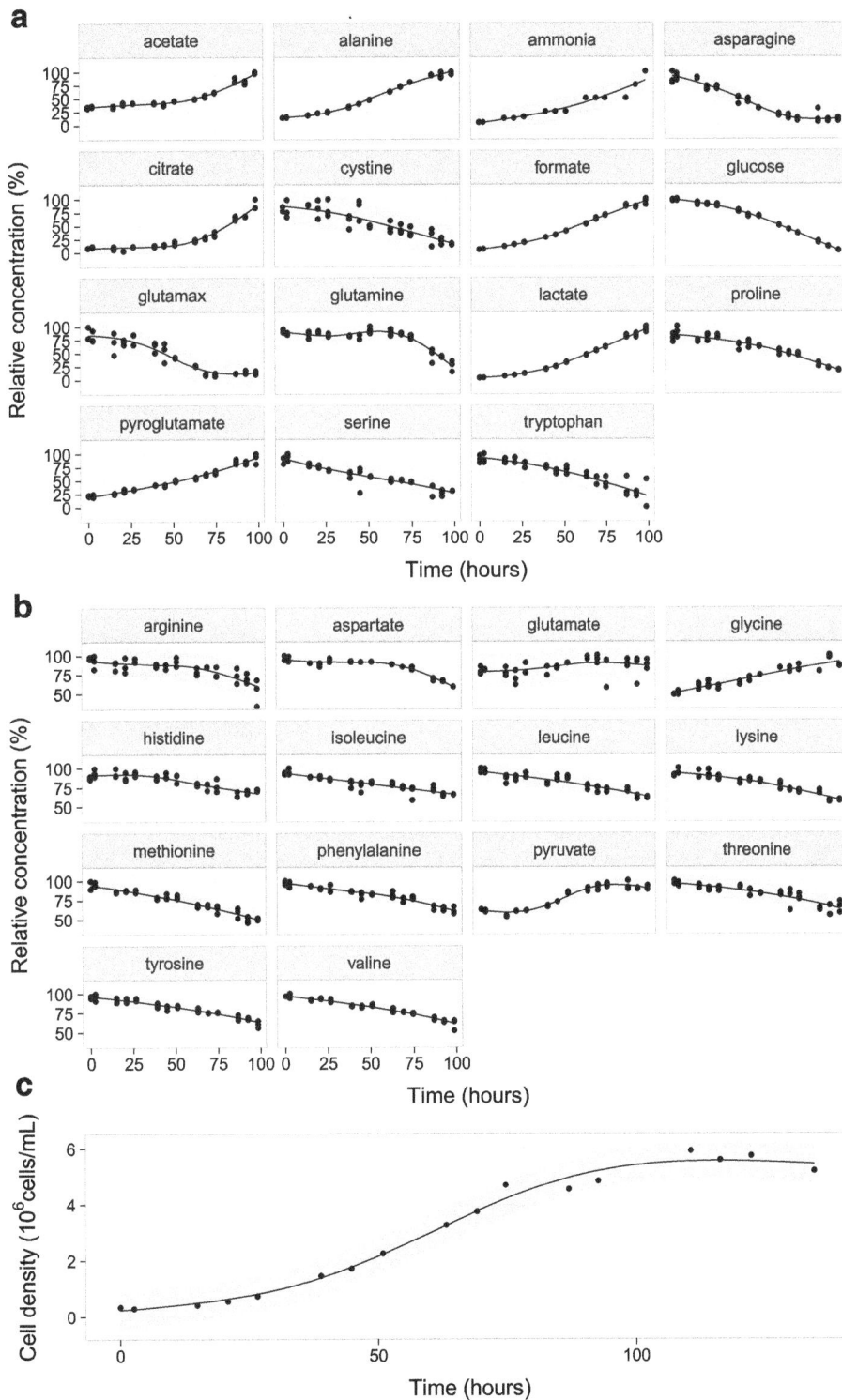

Fig. 1 Observed time-course trends. The presented data depicts all metabolic trends from a CHO cell cultivation carried out in a batch reactor (see Cell culture section of Methods for more detailed information). A single timepoint of 66 hours was chosen for MFA analysis, corresponding to the midpoint of the exponential phase (where the cells are likely to grow under pseudo steady state conditions). Panels depict **a** metabolites that changed by more than 50 % of their maximum concentration, **b** those that changed by less than 50 %, and **c** cell density. All metabolite concentrations are expressed as fractions of their maximum value. Curves were calculated from cubic regression spline fits constrained to 4 basis functions. Grey area designates 99 % prediction interval used for sampling

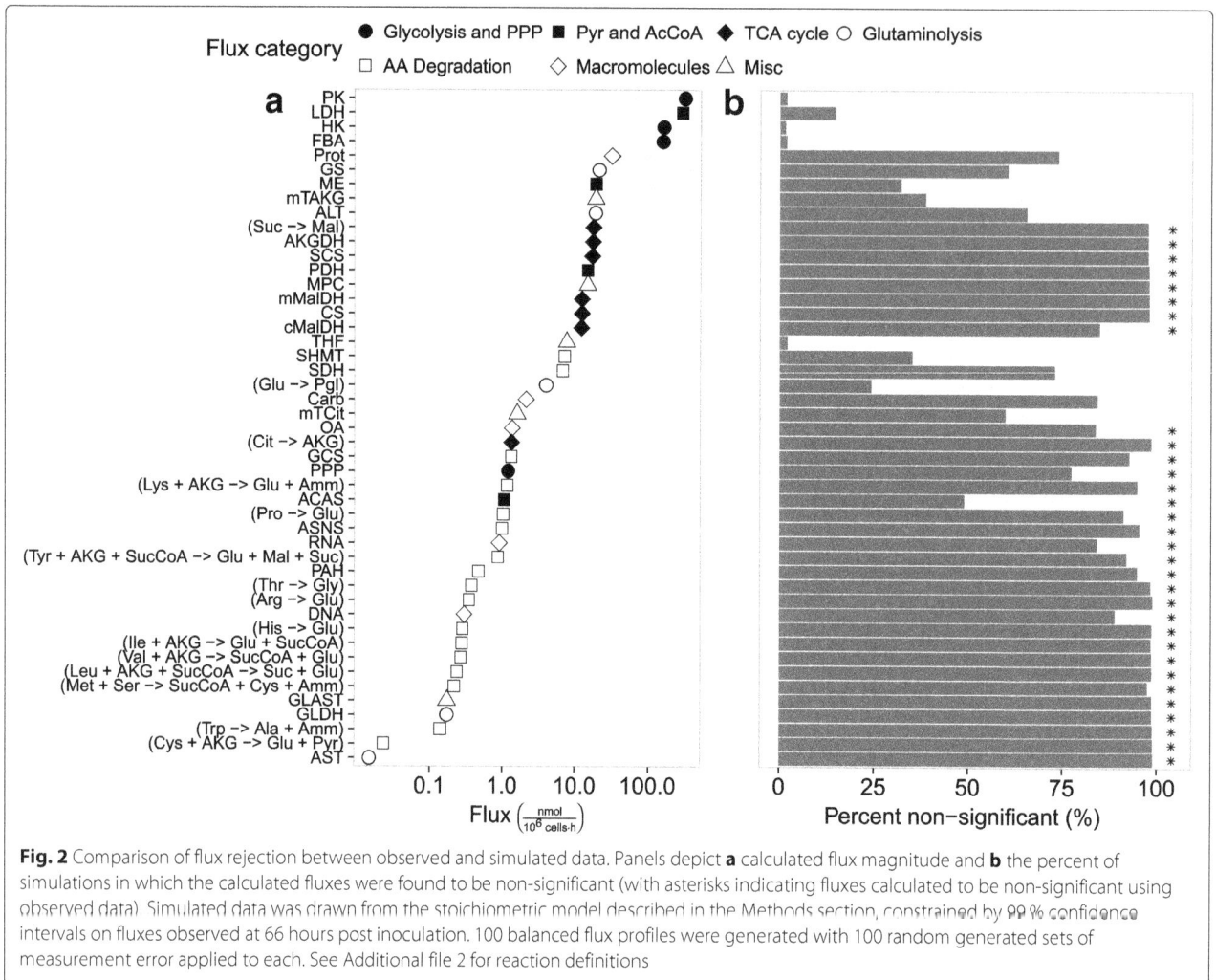

Fig. 2 Comparison of flux rejection between observed and simulated data. Panels depict **a** calculated flux magnitude and **b** the percent of simulations in which the calculated fluxes were found to be non-significant (with asterisks indicating fluxes calculated to be non-significant using observed data). Simulated data was drawn from the stoichiometric model described in the Methods section, constrained by 99 % confidence intervals on fluxes observed at 66 hours post inoculation. 100 balanced flux profiles were generated with 100 random generated sets of measurement error applied to each. See Additional file 2 for reaction definitions

measurements with 5 % coefficient of variation resulted in rejection rates of 50 % or more across practically all fluxes. The TCA fluxes in particular (ranks 36 to 44) were rejected as non-significant 75 % of the time or more (at all levels of measurement variability). The high level of flux rejection at low levels of measurement variability suggested the uncertainty in MFA calculation using observed data was primarily due to model structure rather than the uncertainty of observed data. Despite passing traditional validation tests, the simulation of stoichiometrically balanced fluxes revealed that the model is incapable of explaining observed metabolic profiles with an acceptable degree of confidence.

Effect of model structure

To test the influence of model structure on the significance of calculated fluxes, we simulated the effect of a broken electron transport chain – allowing a closed balanced on NADH and NADPH. Essentially, NADH and NADPH were reintroduced into the model and assumed

to be balanced by the defined stoichiometric relations. Although arbitrary, this assumption is consistent with largely anaerobic metabolism of CHO cells (termed the "Warburg Effect") and allowed the addition of balances around intermediate compounds participating in many reactions. Incorporating the modified model into analysis of the observed fluxes at 66 hours post inoculation revealed no sign of gross measurement error (χ^2 p-value of 0.91) and decreased the number of non-significant fluxes from 32 (of 47) to 16. As before, 10 000 sets of fluxes were simulated from 99 % confidence intervals around the observed measurement fluxes, subject to observed measurement error (Fig. 4). In comparison to Figs. 2b, Fig. 4b reveals a considerable increase in significance across a large number of fluxes, consistent with the idea that model structure plays an important role in uncertainty around calculated fluxes. The impact was particularly drastic for TCA fluxes, most of which changed from entirely non-significant to significant. Despite the improvement in model fit, some model error

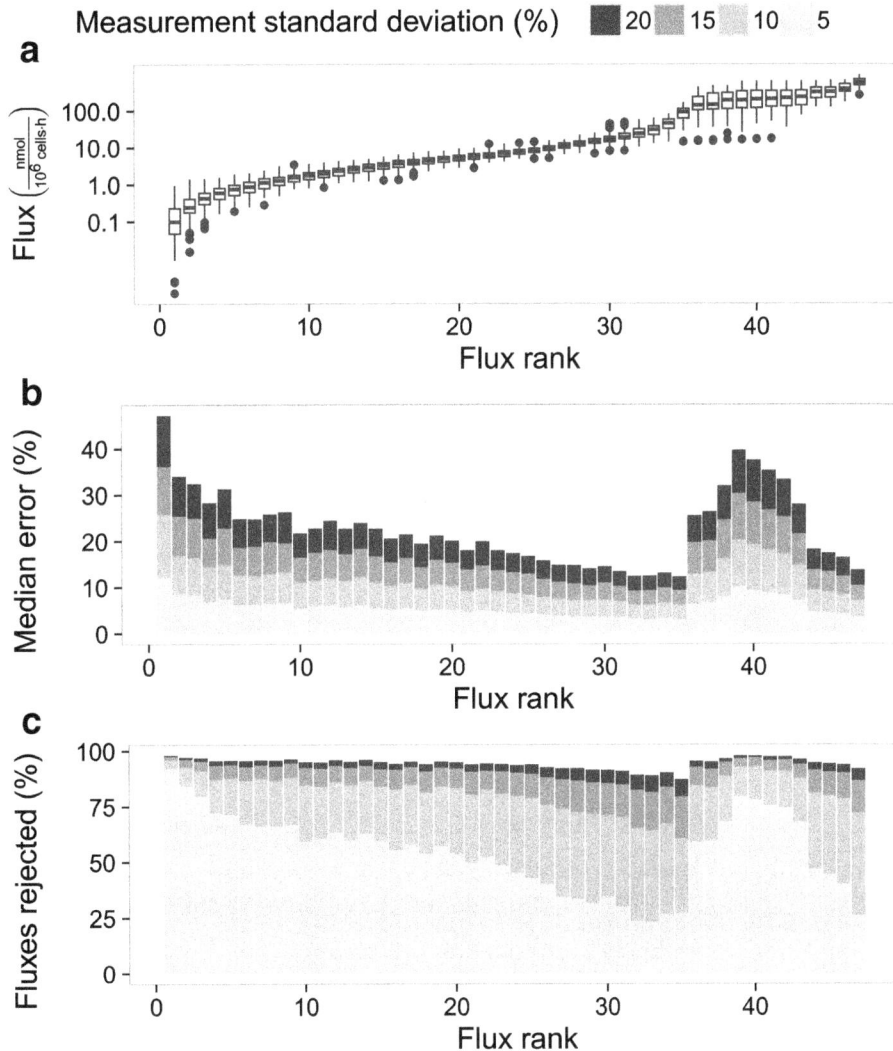

Fig. 3 Comparison of fluxes simulated with different measurement errors. Panels depict **a** flux magnitude, **b** median error, and **c** percent non-significance. Simulated data was drawn from the stoichiometric model described in the Methods section, constrained by 99 % confidence intervals on fluxes observed between 18 and 80 hours post inoculation. 100 balanced flux profiles were generated with 100 random generated sets of measurement error applied to each. Each balanced flux profile was ordered according to increasing absolute flux magnitude to generate an associated rank from 1 to 45

could also be observed – too many of the low magnitude fluxes calculated from observed data were found to be non-significant when compared to the simulated results.

The modified model was also tested with an extended simulation (Fig. 5). As with the original model, 99 % confidence intervals were generated for all fluxes 18–80 hours post inoculation with the minimum and maximum values for each flux used to bound the flux solution space. The most pronounced impact of the modification was on the rate of flux rejection (Fig. 5c). At 5 % measurement variability, approximately two thirds of the fluxes were always significant. The remaining third of the lowest

magnitude fluxes were significant at least 50 % of the time. In comparison, none of the fluxes calculated with the original model were significant for more than 75 % of the simulations. To get a better idea of how the t-test metric related to flux inaccuracy, median errors were separated for significant and non-significant fluxes. At 5 % coefficient of variation, fluxes deemed statistically significant had a constant median error of less than 5 % (with relation to flux rank), while non-significant fluxes had considerably higher errors (Fig. 6). Increasing coefficients of variation resulted in dramatic increases in overall rates of flux rejection (Fig. 5c). However, the median error of statistically significant fluxes also increased, diminishing

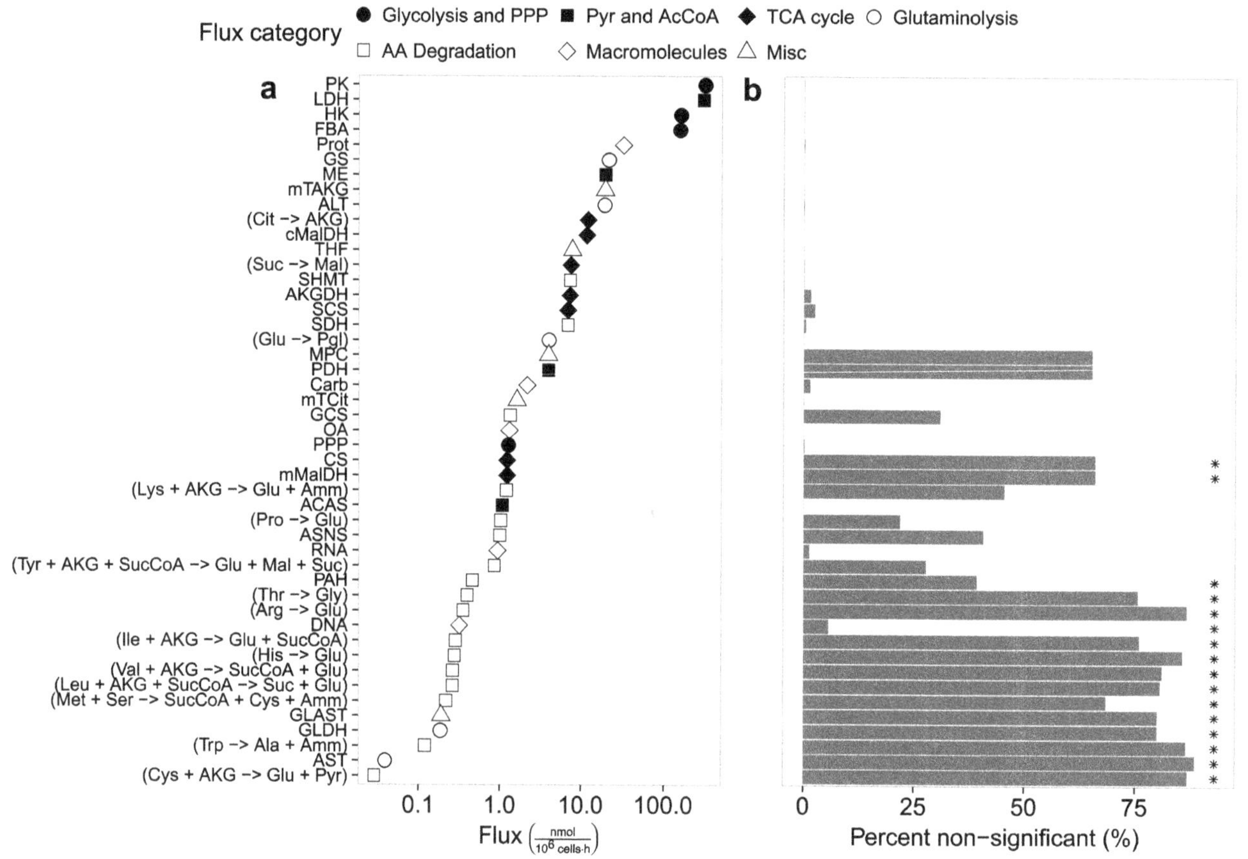

Fig. 4 Comparison of flux rejection between observed and simulated data following model modification. Panels depict **a** calculated flux magnitude and **b** the percent of simulations in which the calculated fluxes were found to be non-significant (with asterisks indicating fluxes calculated to be non-significant using observed data). Simulated data was drawn from a modification of the stoichiometric model described in the Methods section (with balances on NADH and NADPH), constrained by 99 % confidence intervals on fluxes observed at 66 hours post inoculation. 100 balanced flux profiles were generated with 100 random generated sets of measurement error applied to each. See Additional file 2 for reaction definitions

the ability of the t-test metric to identify inaccuracy in higher magnitude fluxes (Fig. 6). In comparison, the typical χ^2 test retained a 5 % rejection rate for all measurement errors (equal to the false positive rate).

Discussion

Taken together, the results of the simulations suggest that both measurement uncertainty and model structure have an impact on MFA results that are not assessed by typical validation methods. The structure of the model may lead to a considerable amount of uncertainty around calculated fluxes despite a high level of measurement precision. Mathematically, this impact can be seen in the $\left(S_c'^T S_c'\right)^{-1}$ term that stems from the variance of estimated regression parameters, i.e., $\mathrm{Cov}(\hat{\beta})$. Less formally, it may be intuitive that a model featuring a balance on important intermediate metabolites such as NADH and NADPH would be able to estimate intracellular fluxes with a greater degree of confidence than a model without the extra information

afforded by the balance. Naturally, the addition of isotopically labelled substrates can add a much greater degree of certainty. Indeed, an important application of the proposed testing and simulation framework is to provide a rigorous assessment of when extra information from sources such as labelled substrate would be essential for accurate flux calculation.

The proposed framework integrates a number of validation steps. While the t-test offers a straightforward post-regression significance test, combining the t-test with balanced flux simulation provides a convenient assessment of practical model identifiability [41, 42]. In addition, comparing the results from simulated and observed values can identify a lack of fit between model and measured data. Model fit is particularly important in the context of overdetermined MFA due to the large degree of simplification involved in model generation. Our findings suggest that the results of such simplification may be poor identifiability and lack of fit. These

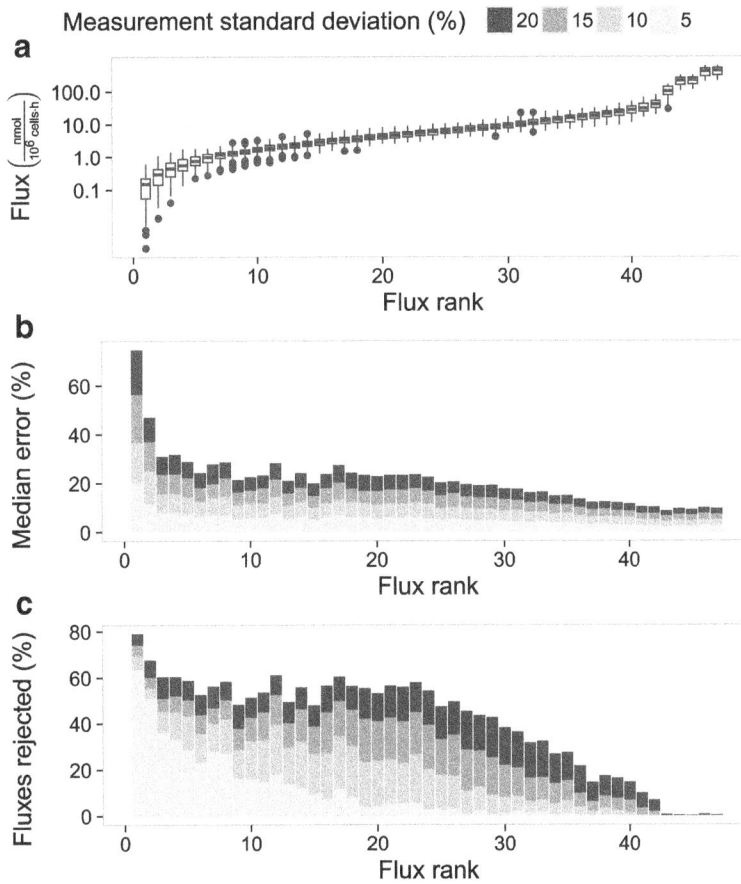

Fig. 5 Comparison of fluxes simulated with different measurement errors following model modification. Panels depict **a** flux magnitude, **b** median error, and **c** percent non-significance of fluxes simulated with different measurement errors. Simulated data was drawn from a modification of the stoichiometric model described in the Methods section (with balances on NADH and NADPH), constrained by 99 % confidence intervals on fluxes observed between 18 and 80 hours post inoculation. 100 balanced flux profiles were generated with 100 random generated sets of measurement error applied to each. Each balanced flux profile was ordered according to increasing absolute flux magnitude to generate an associated rank from 1 to 45

issues are rarely considered outside of "gross measurement error" detection. The combination of t-test validation and balanced flux simulation offers a simple and practical approach that avoids the assumption of model validity in the determination of significance. Although this validation strategy was developed for the analysis of simplified metabolic models, it should be equally useful at larger scales provided that enough observations are available.

It is important to note that the GLS framework for validation is more robust to estimated measurement error than the standard χ^2 test. GLS regression only requires an estimate of relative measurement variance and covariance in the form of V. Residual variance magnitude ($\hat{\sigma}^2$) is still estimated from the model. On the other hand, variance scaling in the χ^2 test allows for large measurement variance to reduce the χ^2 statistic. Effectively, high variability leads to a lower confidence that deviations are

not normally distributed. Given that variance does not factor into any other aspect of validation, assuming a large variance can serve as a way to avoid dealing with lack of fit.

Following the case study presented in this work, we recommend the following validation procedure. Before any experiments are carried out (but after a model of interest has been identified), construct reasonable limits around each observable flux from literature or other available data. Simulate flux profiles from the constrained flux space and perturb them with a range of measurement errors. If the flux space is infeasible, there is considerable disagreement between fluxes and the model that needs to be resolved. Otherwise, generate confidence intervals around the calculated fluxes and calculate the proportion of simulated fluxes that are non-significant. If many high magnitude fluxes are found to be non-significant in the majority of simulations

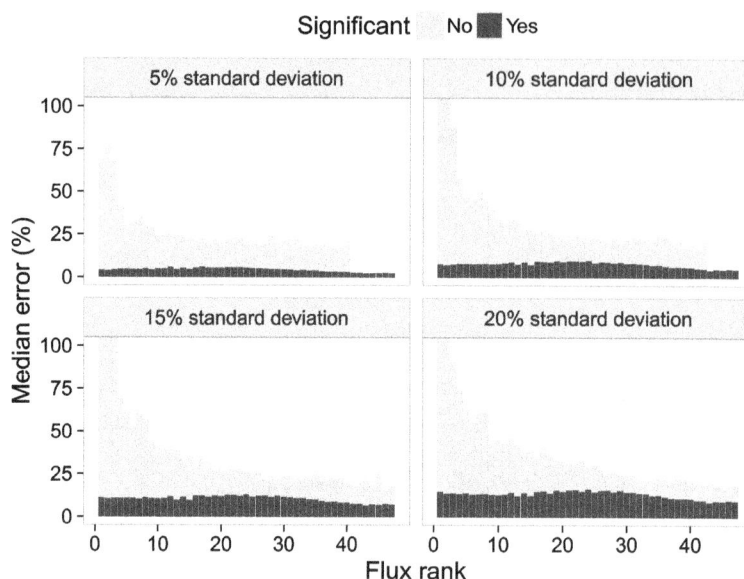

Fig. 6 Comparison of median error of significant and non-significant fluxes (determined by t-test with $\alpha = 0.05$) simulated with different measurement errors. Simulated data was drawn from a modification of the stoichiometric model described in the Methods section (with balances on NADH and NADPH), constrained by 99 % confidence intervals on fluxes observed between 18 and 80 hours post inoculation. 100 balanced flux profiles were generated with 100 random generated sets of measurement error applied to each. Each balanced flux profile was ordered according to increasing absolute flux magnitude to generate an associated rank from 1 to 45

(regardless of measurement error), then the model may have structural issues that need to be resolved. Alternatively, extra flux information may be required. If the model is sound, then experiments can be carried out and collected data analyzed via MFA. Apply the model and generate confidence intervals around calculated fluxes. Construct limits in close vicinity of observed values, simulate flux profiles, and perturb them with estimated measurement error. If the confidence intervals of simulated fluxes are considerably smaller than those of observed fluxes, then the model may have errors resulting in a lack of fit.

Conclusion

The interpretation of MFA through the GLS framework underscores the need for robust validation methods. The mathematical equivalence of MFA and regression suggests that the failure to follow good practices of regression analysis can lead to questionable results. This work highlights the application of simple t-tests for the detection of error due to measurement variability and presents a means to directly assess model error via flux profile simulation. At the same time, we bring attention to the impact of measurement variability on model identifiability, underlining the need for better reporting. Although this work has focused on the validation of a traditional MFA model via t-test analysis, the overall framework is likely

to be just as applicable to other regression validation methods or alternative MFA formulations (such as dynamic MFA).

Endnotes

[1] A more detailed discussion of the theoretical principles, including a worked example and some proofs, is available as an Additional file 1.

[2] It is typically assumed that S_c is sufficient for the estimation of all v_c values. However, failure to observe a key metabolite may result in a case where not all values of v_c can be estimated despite S_c appearing determined or overdetermined. See [30] for details on stoichiometry matrix classification.

Additional files

Additional file 1: Extended theoretical principles. To make the proposed protocol as accessible as possible, the theoretical section has been extended with a number extra details as well as a simplified example model.

Additional file 2: Detailed model description and metabolic map. Definitions of all reactions included in the metabolic model.

Additional file 3: SBML file. An SBML representation of the metabolic model.

Abbreviations

ATP: Adenosine triphosphate; CHO: Chinese hamster ovary; COBRA: Constraint-based reconstruction and analysis; FADH: Flavin adenine dinucleotide; GLS: Generalized least squares; MFA: Metabolic flux analysis; NADH: Nicotinamide adenine dinucleotide; NADPH: Nicotinamide adenine dinucleotide phosphate; NMR: Nuclear magnetic resonance; SBML: Systems biology markup language; TCA: Tricarboxylic acid [cycle]

Acknowledgements
The authors would like to thank Steffen Schulze and Eric J Blondeel for their assistance with bioreactor cultivation.

Funding
The work was supported in part by an NSERC Canada Graduate Scholarship to SS as well as NSERC Discovery (RGPIN 355513-2012) and Mabnet Strategic Network (NETGP 380070-08) grants to MGA. The funding provider did not play a role in the design of the study; collection, analysis, and interpretation of data; or writing the manuscript.

Authors' contributions
SS designed the validation method, performed the analysis, and assisted with data collection. MQ assisted with data collection and performed NMR quantification. MGA assisted in data interpretation and manuscript preparation. All authors read and approved the final manuscript.

Competing interests
The authors declare that they have no competing interests.

References
1. Chen C, Le H, Goudar CT. Integration of systems biology in cell line and process development for biopharmaceutical manufacturing. Biochem Eng J. 2016;107:11–17. doi:10.1016/j.bej.2015.11.013.
2. Stephanopoulos G, Aristidou A, Nielsen J. Metabolic engineering: principles and methodologies. San Diego: Academic Press; 1998.
3. Wang NS, Stephanopoulos G. Application of macroscopic balances to the identification of gross measurement errors. Biotechnol Bioeng. 1983;25(9):2177–08. doi:10.1002/bit.260250906.
4. Bordbar A, Monk JM, King ZA, Palsson BO. Constraint-based models predict metabolic and associated cellular functions. Nat Rev Genet. 2014;15(2):107–20. doi:10.1038/nrg3643.
5. Lewis NE, Nagarajan H, Palsson BO. Constraining the metabolic genotype-phenotype relationship using a phylogeny of in silico methods. Nat Rev Microbiol. 2012;10(4):291–305. doi:10.1038/nrmicro2737.
6. Dauner M. From fluxes and isotope labeling patterns towards in silico cells. Curr Opin Biotechnol. 2010;21(1):55–62. doi:10.1016/j.copbio.2010.01.014.
7. Maertens J, Vanrolleghem PA. Modeling with a view to target identification in metabolic engineering: a critical evaluation of the available tools. Biotechnol Prog. 2010;26(2):313–1. doi:10.1002/btpr.349.
8. Boghigian BA, Seth G, Kiss R, Pfeifer BA. Metabolic flux analysis and pharmaceutical production. Metab Eng. 2010;12(2):81–95. doi:10.1016/j.ymben.2009.10.004.
9. Shlomi T, Eisenberg Y, Sharan R, Ruppin E. A genome-scale computational study of the interplay between transcriptional regulation and metabolism. Mol Syst Biol. 2007;3(101):101. doi:10.1038/msb4100141.
10. Krömer JO, Sorgenfrei O, Klopprogge K, Heinzle E, Wittmann C. In-depth profiling of lysine-producing Corynebacterium glutamicum by combined analysis of the transcriptome, metabolome, and fluxome. J Bacteriol. 2004;186(6):1769–84. doi:10.1128/JB.186.6.1769.

11. Kaas CS, Kristensen C, Betenbaugh MJ, Andersen MR. Sequencing the CHO DXB11 genome reveals regional variations in genomic stability and haploidy. BMC Genomics. 2015;16:160. doi:10.1186/s12864-015-1391-x.
12. Feichtinger J, Hernández I, Fischer C, Hanscho M, Auer N, Hackl M, Jadhav V, Baumann M, Krempl PM, Schmidl C, Farlik M, Schuster M, Merkel A, Sommer A, Heath S, Rico D, Bock C, Thallinger GG, Borth N. Comprehensive genome and epigenome characterization of CHO cells in response to evolutionary pressures and over time. Biotechnol Bioeng. 2016. doi:10.1002/bit.25990.
13. Antoniewicz MR. Methods and advances in metabolic flux analysis: A mini-review. J Ind Microbiol Biotechnol. 2015;42(3):317–25. doi:10.1007/s10295-015-1585-x.
14. Young JD. (13)C metabolic flux analysis of recombinant expression hosts. Curr Opin Biotechnol. 2014;30:238–45. doi:10.1016/j.copbio.2014.10.004.
15. Mashego MR, Rumbold K, De Mey M, Vandamme E, Soetaert W, Heijnen JJ. Microbial metabolomics: past, present and future methodologies. Biotechnol Lett. 2007;29(1):1–16. doi:10.1007/s10529-006-9218-0.
16. Quek LE, Dietmair S, Krömer JO, Nielsen LK. Metabolic flux analysis in mammalian cell culture. Metab Eng. 2010;12(2):161–71. doi:10.1016/j.ymben.2009.09.002.
17. Quek LE, Dietmair S, Hanscho M, Martínez VS, Borth N, Nielsen LK. Reducing recon 2 for steady-state flux analysis of HEK cell culture. J Biotechnol. 2014;184:172–8. doi:10.1016/j.jbiotec.2014.05.021.
18. Klamt S, Saez-Rodriguez J, Gilles ED. Structural and functional analysis of cellular networks with cellnetanalyzer. BMC Syst Biol. 2007;1(1):1–13. doi:10.1186/1752-0509-1-2.
19. Klamt S, von Kamp A. An application programming interface for CellNetAnalyzer. BioSystems. 2011;105(2):162–8. doi:10.1016/j.biosystems.2011.02.002.
20. Erdrich P, Steuer R, Klamt S. An algorithm for the reduction of genome-scale metabolic network models to meaningful core models. BMC Syst Biol. 2015;9:48. doi:10.1186/s12918-015-0191-x.
21. Xing Z, Kenty B, Koyrakh I, Borys M, Pan SH, Li Z. Optimizing amino acid composition of CHO cell culture media for a fusion protein production. Process Biochem. 2011;46(7):1423–9. doi:10.1016/j.procbio.2011.03.014.
22. Niklas J, Schräder E, Sandig V, Noll T, Heinzle E. Quantitative characterization of metabolism and metabolic shifts during growth of the new human cell line AGE1.HN using time resolved metabolic flux analysis. Bioprocess Biosyst Eng. 2011;34(5):533–45. doi:10.1007/s00449-010-0502-y.
23. Niklas J, Priesnitz C, Rose T, Sandig V, Heinzle E. Primary metabolism in the new human cell line AGE1.HN at various substrate levels: increased metabolic efficiency and α1-antitrypsin production at reduced pyruvate load. Appl Microbiol Biotechnol. 2012;93(4):1637–50. doi:10.1007/s00253-011-3526-6.
24. Priesnitz C, Niklas J, Rose T, Sandig V, Heinzle E. Metabolic flux rearrangement in the amino acid metabolism reduces ammonia stress in the α1-antitrypsin producing human AGE1.HN cell line. Metab Eng. 2012;14(2):128–37. doi:10.1016/j.ymben.2012.01.001.
25. Bernal V, Carinhas N, Yokomizo AY, Carrondo MJT, Alves PM. Cell density effect in the baculovirus-insect cells system: A quantitative analysis of energetic metabolism. Biotechnol Bioeng. 2009;104(1):162–80. doi:10.1002/bit.22364.
26. Carinhas N, Bernal V, Monteiro F, Carrondo MJT, Oliveira R, Alves PM. Improving baculovirus production at high cell density through manipulation of energy metabolism. Metab Eng. 2010;12(1):39–52. doi:10.1016/j.ymben.2009.08.008.
27. Carinhas N, Duarte TM, Barreiro LC, Carrondo MJT, Alves PM, Teixeira AP. Metabolic signatures of GS-CHO cell clones associated with butyrate treatment and culture phase transition. Biotechnol Bioeng. 2013;110(12):3244–57. doi:10.1002/bit.24983.
28. Vallino JJ, Stephanopoulos GN. Flux determination in cellular bioreaction networks: Applications to lysine fermentations. In: Sikdar SK, Bier M, editors. Frontiers in Bioprocessing. Boulder, Colorado: CRC Press; 1990. p. 205–19.
29. Goudar CT, Biener R, Konstantinov KB, Piret JM. Error propagation from prime variables into specific rates and metabolic fluxes for mammalian cells in perfusion culture. Biotechnol Prog. 2009;25(4):986–8. doi:10.1002/btpr.155.

30. van der Heijden RT, Romein B, Heijnen JJ, Hellinga C, Luyben KC. Linear constraint relations in biochemical reaction systems: II, Diagnosis and estimation of gross errors. Biotechnol Bioeng. 1994;43(1):11–20. doi:10.1002/bit.260430104.

31. Leighty RW, Antoniewicz MR. Dynamic metabolic flux analysis (DMFA): A framework for determining fluxes at metabolic non-steady state. Metab Eng. 2011;13(6):745–55. doi:10.1016/j.ymben.2011.09.010.

32. Antoniewicz MR, Kelleher JK, Stephanopoulos G. Determination of confidence intervals of metabolic fluxes estimated from stable isotope measurements. Metab Eng. 2006;8(4):324–7. doi:10.1016/j.ymben.2006.01.004.

33. Sokolenko S, Aucoin MG. A correction method for systematic error in (1)H-NMR time-course data validated through stochastic cell culture simulation. BMC Syst Biol. 2015;9:51. doi:10.1186/s12918-015-0197-4.

34. Altamirano C, Illanes A, Casablancas A, Gámez X, Cairó JJ, Gòdia C. Analysis of CHO cells metabolic redistribution in a glutamate-based defined medium in continuous culture. Biotechnol Prog. 2001;17(6):1032–41. doi:10.1021/bp0100981.

35. Kumar A, Bachhawat AK. Pyroglutamic acid: Throwing light on a lightly studied metabolite. Curr Sci. 2012;102(2):288–97.

36. Wood SN. Fast stable restricted maximum likelihood and marginal likelihood estimation of semiparametric generalized linear models. J R Stat Soc B. 2011;73(1):3–36. doi:10.1111/j.1467-9868.2010.00749.x.

37. R Core Team. R: A Language and Environment for Statistical Computing. Vienna, Austria: R Foundation for Statistical Computing; 2016.

38. Smith R. Efficient Monte Carlo procedures for generating points uniformly distributed over bounded regions. Oper Res. 1984;32(6):1296–308.

39. Van den Meersche K, Soetaert K, Van Oevelen D. xsample(): An R function for sampling linear inverse problems. J Stat Softw. 2009;30(1):1296–308.

40. Sokolenko S, Blondeel EJM, Azlah N, George B, Schulze S, Chang D, Aucoin MG. Profiling convoluted single-dimension proton NMR spectra: A Plackett-Burman approach for assessing quantification error of metabolites in complex mixtures with application to cell culture. Anal Chem. 2014;86(7):3330–7. doi:10.1021/ac4033966.

41. Banga JR, Balsa-Canto E. Parameter estimation and optimal experimental design. Essays Biochem. 2008;45:195–209. doi:10.1042/BSE0450195.

42. Jaqaman K, Danuser G. Linking data to models: data regression. Nat Rev Mol Cell Biol. 2006;7(11):813–9. doi:10.1038/nrm2030.

7

Quantifying the roles of random motility and directed motility using advection-diffusion theory for a 3T3 fibroblast cell migration assay stimulated with an electric field

Matthew J. Simpson[1]* (iD), Kai-Yin Lo[2] and Yung-Shin Sun[3]

Abstract

Background: Directed cell migration can be driven by a range of external stimuli, such as spatial gradients of: chemical signals (chemotaxis); adhesion sites (haptotaxis); or temperature (thermotaxis). Continuum models of cell migration typically include a diffusion term to capture the undirected component of cell motility and an advection term to capture the directed component of cell motility. However, there is no consensus in the literature about the form that the advection term takes. Some theoretical studies suggest that the advection term ought to include receptor saturation effects. However, others adopt a much simpler constant coefficient. One of the limitations of including receptor saturation effects is that it introduces several additional unknown parameters into the model. Therefore, a relevant research question is to investigate whether directed cell migration is best described by a simple constant tactic coefficient or a more complicated model incorporating saturation effects.

Results: We study directed cell migration using an experimental device in which the directed component of the cell motility is driven by a spatial gradient of electric potential, which is known as electrotaxis. The electric field (EF) is proportional to the spatial gradient of the electric potential. The spatial variation of electric potential across the experimental device varies in such a way that there are several subregions on the device in which the EF takes on different values that are approximately constant within those subregions. We use cell trajectory data to quantify the motion of 3T3 fibroblast cells at different locations on the device to examine how different values of the EF influences cell motility. The undirected (random) motility of the cells is quantified in terms of the cell diffusivity, D, and the directed motility is quantified in terms of a cell drift velocity, v. Estimates D and v are obtained under a range of four different EF conditions, which correspond to normal physiological conditions. Our results suggest that there is no anisotropy in D, and that D appears to be approximately independent of the EF and the electric potential. The drift velocity increases approximately linearly with the EF, suggesting that the simplest linear advection term, with no additional saturation parameters, provides a good explanation of these physiologically relevant data.

Conclusions: We find that the simplest linear advection term in a continuum model of directed cell motility is sufficient to describe a range of different electrotaxis experiments for 3T3 fibroblast cells subject to normal physiological values of the electric field. This is useful information because alternative models that include saturation effects involve additional parameters that need to be estimated before a partial differential equation model can be applied to interpret or predict a cell migration experiment.

Keywords: Cell migration, Random motility, Directed motility, Electrotaxis, Partial differential equation, Keller-Segal model

*Correspondence: matthew.simpson@qut.edu.au
[1] School of Mathematical Sciences, Queensland University of Technology (QUT), Brisbane, Australia
Full list of author information is available at the end of the article

Background

Continuum models are used to describe cell migration in a number of contexts including wound repair [1–3] and malignant invasion [4, 5]. Here, we consider a continuum partial differential equation to describe the motion of a population of cells, with cell density $C(x, y, t)$, where x and y are the Cartesian coordinates, and t is time. The continuum model allows the cell migration mechanism to involve an undirected (diffusive) and directed (tactic) component. Conservation arguments lead to

$$\frac{\partial C}{\partial t} = \frac{\partial}{\partial x}\left(D(S)\frac{\partial C}{\partial x}\right) + \frac{\partial}{\partial y}\left(D(S)\frac{\partial C}{\partial y}\right) \\ - \frac{\partial}{\partial x}\left(\chi(S)\frac{\partial S}{\partial x}C\right) - \frac{\partial}{\partial y}\left(\chi(S)\frac{\partial S}{\partial y}C\right), \tag{1}$$

where $D(S) > 0$ is the cell diffusivity, and $\chi(S)$ is the tactic sensitivity function. In this Keller-Segel [6] type model, the tactic flux is proportional to the gradient of some signal, $S(x, y, t)$, and the strength of the tactic response is governed by the tactic sensitivity function, $\chi(S)$ [6, 7]. Setting $\chi(S) > 0$ represents attraction, since the directed component of the cell flux is in the direction of increasing S. Alternatively, setting $\chi(S) < 0$ represents repulsion. To maintain generality, the cell diffusivity $D(S) > 0$ is also written as a function of the signal, S [1, 8, 9]. If $D(S)$ is increasing, this model represents an increase in undirected motility with the signal, as in the case of chemokinesis [10]. Since there is no source/sink term in Eq. (1) we are focusing on cell migration processes on short time scales so that cell proliferation and cell death have a negligible impact on the cell density.

Directed cell migration can occur in response to various types of external spatial gradients. In Eq. (1) we have not specified the physical interpretation of S. In a model of chemotaxis S would represent the concentration of a chemical signal, whereas in a model of thermotaxis S would represent the temperature. In a model of electrotaxis S represents the electric potential. In this work we focus on stimulating directed cell migration in an electric field.

Electrotaxis plays an important role in guiding epithelial and corneal wound healing processes, and could potentially be used to design novel therapies [11–16]. While the precise molecular-level mechanisms behind electrotaxis remain unresolved, a common hypothesis is that exposing cells to an electric field leads to changes in plasma membrane potentials [11, 12] with the membrane facing the cathode becoming depolarized, and the membrane facing the anode becoming hyperpolarized [11, 12]. In a cell with negligible voltage-gated conductance, the hyperpolarized membrane attracts calcium ions, leading to a contraction of this side of the cell which propels the cell toward the cathode [11, 12]. In a cell with voltage-gated calcium channels, the channels near the depolarized side

open to allow an influx of calcium ions leading to a rise in the intracellular calcium ion level throughout such a cell. The direction of cell movement in this situation will depend on the balance between the opposing contractile forces [11, 12].

A key question in applying Eq. (1) is to determine the functional forms of $D(S)$ and $\chi(S)$. In many theoretical studies focusing on directed cell movement, an explicit relationship between the tactic response function and the signal, S, is emphasized. Often, particularly in more theoretical studies, an argument about saturation of receptor cites on cells is made to suggest that $\chi(S)$ ought to be a decreasing function of S, so that $d\chi/dS < 0$ [6, 7]. Several putative functional forms have been put forward. For example, relationships such as $\chi(S) = \chi_0/S$ and $\chi(S) = \chi_0 K / (K + S)^2$, and several others, have been suggested [13, 18–23]. In contrast, other studies simply adopt a constant $\chi(S) = \chi$ [24–33]. Under the situation where we treat $D(S)$ and $\chi(S)$ as constants, Eq. (1) simplifies to

$$\frac{\partial C}{\partial t} = D\left[\frac{\partial^2 C}{\partial x^2} + \frac{\partial^2 C}{\partial y^2}\right] - \chi\left[\frac{\partial}{\partial x}\left(\frac{\partial S}{\partial x}C\right) + \frac{\partial}{\partial y}\left(\frac{\partial S}{\partial y}C\right)\right]. \tag{2}$$

The advantage of working with Eq. (2) compared to Eq. (1), is that there are just two unknown parameters in Eq. (2), χ and D. In contrast, the more complicated models involving receptor saturation effects can involve six or more unknown parameters [13, 18–23].

Making a distinction between choosing models where the tactic sensitivity incorporates receptor saturation effect (Eq. (1)) and a simpler model where the tactic sensitivity coefficient is constant (Eq. (2)) is not obvious unless we are guided by a reasonable quantity of experimental data. From a theoretical point of view, it might be attractive to incorporate receptor saturation dynamics into a mathematical model, but this comes with the trade off that this is typically achieved by introducing a complicated relationship between the tactic sensitivity coefficient and the attractant concentration, which can introduce several unknown parameters into the mathematical model thereby over complicating the process of model calibration [17]. To provide some insight into this question, here we analyze a suite of cell migration data. The data we analyze comes from an electrotaxis experiment where the strength of the attraction gradient is carefully varied so that we can analyze both the random component of the cell migration as well as the directed component over a range of applied gradients.

Results

Qualitative assessment of trajectory data

Cell trajectory data, describing the motion of 80 randomly-chosen 3T3 fibroblast cells [34] (Fig. 1b) under

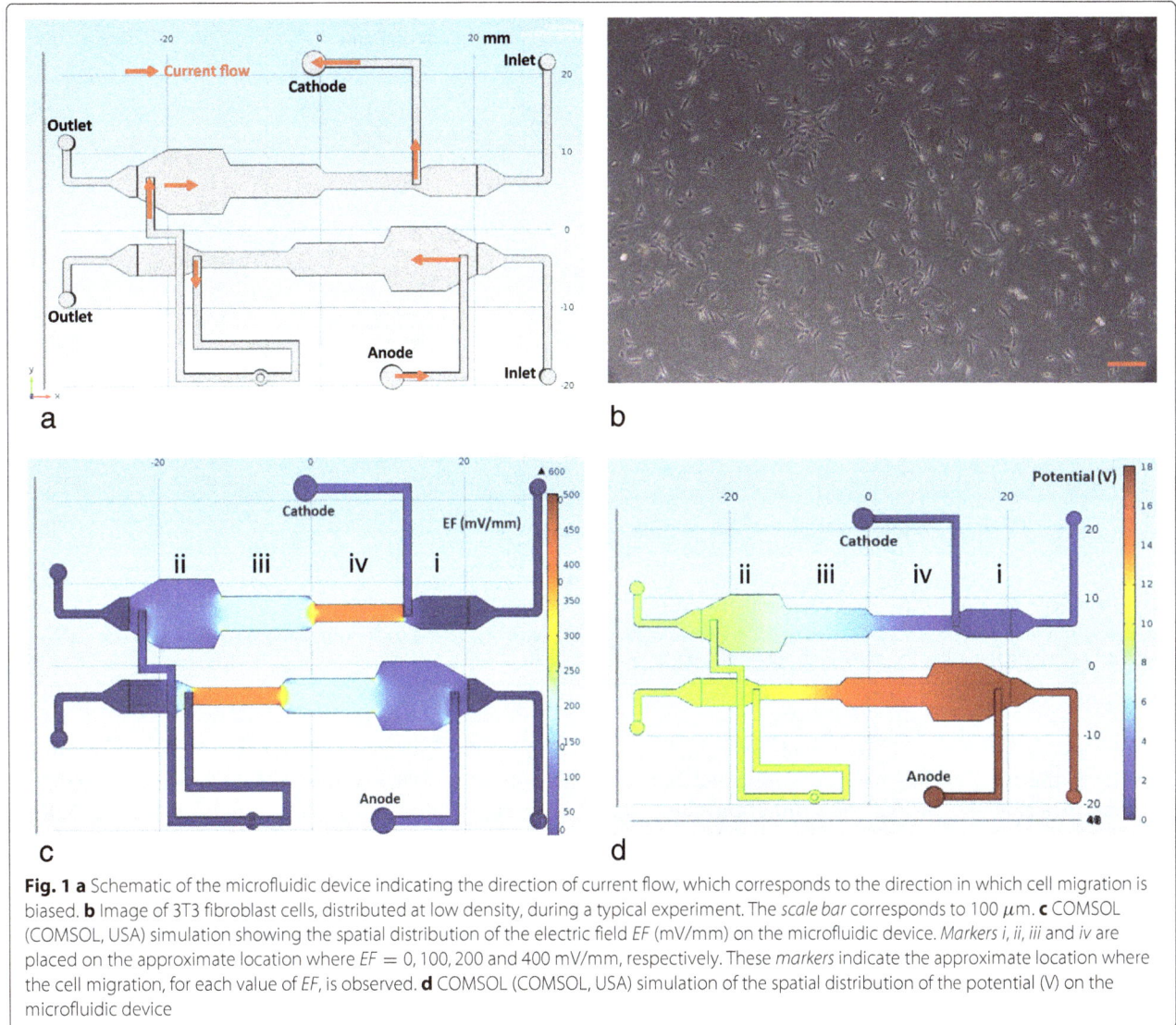

Fig. 1 a Schematic of the microfluidic device indicating the direction of current flow, which corresponds to the direction in which cell migration is biased. **b** Image of 3T3 fibroblast cells, distributed at low density, during a typical experiment. The *scale bar* corresponds to 100 μm. **c** COMSOL (COMSOL, USA) simulation showing the spatial distribution of the electric field *EF* (mV/mm) on the microfluidic device. *Markers i, ii, iii* and *iv* are placed on the approximate location where *EF* = 0, 100, 200 and 400 mV/mm, respectively. These *markers* indicate the approximate location where the cell migration, for each value of *EF*, is observed. **d** COMSOL (COMSOL, USA) simulation of the spatial distribution of the potential (V) on the microfluidic device

a range of gradients, $EF = 0, 100, 200$ and 400 mV/mm, within the experimental apparatus (Fig. 1a, c-d) are analysed [35]. Since 3T3 fibroblast cells are known to migrate towards the cathode in these types of experiments [35], the Cartesian coordinate axes are aligned so that the positive x-direction points towards the cathode (Fig. 1c-d). We note that there is no gradient in the y-direction (Fig. 1c-d).

The data involves recording the initial position of each trajectory, $(x'(0), y'(0))$ and the position of each cell every half-hour over a two hour interval, giving: $(x'(0.5), y'(0.5))$, $(x'(1), y'(1))$, $(x'(1.5), y'(1.5))$ and $(x'(2), y'(2))$. Using this data, we shift the coordinate system for each trajectory so that the initial location of the cell is at the origin, giving $(x(t), y(t)) = (x'(t) - x'(0), y'(t) - y'(0))$. Plots showing $(x(2), y(2))$ for 80 trajectories under four different gradients are shown in Fig. 2. The scatter plot in Fig. 2a, under the action of no gradient,

shows an approximately symmetric distribution of the end points of the trajectories. In this case the trajectories extend no further than approximately 40 μm away from the origin. Since these trajectories appear to follow no particular preferred direction, this cells seem to undergo an unbiased migration process. In comparison, the scatter plot in Fig. 2b shows that there is some drift in the positive x-direction when the cells move under the action of a gradient. Despite the fact that there is an obvious drift in the positive x-direction in Fig. 2b, there remains some randomness in the distribution of $(x(2), y(2))$. Therefore, under the action of the electric field, these 3T3 fibroblast cells move with both a directed and an undirected component. Comparing results in Fig. 2b-d confirms that the drift in the positive x-direction increases with the increasing electric field, and there appears to be some randomness in the distribution of cells regardless of the

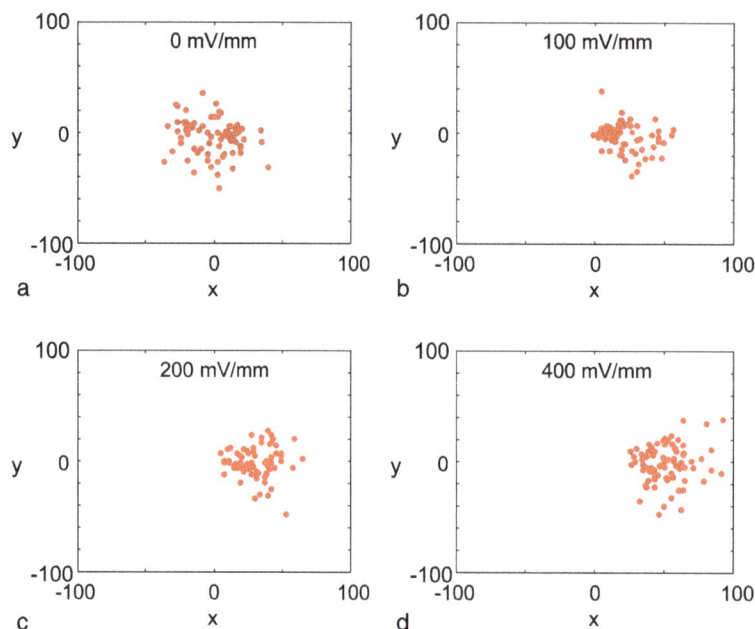

Fig. 2 End points of cell trajectories under different experimental conditions. Results correspond to: **a** *EF* = 0; **b** *EF* = 100; **c** *EF* = 200, and **d** *EF* = 400 mV/mm. All trajectories are shifted so that the initial location of the trajectory is at the origin. In each subfigure there are 80 *red dots*, each corresponding to the location of the each cell after a duration of two hours

strength of the electric field. To provide more information about the roles of directed and undirected motion in these experiments, we will now interpret this data using a biased random walk model that is related to an advection-diffusion equation.

Quantitative assessment of trajectory data

We first quantify the directed component of the motility depicted in Fig. 2. Estimates of the drift velocity are obtained, in both the x and y directions, for each of the 80 trajectories, under the four different gradient conditions. These data are presented as histograms in Fig. 3. Results in Fig. 3a-b characterize the estimates of v_x and v_y when there is no gradient, and averaging these 80 estimates gives us an approximation of the average drift velocity in each direction. This gives $\langle v_x \rangle = -1$ μm/h and $\langle v_y \rangle = -1$ μm/h. Therefore, the average drift velocity in both directions is approximately zero, as we anticipate intuitively by inspecting the data in Fig. 2a. Results in Fig. 3c-h show estimates of v_x and v_y for $EF = 100, 200$ and 400 mV/mm, respectively. In each case we see that $\langle v_y \rangle \approx 0$ μm/h, which is consistent with the experimental design since there is no gradient in the y direction (Fig. 1c-d). In contrast, estimates of $\langle v_x \rangle$ increase with EF, as we have $\langle v_x \rangle = -1, 9, 14$ and 25 μm/h when $EF = 0, 100, 200$ and 400 mV/mm, respectively. In addition to characterizing the mean drift velocities, $\langle v_x \rangle$ and $\langle v_y \rangle$, the data in the histograms in Fig. 3a-h show how the individual estimates of v_x and v_y are distributed for each of the 80 trajectories

considered. A qualitative assessment of these distributions indicates that, for each value of the *EF*, estimates of v_x and v_y are approximately symmetrically distributed about the mean. Furthermore, the spread about the mean appears to be approximately constant for each value of the *EF*.

Given our estimates of $\langle v_x \rangle$ and $\langle v_y \rangle$ (Fig. 3a-h), we now estimate the diffusivity coefficients, D_x and D_y, for each experiment. Results showing estimates of D_x and D_y under the application of no gradient are summarised in Fig. 3i-j. Averaging our estimates across the 80 trajectories we obtain $\langle D_x \rangle = 59$ μm^2/h and $\langle D_y \rangle = 50$ μm^2/h for the experiments in which there is no gradient. The magnitude of these estimates of cell diffusivity are consistent with previous estimates 3T3 fibroblast cells obtained using single cell trajectory data [36, 37]. Additional estimates of D_x and D_y, and $\langle D_x \rangle$ and $\langle D_y \rangle$ are shown in Fig. 3k-p for cell migration under the influence of gradients of 100, 200 and 400 mV/mm, respectively. For each of these data sets we have $\langle D_x \rangle \approx \langle D_y \rangle$, indicating that the random motility coefficient is isotropic. Furthermore, unlike our estimates of $\langle v_x \rangle$, our estimates of $\langle D_x \rangle$ and $\langle D_y \rangle$ appear not to depend on the electric field.

Relationship between the applied gradient, cell diffusivities and drift velocities

To further explore the relationships between D_x, D_y, v_x, v_y and the applied gradient, we calculate the sample mean and sample standard deviation for each of the 16 histograms in Fig. 3. Results in Fig. 3 show $\langle v_x \rangle$, $\langle v_y \rangle$, $\langle D_x \rangle$

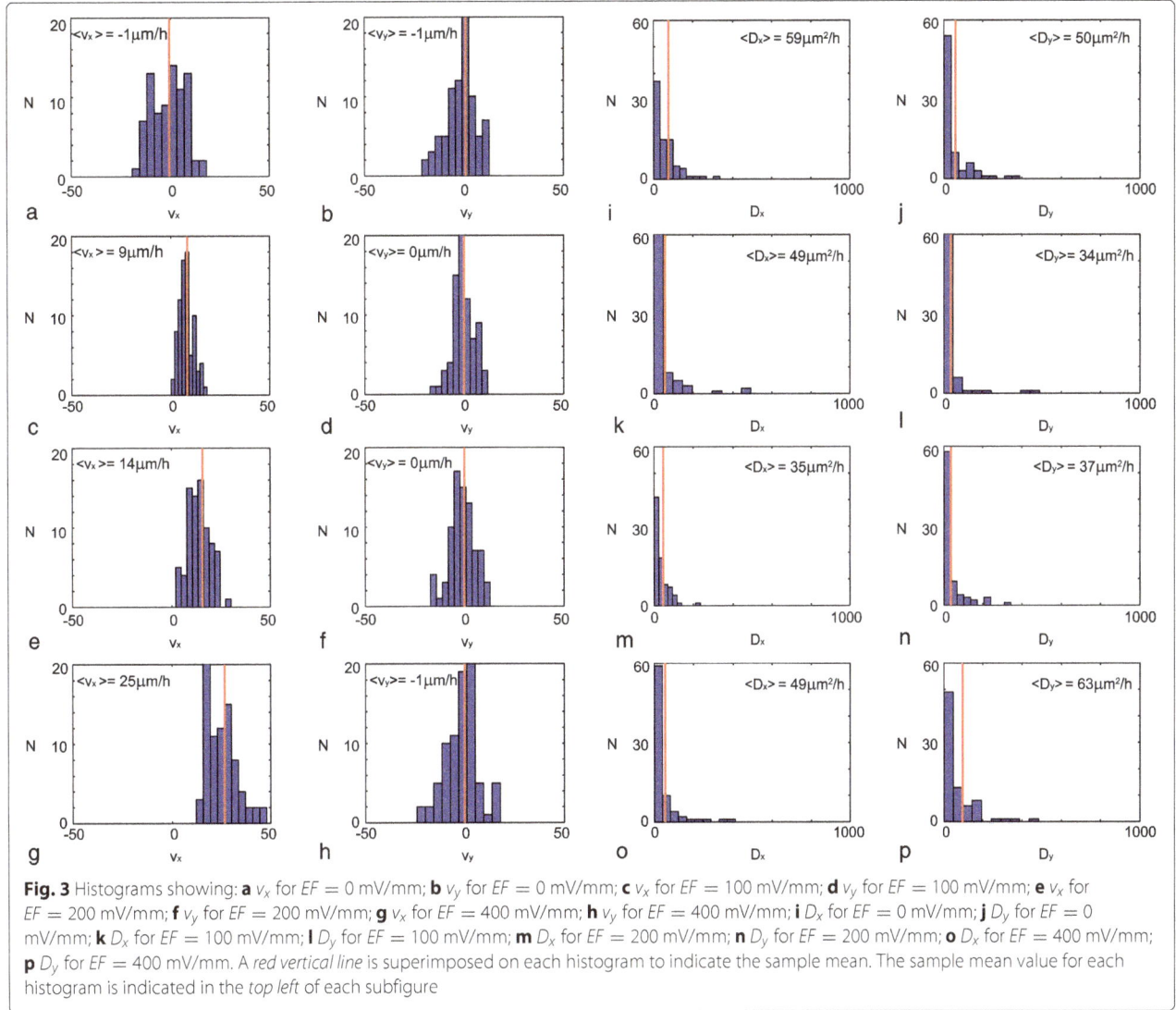

Fig. 3 Histograms showing: **a** v_x for $EF = 0$ mV/mm; **b** v_y for $EF = 0$ mV/mm; **c** v_x for $EF = 100$ mV/mm; **d** v_y for $EF = 100$ mV/mm; **e** v_x for $EF = 200$ mV/mm; **f** v_y for $EF = 200$ mV/mm; **g** v_x for $EF = 400$ mV/mm; **h** v_y for $EF = 400$ mV/mm; **i** D_x for $EF = 0$ mV/mm; **j** D_y for $EF = 0$ mV/mm; **k** D_x for $EF = 100$ mV/mm; **l** D_y for $EF = 100$ mV/mm; **m** D_x for $EF = 200$ mV/mm; **n** D_y for $EF = 200$ mV/mm; **o** D_x for $EF = 400$ mV/mm; **p** D_y for $EF = 400$ mV/mm. A *red vertical line* is superimposed on each histogram to indicate the sample mean. The sample mean value for each histogram is indicated in the *top left* of each subfigure

and $\langle D_y \rangle$, each plotted as a function of the electric field. The plots show the variation in the average transport coefficients with the EF. In addition, the variability in the estimates of the average transport coefficients is indicated by the error bars. The error bars indicate the sample mean plus or minus one sample standard deviation.

Results in Fig. 3a-b show $\langle v_x \rangle$ and $\langle v_y \rangle$ as a function of the EF. As we anticipate, $\langle v_x \rangle$ increases with EF whereas $\langle v_y \rangle \approx 0$ for all EF considered. To examine the putative relationship between $\langle v_x \rangle$ and EF, and between $\langle v_y \rangle$ and EF, we perform an unconstrained linear regression. The coefficient of determination for the $\langle v_x \rangle$ data is very high, $r^2 = 0.98$, suggesting that the linear relationship between $\langle v_x \rangle$ and EF provides a good explanation of the variability. In contrast, the coefficient of determination for $\langle v_y \rangle$ is very low, $r^2 = 0.00$, suggesting that the null hypothesis is valid and there is no relationship between $\langle v_y \rangle$ and EF. In summary, these results imply that a linear relationship

between $\langle v_x \rangle$ and EF is consistent with the observed data. To match the drift term in Eq. (1) with the advection-diffusion (Eq. (6)) we require that $v_x = \chi(S)\partial S/\partial x$. Since our data is consistent with a linear relationship between v_x and the applied gradient, $\partial S/\partial x$, it appears that a constant tactic sensitivity function, $\chi(S) = \chi$, provides the simplest explanation of our experimental results.

Results in Fig. 3c-d show $\langle D_x \rangle$ and $\langle D_y \rangle$ as a function of EF. Visually, we see no discernible trend in the data for different values of EF. This visual interpretation is consistent with the fact that we obtain a small coefficient for each of the linear regressions in Fig. 3c-d. Therefore, it is reasonable to assume that the cell diffusivities appear to be independent of the electric field. If we accept this assumption and further average the data in Fig. 3i-p in each direction we obtain overall estimates of $\langle D_x \rangle = 48$ μm^2/h and $\langle D_y \rangle = 46$ μm^2/h. Again, this suggests that the diffusion of 3T3 fibroblast cells is approximately isotropic since

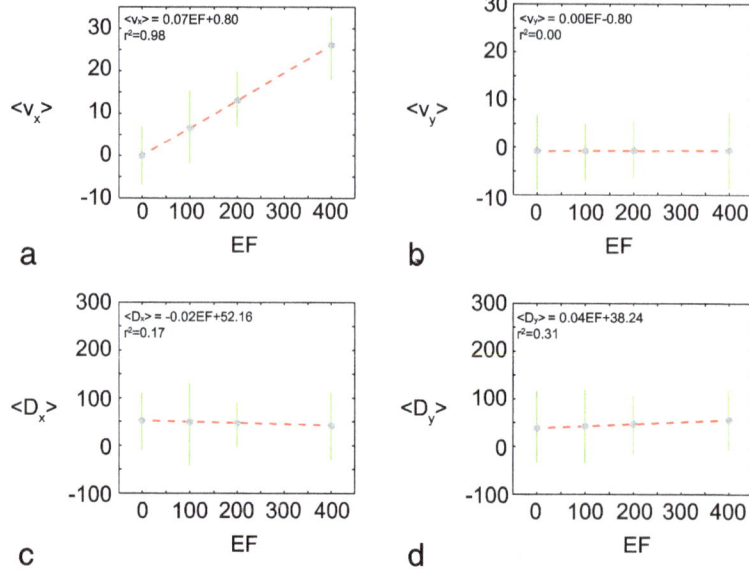

Fig. 4 Summary of the average transport coefficients as a function of the applied gradient. **a** $\langle v_x \rangle$ as a function of EF; **b** $\langle v_y \rangle$ as a function of EF; **c** $\langle D_x \rangle$ as a function of EF; and **d** $\langle D_y \rangle$ as a function of EF. In each *plot* the sample mean is shown (*grey circle*), and the *error bars* indicate the variability. In this case the variability is reported as the one sample standard deviation about the mean. In each case an unconstrained linear regression is superimposed in each subfigure

we have $D_x \approx D_y$, across all the experimental conditions considered.

Now that we have summarised the estimates of the directed and undirected components of cell migration in the experiments, we can quantify the relative roles in terms of the dimensionless Peclet number [38],

$$Pe = \frac{vL}{D}, \tag{3}$$

where v is the drift velocity, D is the diffusivity and L is a relevant length-scale, which here we will take to be the cell diameter of fibroblast cells, $L \approx 25 \ \mu$m [37]. The Peclet number is a measure of the time scale of advection to the time scale of diffusion [38]. When $Pe \ll 1$, undirected diffusive transport dominates, when $Pe \gg 1$, directed transport dominates, and when $Pe \approx 1$ to two mechanisms are in balance. Comparing estimates of the drift velocity and the diffusivity in the x-direction suggests that our experiments deal with a range of Peclet numbers from $Pe \approx 0$ when EF=0 mV/mm to $Pe \approx 10$ when EF=400 mV/mm. Therefore, our experimental data covers a wide range of transport conditions ranging from purely undirected, diffusive transport to highly directed, advection-dominant conditions.

To summarise our findings, results in Fig. 3 suggest that $\langle v_x \rangle$ increases linearly with EF, whereas the data suggests that the other transport coefficients, $\langle v_y \rangle$, $\langle D_x \rangle$ and $\langle D_y \rangle$, appear to be independent of EF. Guided by these results, we assume that $\langle v_x \rangle$ increases linearly with EF,

and that the other transport coefficients are independent of EF. Comparing the results in Fig. 1c and d also allows us to also consider whether there is any possible relationship between the transport coefficients and the electric potential. Repeating the process of plotting our estimates of the four transport coefficients as a function of the electric potential (not shown) suggests that there is no obvious trends in the data. Furthermore, linear regressions between each transport coefficient and the associated value of the electric potential reveals a low coefficient of determination, $r^2 < 1$. Therefore, based on the data, we assume that the transport coefficients appear to be independent of the electric potential in these experiments.

Discussion

Our results indicate that when we quantify the roles of directed and undirected migration of 3T3 fibroblast cells under the influence of an applied electric field, the undirected component of the migration appears to be independent of the EF, and the directed migration appears to increase linearly with EF. Furthermore, we observe no consistent differences in the cell diffusivity estimates in the x and y Cartesian directions, implying that the undirected migration is isotropic. The simplest way to explain these results in terms of a Keller-Segel-type continuum model (Eq. (1)) is that we have a constant diffusivity, $D(S) = D$, and a constant chemotactic sensitivity function, $\chi(S) = \chi$. While the assumption that the chemotactic sensitivity function can be treated as a constant is widely invoked [24–33], this assumption is infrequently tested using experimental data collected under a range of gradient conditions. The question of whether the tactic sensitivity function ought to be treated as a constant or a more complicated expression is of interest because many theoretical models incorporate these kinds of details, such as receptor saturation, without necessarily being guided by experimental observations [6, 7, 13, 18–23].

Conclusion

By examining trajectories of 3T3 fibroblast cells under a range of physiologically-relevant electric gradients [11, 14], we quantify the roles of directed and undirected migration. In summary we find that the undirected migration is isotropic and the cell diffusivity is approximately 50 $\mu m^2/h$, and that the drift velocity increases approximately linearly with the applied electric field, suggesting that the tactic sensitivity function is a constant.

Although our results apply to 3T3 fibroblast cells, we anticipate that repeating the experiments and analysis outlined here for different cell lines would provide insight into the roles of directed and undirected motility for any cell line of interest. Although we have found that the drift of fibroblasts to increase approximately linearly with the electric field in the range of $EF = 0 - 400$ mV/mm, it is possible that we may observe a different response for different cell lines, or we may observe a different response for the same cell line when we apply a stronger electric field. However, here we deal only with gradients in the range of 0-400 mV/mm because this is a physiologically relevant range [11, 14].

Methods

Experimental methods

As shown in Fig. 1a, we use a specifically designed and fabricated microfluidic chip to study the electrotaxis of NIH 3T3 fibroblasts. A CO_2 laser scriber (ILS2, Laser Tools & Technics Corp, Taiwan) is used to ablate desired patterns on polymethylmethacrylate (PMMA) substrates [39–41]. Four layers of PMMA sheets are thermally bonded to form the fluidic channel, which is then attached to a cover glass to act as the cell culture area. The thickness of the fluidic channel is 1 mm, and the widths of the four culture areas (two copies) are 4.00, 8.28, 4.14 and 2.07 mm, respectively. By applying a direct current (dc) of 80 μA, the EF inside these areas are calculated to be 0, 100, 200, and 400 mV/mm, respectively, based on Ohm's law [35]. Numerical simulations of the EF and the potential inside the microfluidic chip is simulated using the commercial software package COMSOL Multiphysics (COMSOL, USA) to confirm these calculations (see Fig. 1c and d).

The NIH 3T3 fibroblast cell line, purchased from Bioresource Collection and Research Center (BCRC, Taiwan), is cultured in a complete medium composed of Dulbecco's modified Eagle medium (Gibco, USA) and 10% calf serum (Invitrogen, USA). 10^6 cells are injected into the chip and the temperature is maintained at 37 $\pm$ 0.5 °C using a customized temperature controller. Different EF strengths are introduced by connecting the Ag(anode)/AgCl(cathode) electrodes (see Fig. 1a) to a dc power supply (GWInstek, Taiwan) set at the constant-current model [42]. The microfluidic chip is mounted on a motorized, bright-field inverted microscope (CKX41, Olympus, USA) to observe cell migration. Figure 1b shows an image of the cells in one culture area. For each culture area, corresponding to a different EF, images are taken over a period of 2 h. In each area, at least 80 cells were selected at random for data analysis.

Modelling methods

Since we are dealing with trajectory data over a finite period of time for which no trajectory touches any physical boundary, we model the system as a random walk on $\Omega = \{(x,y) : -\infty < x < \infty, -\infty < y < \infty\}$. For the analysis we denote the position of a cell at time t, relative to the position at $t = 0$ as a random vector $(x(t), y(t))$, where $x(t)$ and $y(t)$ are the Cartesian coordinates at time t. These coordinates are related to a probability density function, $p(x,y,t)$ so that

$$\mathbb{P}\{(x(t), y(t)) \in A\} = \iint_A p(x,y,t) \, \mathrm{d}x \, \mathrm{d}y, \qquad (4)$$

where A is a plane region that is a subset of Ω.

We take the simplest possible, standard approach by setting the transport coefficients, D and χ, to be constants [36, 43]. Furthermore, we also make use of the fact that the spatial gradient of electric field $(\partial S/\partial x)$ is approximately constant across several subregions on the experimental device. However, at this stage we allow for the transport coefficients to potentially take on different values in different directions. These simplifications allow us to work with an anisotropic analogue of the linear advection-diffusion equation in a two-dimensional Cartesian geometry, which can be written as

$$\frac{\partial p}{\partial t} = D_x \frac{\partial^2 p}{\partial x^2} + D_y \frac{\partial^2 p}{\partial y^2} - v_x \frac{\partial p}{\partial x} - v_y \frac{\partial p}{\partial y}, \qquad (5)$$

on Ω. Since the distribution of cells on the experimental device is deliberately kept low so that the density of cells is well below carrying capacity, we deal with a linear model which is appropriate for cell migration under low cell density conditions where cell-to-cell collisions are relatively infrequent [44, 45]. If we consider the initial condition $p(x,y,0) = \delta(x)\delta(y)$, which is relevant to following the motion of a single agent in the random walk starting from the origin [36], the solution of Eq. (5) is [43]

$$p(x,y,t) = \frac{1}{4\pi t\sqrt{D_x D_y}} \exp\left[-\left(\frac{(x - v_x t)^2}{4D_x t} + \frac{(y - v_y t)^2}{4D_y t}\right)\right].$$
(6)

To interpret the random walk data in terms of this model we will deal with a series of individual trajectory data, $(x(t), y(t))$ with $(x(0), y(0)) = (0, 0)$. We analyze each spatial component of the shifted trajectory separately. To achieve this we consider the marginal probability density functions for each spatial component,

$$p_x(x,t) = \int_{-\infty}^{\infty} p(x,y,t)\, dy, \qquad p_y(y,t) = \int_{-\infty}^{\infty} p(x,y,t)\, dx,$$

and we evaluate the first two positive moments, the mean and variance, of $x(t)$ and $y(t)$, respectively. The first moments of the marginal probability density functions are given by

$$\langle x^1(t)\rangle = \int_{-\infty}^{\infty} x\, p_x(x,t)\, dx, \qquad \langle y^1(t)\rangle = \int_{-\infty}^{\infty} y\, p_y(y,t)\, dy.$$

Using Eq. (6) we obtain

$$\langle x^1(t)\rangle = v_x t, \qquad \langle y^1(t)\rangle = v_y t.$$

Therefore, for each trajectory, $(x(t), y(t))$, we can obtain separate estimates of v_x and v_y. Fitting a series of straight lines constrained to pass through the origin gives us an estimate of v_x and v_y for each trajectory. Since we have 80 trajectories for each gradient condition, we obtain 80 estimates of v_x and 80 estimates of v_y. The variability amongst these estimates can be observed by plotting the results as a histogram. Furthermore, we can characterise the average coefficients by evaluating the sample mean and sample standard deviation of these 80 estimates. We will denote the sample mean as $\langle v_x\rangle$ and $\langle v_y\rangle$, respectively.

To provide information about the diffusivity, we will make use of the second moments of the marginal probability density functions are given by

$$\langle x^2(t)\rangle = \int_{-\infty}^{\infty} \left(x - \langle x^1(t)\rangle\right)^2 p_x(x,t)\, dx,$$

$$\times\ \langle y^2(t)\rangle = \int_{-\infty}^{\infty} \left(y - \langle y^1(t)\rangle\right)^2 p_y(y,t)\, dy.$$

Using Eq. (6) we obtain

$$\langle x^2(t)\rangle = 2D_x t, \qquad \langle y^2(t)\rangle = 2D_y t.$$

Therefore, given our previous estimates of the average drift velocity in each direction $\langle v_x\rangle$ and $\langle v_y\rangle$, for each trajectory we can obtain separate estimates of D_x and D_y. Fitting a series of straight lines constrained to pass through the origin give us estimates of D_x and D_y for each trajectory. Since we have 80 trajectories for each gradient condition, we obtain 80 estimates of D_x and 80

estimates of D_y. The variability amongst these estimates can be observed by plotting the results as a histogram. Furthermore, we can characterise the average coefficients by evaluating the sample mean and sample standard deviation of these 80 estimates. We will denote the sample mean as $\langle D_x\rangle$ and $\langle D_y\rangle$, respectively.

Abbreviation
EF: electric field

Acknowledgments
The authors would like to thank Dr. Ji-Yen Cheng for help in fabricating microfluidic chips, and the Center for Emerging Material and Advanced Devices, National Taiwan University, for the cell culture room facility. We appreciate the helpful comments from three referees and the BMC Systems Biology editor.

Funding
This work is supported by the Australian Research Council (DP140100249, DP170100474). This work is financially supported by the Ministry of Science and Technology of Taiwan under Contract No. MOST 104-2311-B-002-026 (K. Y. Lo), No. MOST 104-2112-M-030-002 (Y. S. Sun), and the National Taiwan University Career Development Project (103R7888) (K. Y. Lo).

Authors' contributions
MJS, K-YL and Y-SS conceived the study and designed the experiments. K-YL and Y-SS performed the experiments. MJS analysed the data. MJS wrote the manuscript. All authors read and approved the final manuscript.

Competing interests
The authors declare that they have no competing interests.

Author details
[1] School of Mathematical Sciences, Queensland University of Technology (QUT), Brisbane, Australia. [2] Department of Agricultural Chemistry, National Taiwan University, 10617 Taipei, Taiwan. [3] Department of Physics, Fu-Jen Catholic University, 24205 New Taipei City, Taiwan.

References
1. Sherratt JA, Murray JD. Models of epidermal wound healing. Proc R Soc Lond B. 1990;241:29–36.
2. Maini PK, McElwain DLS, Leavesley DI. Traveling wave model to interpret a wound-healing cell migration assay for human peritoneal mesothelial cells. Tissue Eng. 2004;10:475–82.
3. Treloar KK, Simpson MJ, McElwain DLS, Baker RE. Are in vitro estimates of cell diffusivity and cell proliferation rate sensitive to assay geometry? J Theor Biol. 2014;356:71–84.
4. Gatenby RA, Gawlinski ET. A reaction-diffusion model of cancer invasion. Cancer Res. 1996;56:5745–53.

5. Johnston ST, Shah ET, Chopin LK, McElwain DLS, Simpson MJ. Estimating cell diffusivity and cell proliferation rate by interpreting IncuCyte ZOOM^{TM} assay data using the Fisher-Kolmogorov model. BMC Syst Biol. 2015;9:38.

6. Keller EF, Segel LA. Model for chemotaxis. J Theor Biol. 1971;30:225–34.

7. Murray JD. Mathematical biology: i an introduction. Springer-Verlag Berlin Heidelberg: Springer; 2002.

8. Painter KJ, Sherratt JA. Modelling the movement of interacting cell populations. J Theor Biol. 2003;225:327–39.

9. Cai AQ, Landman KA, Hughes BD. Modelling directional guidance and motility regulation in cell migration. Bull Math Biol. 2006;68:25–52.

10. Simpson MJ, Landman KA, Hughes BD, Newgreen DF. Looking inside an invasion wave of cells using continuum models: proliferation is the key. J Theor Biol. 2007;243:343–60.

11. Robinson KR. The responses of cells to electrical fields: a review. J Cell Biol. 1985;101:2023–7.

12. Mycielska ME, Djamgoz MBA. Cellular mechanisms of direct-current electric field effects: galvanotaxis and metastatis diseasse. J Cell Sci. 2004;117:1631–9.

13. Vanegas-Acosta JC, Garzon-Alvarado DA, Zwamborn APM. Mathematical model of electrotaxis in osetoblastic cells. Bioelectrochemistry. 2012;88: 134–43.

14. Nuccitelli R. A role for endogenous electric fields in wound healing. Curr Top Dev Biol. 2003;58:1–26.

15. Farboud B, Nuccitelli R, Schwab IR, Isseroff RR. DC electric fields induce rapid directional migration in cultured human corneal epithelial cells. Exp Eye Res. 2000;70:667–73.

16. Zhao M. Electrical fields in wound healing - an overriding signal that directs cell migration. Semin Cell Dev Biol. 2009;20:674–82.

17. Jin W, Shah ET, Penington CJ, McCue SW, Chopin LK, Simpson MJ. Reproducibility of scratch assays is affected by the initial degree of confluence: experiments, modelling and model selection. J Theor Biol. 2016;390:136–45.

18. Hughes-Alford SK, Lauffenburger DA. Quantitative analysis of gradient sensing: towards building predictive models of chemotaxis in cancer. Curr Opin Cell Biol. 2012;24:284–91.

19. Tranquillo RT, Lauffenburger DA, Zigmond SH. A stochastic model for leukocyte random motility and chemotaxis based on receptor binding fluctuations. J Cell Biol. 1988;106:303–09.

20. Tranquillo RT, Zigmond SH, Lauffenburger DA. Measurement of the chemotaxis coefficient for human neutrophils in the under-agarose migration assay. Cell Motil Cytoskel. 1988;11:1–15.

21. Marchant BP, Norbury J, Byrne HM. Biphasic behaviour in malignant invasion. Math Med Biol. 2006;23:173–96.

22. Landman KA, Simpson MJ, Pettet GJ. Tactically-driven nonmonotone travelling waves. Physica D. 2008;237:678–91.

23. Wu D, Lin F. A receptor-electromigration-based model for cellular electrostatic sensing and migration. Biochem Bioph Res Co. 2011;411: 695–701.

24. Perumpanani AJ, Sherratt JA, Norbury J, Byrne HM. A two parameter family of travelling waves with a singular barrier arising from the modelling of extracellular matrix mediated cellular invasion. Physica D. 1999;126:145–59.

25. Ford RM, Phillips BR, Quinn JA, Lauffenburger DA. Measurement of bacterial random motility and chemotaxis coefficients: 1 stopped-flow diffusion chamber assay. Biotechnol Bioeng. 1991;37:647–60.

26. Ford RM, Lauffenburger DA. Measurement of bacterial random motility and chemotaxis coefficients: II application of single-cell-based mathematical model. Biotechnol Bioeng. 1991;37:661–72.

27. Stokes CL, Lauffenburger DA, Williams SK. Migration of individual microvessel enthothelial cells: stochastic model and parameter measurement. J Cell Sci. 1991;99:419–30.

28. Stokes CL, Lauffenburger DA. Analysis of the roles of microvessel endothelial cell random motility and chemotaxis in angiogenesis. J Theor Biol. 1991;152:377–403.

29. Pettet GJ, McElwain DLS, Norbury J. Lotka-Volterra equations with chemotaxis: walls barriers and travelling waves. Math Med Biol. 2000;17: 395–413.

30. Wechselberger M, Pettet GJ. Folds, canards and shocks in advection-reaction-diffusion models. Nonlinearity. 2010;23:1949–69.

31. Harley K, van Heijster P, Marangell R, Pettet GJ, Weschelberger M. Existence of traveling wave solutions for a model of tumour invasion. SIAM J Appl Dyn Syst. 2014;13:366–96.

32. Charteris N, Khain E. Modeling chemotaxis of adhesive cells: stochastic lattice approach and continuum description. New J Phys. 2014;16:025002.

33. Irons C, Plank MJ, Simpson MJ. Lattice-free models of directed cell motility. Physica A. 2016;442:110–21.

34. Todaro GJ, Green H. Quantitative studies of the growth of mouse embryo cells in culture and their development into established lines. J Cell Biol. 1963;17:299–313.

35. Wu SY, Hou HS, Sun YS, Cheng JY, Lo KY. Correlation between cell migration and reactive oxygen species under electric field stimulation. Biomicrofluidics. 2015;9:054120.

36. Cai AQ, Landman KA, Hughes BD. Multi-scale modeling of a wound-healing cell migration assay. J Theor Biol. 2007;245:576–94.

37. Simpson MJ, Binder BJ, Haridas P, Wood BK, Treloar KK, McElwain DLS, Baker RE. Experimental and modelling investigation of monolayer development with clustering. Bull Math Model. 2013;75:871–89.

38. Bird RB, Stewart WE, Lightfood EN. Transport phenomena. Wiley Singapore: Wiley; 2005.

39. Cheng JY, Wei CW, Hsu KH, Young TH. Direct-write laser micromachining and universal surface modification of PMMA for device development. Sensor Actuat B-Chem. 2004;99:186–96.

40. Cheng JY, Yen MH, Wei CW, Chuang YC, Young TH. Crack-free direct-writing on glass using a low-power UV laser in the manufacture of a microfluidic chip. J Micromech Microeng. 2005;15:1147–56.

41. Cheng JY, Yen MH, Kuo CT, Young TH. A transparent cell-culture microchamber with a variably controlled concentration gradient generator and flow field rectifier. Biomicrofluidics. 2008;2:24105.

42. Hou HS, Tsai HF, Chiu HT, Cheng JY. Simultaneous chemical and electrical stimulation on lung cancer cells using a multichannel-dual-electric-field chip. Biomicrofluidics. 2014;8:052007.

43. Codling EA, Plank MJ, Benhamou S. Random walk models in biology. J R Soc Interface. 2008;5:813–34.

44. Simpson MJ, Landman KA, Hughes BD. Multi-species simple exclusion processes. Physica A. 2009;388:299–406.

45. Simpson MJ, Treloar KK, Binder BJ, Haridas P, Manton KJ, Leavesley DI, McElwain DLS, Baker RE. Quantifying the roles of cell motility and cell proliferation in a circular barrier assay. J R Soc Interface. 2013;10:130007.

8

Synthetic circuits that process multiple light and chemical signal inputs

Lizhong Liu[1,2], Wei Huang[4] and Jian-Dong Huang[1,2,3]*

Abstract

Background: Multi-signal processing circuits are essential for rational design of sophisticated synthetic systems with good controllability and modularity, therefore, enable construction of high-level networks. Moreover, light-inducible systems provide fast and reversible means for spatiotemporal control of gene expression.

Results: Here, in HEK 293 cells, we present combinatory genetic circuits responding to light and chemical signals, simultaneously. We first constructed a dual input circuit converting different light intensities into varying of the sensitivity of the promoter to a chemical inducer (doxycycline). Next, we generated a ternary input circuit, which responded to light, doxycycline and cumate. This circuit allowed us to use different combinations of blue light and the two chemical inducers to generate gradual output values over two orders of magnitude.

Conclusions: Overall, in this study, we devise genetic circuits sensing and processing light and chemical inducers. Our work may provide insights into bio-computation and fine-tuning expression of the transgene.

Keywords: Synthetic circuit, Multi-input, Signal integration, Light-inducible

Background

Synthetic biology adopts the concepts of engineering and computational science into biological systems, aiming to generate artificial genetic circuits and systems with desirable functions. It offers a promising way to address global challenges, for example, clean energy, environment restoration, and increasing medical needs. During the past decade, a remarkable development of synthetic biology has been achieved. Various synthetic devices and systems have been established, including biological oscillators, switches, counters, as well as logic gates [1–4]. However, it is still challenging to generate complex synthetic systems with good controllability and programmability. For instance, like electronic or mechanical systems that can be fine-tuned by various inputs and produce predictable outputs. Therefore, one can program the systems to act in a desirable way by altering input information. Development and characterization of standard modules, which sense and convert multiple input signals into cellular responses will help address the challenges [5, 6].

In natural biological systems, multi-signal processing is a fundamental aspect. For example, the bow-tie (also called hourglass) architecture, which refers to systems that receive a diversity of inputs and convert the input signals through an intermediate "core", and finally generate a variety of outputs. Since the intermediate "core" is composed of relatively few universal components, the overall structure of the system resembles a bow-tie or hourglass [7]. For instance, in metabolic networks, multiple input nutrients are converted into multiple biomass components by a small number of mediator factors [7]. Previous work suggests that the recurrence of bow-tie architecture in various biological systems indicates its significance on enhancing the robustness of the biological systems [8]. In the counterpart electronic systems, modules for multi-input integration are also widely used, for example, a module called "digital-to-analog" converter (DAC) is commonly used in audio or video devices for converting multiple digital-input signals into the analog output signals [9].

Previous work reported some chemically-inducible expression systems in mammalian cells [10, 11]. Recently, Optogenetics has demonstrated that light is an ideal

* Correspondence: jdhuang@hku.hk
[1]School of Biomedical Sciences, Li Ka Shing Faculty of Medicine, University of Hong Kong, Pok Fu Lam, Hong Kong, People's Republic of China
[2]Shenzhen Institute of Research and Innovation, University of Hong Kong, Shenzhen 518057, People's Republic of China
Full list of author information is available at the end of the article

source of signal for spatiotemporal control of gene expression [12–15]. The combination of chemical inducer and light inducer, for example by generating chimeric promoters that consist chemical-responsive and light-responsive elements, can achieve spatial and stringent control of transgenes [16]. Using light as inducer can avoid drawbacks of using chemical inducers. For instance, the chemical inducers are needed to be transported into cells by passive or active manner before they encounter the sensors, which causes a delay of target gene expression. The delay may lead to undesirable cell-to-cell variation. However, using light does not result in this problem. In addition, recent work demonstrated that light can be used as a communication signal between the computer and modified *E. coli* cells [17, 18]. Connecting synthetic biology systems with a computer, and then monitor and control the behaviors of the circuits by a computer program can tremendously increase programmability of synthetic systems.

In this study, we first developed a 2-input circuit that exhibited different sensitivity to doxycycline (Dox) upon different doses of blue light illumination. Specifically, a blue light-inducible system, called LightOn system [14], was used to control the expression level of a transcriptional repressor TetR. A reporter GFP was driven by TetR-repressible promoter. The repression of TetR can be relieved by adding Dox. Therefore, light and Dox acted as inducers of this circuit. Next, we generated a 3-input circuit for conversion of the binary input sequence, consisting of light and chemicals, into graded output promoter activities. Specifically, this circuit was composed of a cumate-inducible promoter driving a modified rtTA (hereafter, rtTAm) [19], a light-inducible promoter driving the TetR co-repression peptides (hereafter, TCP) [20]. TCP-rtTAm complex activates the output TRE3G promoter. Therefore, Light-inducible system and cumate-switch system form an AND-gate. On the other hand, Dox also can trigger the DNA-binding of rtTAm. Therefore, Dox-inducible system and cumate-switch system also compose an AND-gate. Previous work suggests that short peptide inducer may be less efficient than Dox [21]. Moreover, it has been reported that the peptide competes with Dox for the tc-binding pocket of TetR [20]. Thereby, the potency of TCP fusion protein might be much lower than Dox as rtTAm inducer, and the presence of TCP fusion protein could inhibit Dox inducing ability.

Results
The dual input circuit converting illumination dose into sensitivity variations of a promoter to Dox
In this circuit, LightOn system was used to control the expression of TetR. And a reporter GFP was driven by the CMV(tetO2) promoter (Life Technologies, T-REx

system, and Additional file 1: Supplementary note) containing two copies of tet operator 2 (Fig. 1a). The LightOn system comprises a synthetic photoactive transactivator GAVPO and its cognate synthetic promoter U5 [14]. GAVPO monomers form a homodimer upon blue light illumination. The GAVPO dimer then binds to the UAS_G element in the U5 promoter to recruit general transcription factors and coactivators to bind to the U5 promoter.

Our data indicated the expression level of the TetR::mCherry::NLS fusion could be tuned by adjusting the exposure of blue light (Fig. 1b-d). We examined the spatial resolution of this circuit. Specifically, we illuminated a small square area of the dish, while the other area of the dish was kept in dark. The result showed that cells in the illuminated area were TetR::mCherry::NLS positive and GFP low, while cells in the adjacent dark area were TetR::mCherry::NLS negative and GFP high. The light-induced TetR::mCherry::NLS suppressed the expression of GFP, and addition of Dox relieved this repression (Fig. 1c). Furthermore, we illuminated the cells with different doses of blue light before Dox treatment. Cells exposed to various amounts of light showed different activation thresholds of Dox induction. The increase in the level of the repressor resulted in an increase in the Dox threshold. Whilst, the dynamic range of the promoter was not affected (Fig. 1d).

Design and construction of a multi-input circuit for conversion of light and chemical binary information into different promoter activities
Next, we attempted to design and construct a circuit for integration and conversion binary combinations of light and chemical signals into graded output values (Fig. 2a).

The circuit consists of two AND-gates that response to blue light, cumate and Dox, respectively (Fig. 2a, b). However, the output strengths of the two AND-gates were not equivalent. The addition of saturated cumate and Dox resulted in strong output while the addition of cumate together with light illumination led to moderate output (Fig. 5).

The components composing the circuit are as following: LightOn system which was introduced above; the cumate-inducible system consists of a transcription repressor CymR and the repressible promoter CMV5CuO promoter (pCMV5CuO). CymR binds to the operator sequence (CuO) downstream of a strong promoter CMV5 and inhibits transcription. The addition of a cumate relieve the repression by CymR [19]; and a TetR co-repression peptide (TCP) -inducible system. Specifically, the mCherry-TCP fusion (Additional file 1: Supplementary note) can aid in DNA-binding of rtTAm. The rtTAm is a chimeric protein composed of a reverse TetR variant and 3 copies of VP16 activation domain

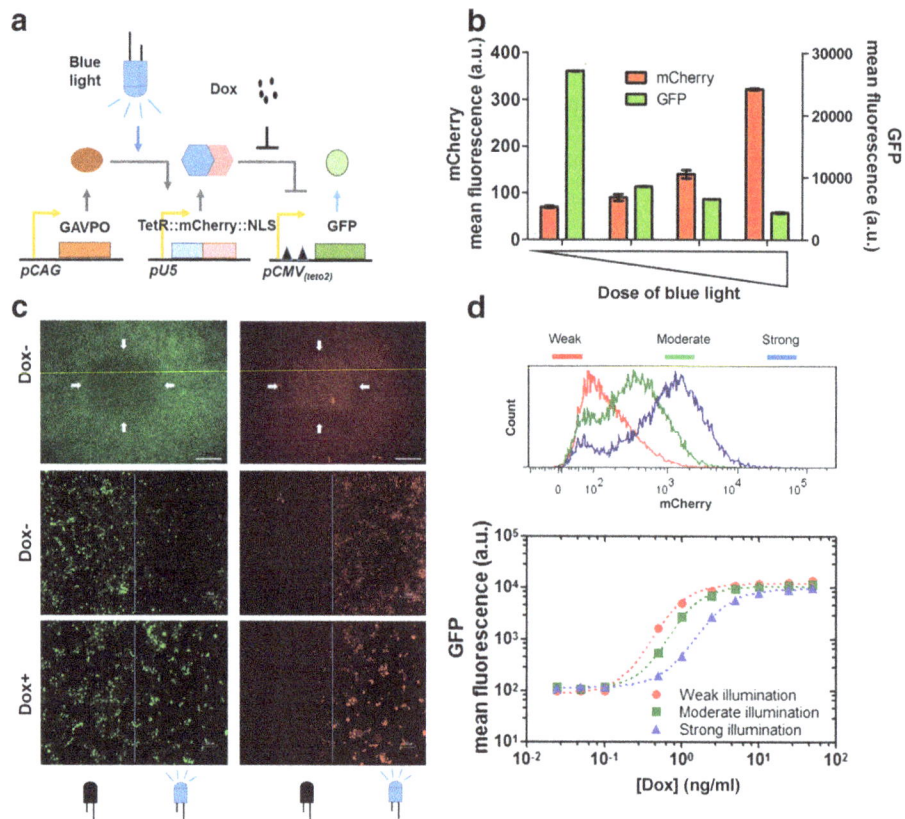

Fig. 1 Light-switchable synthetic circuit with tunable activation threshold and spatial resolution. (**a**) Schematic diagram of the circuit. The CAG promoter is constitutively expressing the photoactive transactivator GAVPO. Upon blue light illumination GAVPO forms a homodimer, which then initiates the transcription of TetR::mCherry::NLS from the pU5 promoter. GFP is under the control of TetR::mCherry::NLS-repressible promoter CMV(tetO2). Dox can release the repression. (**b**) Cells were illuminated with blue light (1.25 W m^{-2}) for different durations (dark, 10 min, 30 min, and 3 h) in the absence of Dox, followed by 24 h incubation in dark. Data are presented as mean ± SEM ($n = 3$). (**c**) A square area, which is indicated by white arrows in the upper panel, was illuminated by blue light (1.25 W m^{-2}) for 24 h. In the middle and low panel, the boundary between illuminated and dark area was indicated by the blue line. The right part of each picture is the illuminated area, while the left part is the dark area. Cells shown in the low panel were treated with 1 μg/ml of Dox. Scale bar is 2 mm in the upper panel, and 100 μm in the middle and low panel. (**d**) Cells with different levels of TetR::mCherry::NLS differentially responded to Dox. The upper panel shows the mCherry intensity of the cells illuminated with blue light (1.25 W m^{-2}) for 1 h (weak, red line), 5 h (moderate, green line), or 20 h (strong, blue line), respectively. The lower panel shows GFP intensity of cells treated with different concentration of Dox after illumination. The data are presented as mean ± SEM ($n = 3$). Data are fitted to a modified Hill equation (dashed lines). The EC50s for the three curves are 2.70 ± 0.15 ng/ml (red), 4.74 ± 0.13 ng/ml (green), and 35.81 ± 1.03 ng/ml (blue). The Hill coefficients for the three curves are 1.81 ± 0.15 (red), 1.77 ± 0.11 (green), and 1.67 ± 0.06 (blue)

(Additional file 1: Supplementary note). The TCP binds to the tc-binding pocket of the reverse TetR variant and triggers allosteric conformational change in the reverse TetR variant, leading to binding of the latter to its cognate DNA [20]. The responsive TRE3G promoter contains seven repeats of tet operator site (tetO) upstream of a CMV minimal promoter (Fig. 2b).

The circuit can be divided into three layers. The first layer receives light and cumate signals. In this layer. constitutively expressed GAVOP (it does not appear in the scheme) activates expression of mCherry-TCP upon blue light illumination. Constitutively expressed CymR (it does not appear in the scheme) suppresses the expression of rtTAm. Cumate is required to switch on the expression of rtTAm. The second layer is the

information integration node. Specifically, the rtTAm, representing the presence of cumate, and mCherry-TCP, representing the presence of blue light, interact and form a protein complex, which can activate the output promoter. On the other hand, Dox also can aid binding of rtTAm to output promoter. The third layer is the responding (output) node, in which there is a luciferase gene driven by the TRE3G promoter (Fig. 2b).

Characterization of the ternary input circuit

We first characterized each inducible expression node, separately. And identified the saturation dose for each inducer. Then we characterized the complete circuit.

To our knowledge, there is no demonstration of induction of rtTA (in our case, it is rtTAm) by TCP in

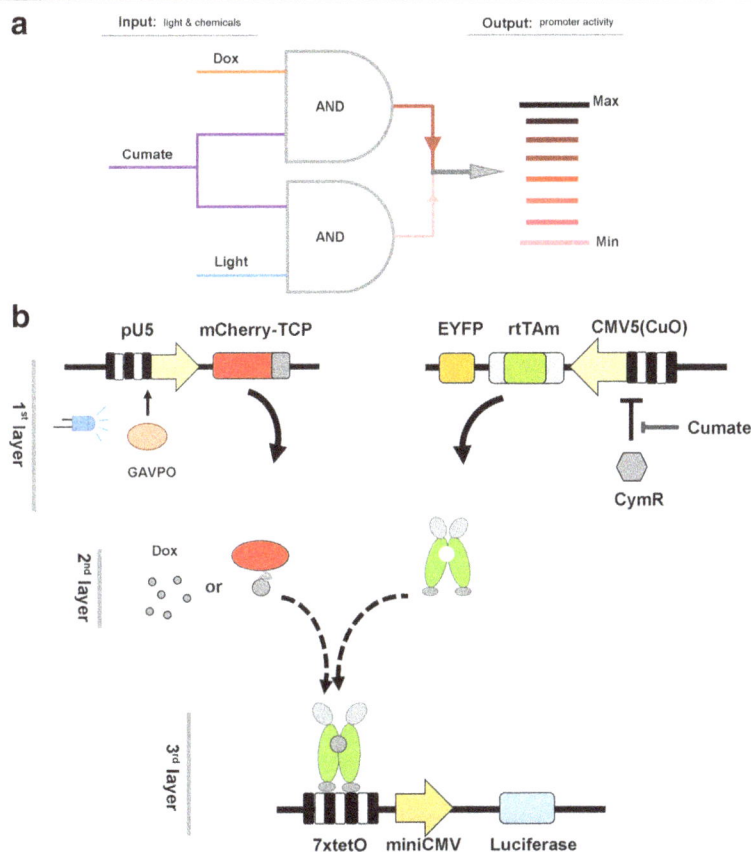

Fig. 2 The design of the 3-input circuit. (**a**) Illustration of the conversion of input signals including Dox, cumate and light into graded output signals, which are promoter activities. The system consists of three inducible systems formed Boolean logic gates. (**b**) The scheme of the circuit. It can be divided into three layers. In the first layer, LightOn system controls the expression of mCherry-TCP fusion; cumate-switch system controls the expression of rtTAm (co-expressed with EYFP linked by 2A peptide). In the second layer, Dox or mCherry-TCP fusion protein can serve as an inducer to trigger the binding of rtTAm to the cognate TRE3G promoter. In the third layer, rtTAm serves as transcription activator to initiate the transcription of luciferase from TRE3G promoter

mammalian cells, to date. We first attempted to test whether intracellularly expressed TCP fusion protein, i.e., mCherry-TCP fusion, could induce rtTAm-dependent expression from the TRE3G promoter. To this end, we used CMV promoter to control the expression of mCherry-TCP and rtTAm. And used TRE3G promoter to drive a GFP gene (Fig. 3a). We introduced the circuit DNA into HEK293 cells by transfection and observed the transfected cells by fluorescent microscope, 72 h after transfection. In the control group, the mCherry-TCP fusion was replaced by mCherry. We observed GFP signal in the mCherry-TCP cells but not in the mCherry cells (Fig. 3b and Additional file 2: Figure S1).

To characterize the LightOn system regulated expression of mCherry-TCP, we generated a stable cell line integrated with a modified circuit without expression of CymR. However, the other components were the same as the complete circuit (Fig. 4a). We illuminated the cells with blue LED (1.25 W m^{-2}) for 24 h and then put the cells back in a dark environment for 0 h, 24 h, or 48 h (the total incubation time was 24 h, 48 h or 72 h, respectively). Next, we examined the mCherry-TCP expression levels by flow cytometry (Fig. 4b). At the meantime, we examined luciferase expression induced by mCherry-TCP. The data suggested that after 24 h illumination, the expression of mCherry-TCP reached the highest level, and then it start to decrease. However, the maximum expression level of luciferase was observed 48 h after the illumination started (Fig. 4c). In another experiment, we examined the kinetics of the circuit at earlier phases after illumination with constitutively expressed rtTAm. The results show that 10 h after illumination a moderate increase of luciferase activity was detected (Additional file 3: Figure S2A). And the response of the circuit to blue light was slower than the response to Dox (Additional file 3: Figure S2B).

Also, we characterized the cumate-inducible node in the circuit (Fig. 4d). We varied cumate concentrations,

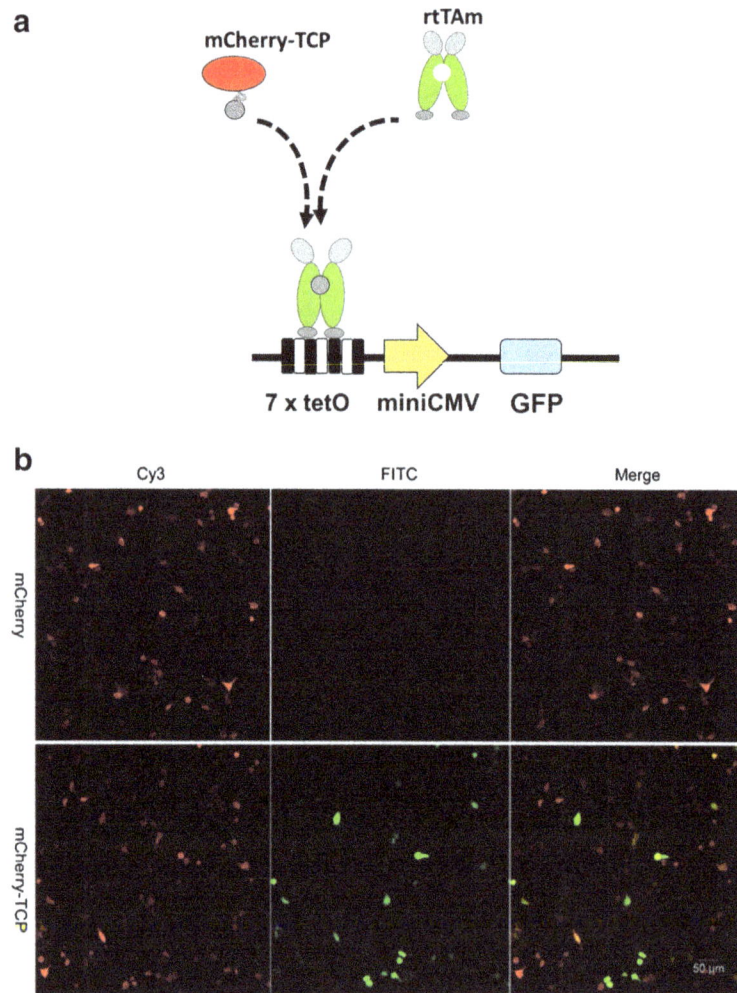

Fig. 3 TCP induces rtTAm. (**a**) Scheme of TCP induced rtTAm binding to TRE3G promoter. Constitutively expressed mCherry-TCP and rtTAm (both driven by CMV promoter) interact with each other, and then the complex bind to the tetO elements within TRE3G promoter (consist of 7 × tetO elements and a minimal CMV promoter). A GFP gene is placed downstream of TRE3G promoter. (**b**) mCherry-TCP induced the rtTAm-dependent expression of GFP. GFP-positive cells were observed in cells co-transfected with mCherry-TCP and rtTAm coding plasmids, but not in cells co-transfected with mCherry and rtTAm coding plasmids. The scale bar is 50 µm

and then examined the expression of EYFP. The data suggested that addition of 30 µg/ml of cumate in the medium was enough to produce the maximal expression level of rtTAm and EYFP from pCMV5CuO (Fig. 4e). The kinetics of the circuit responding to cumate induction with constitutive mCherry-TCP expression was also characterized. In the experiment, the cells were continuously illuminated, meanwhile were treated with 30 µg/ml of cumate for different time durations. 10 h after cumate addition, a moderate increase of luciferase activity was detected (Additional file 3: Figure S2C).

Finally, we characterized the complete circuit. We applied all the eight combinations of inputs to the cells and measured the output values, i.e., the luciferase activities. Expression levels of mCherry-TCP and rtTAm

(indicated by EYFP) were also examined. The data showed that the circuit responded to different input combinations and generated different output values, which evenly distributed in a range of two orders of magnitude. The output signal induced by TCP was weaker than the signal induced by Dox. In agreement with previous work [20], TCP inhibited the rtTAm-binding of Dox, which might explain that the input combination of "Light +, Cumate +, Dox +" induced lower luciferase level than the combination of "Light -, Cumate +, Dox+" (Fig. 5). We also introduced a conditional positive feedback loop into the third layer. A trans-activator, i.e., tTA was placed downstream of TRE3G promoter. The binding of tTA to TRE3G promoter can be blocked by Dox, but not by TCP (data not shown). The result suggested that the conditional positive

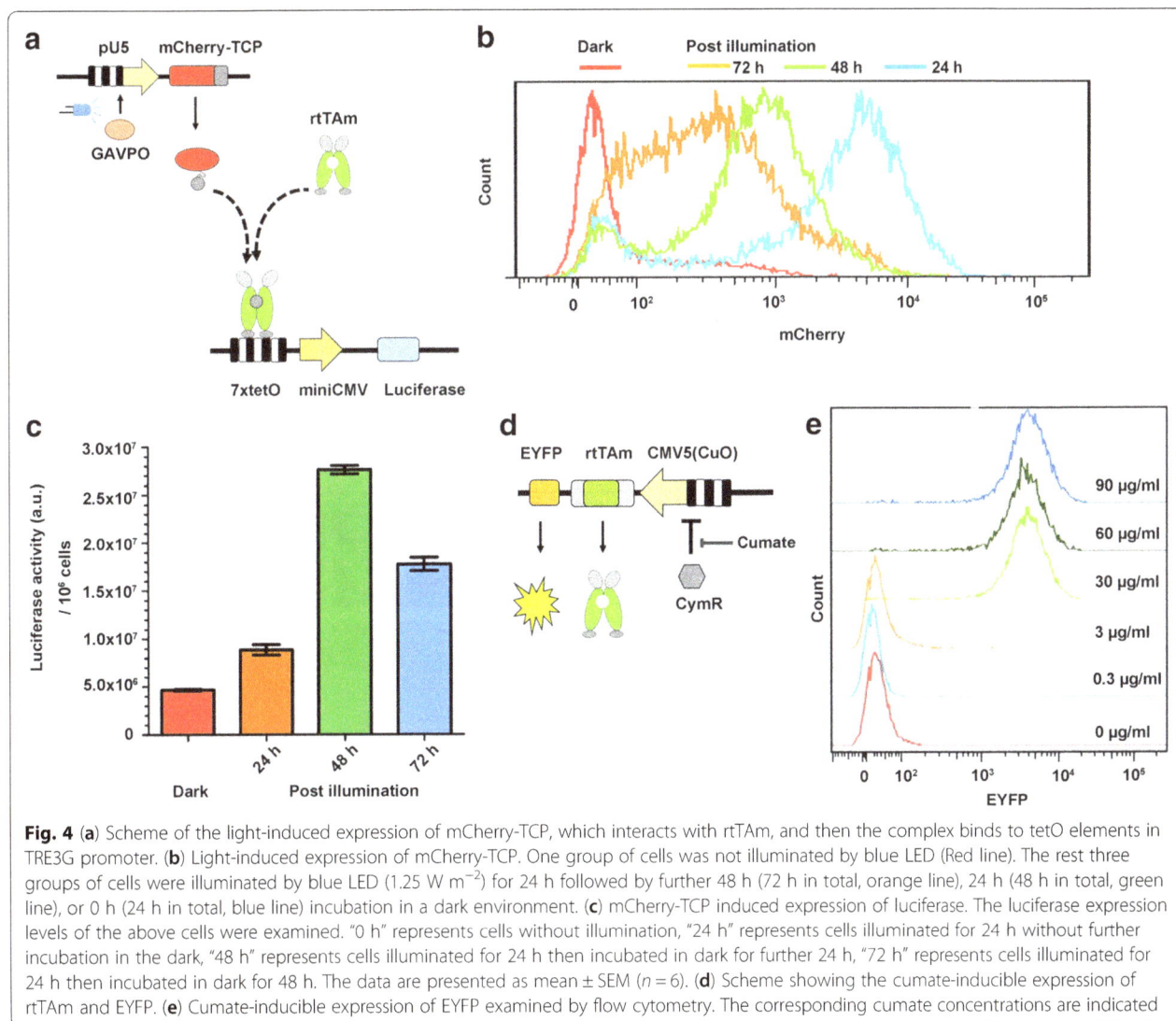

Fig. 4 (**a**) Scheme of the light-induced expression of mCherry-TCP, which interacts with rtTAm, and then the complex binds to tetO elements in TRE3G promoter. (**b**) Light-induced expression of mCherry-TCP. One group of cells was not illuminated by blue LED (Red line). The rest three groups of cells were illuminated by blue LED (1.25 W m^{-2}) for 24 h followed by further 48 h (72 h in total, orange line), 24 h (48 h in total, green line), or 0 h (24 h in total, blue line) incubation in a dark environment. (**c**) mCherry-TCP induced expression of luciferase. The luciferase expression levels of the above cells were examined. "0 h" represents cells without illumination, "24 h" represents cells illuminated for 24 h without further incubation in the dark, "48 h" represents cells illuminated for 24 h then incubated in dark for further 24 h, "72 h" represents cells illuminated for 24 h then incubated in dark for 48 h. The data are presented as mean ± SEM ($n = 6$). (**d**) Scheme showing the cumate-inducible expression of rtTAm and EYFP. (**e**) Cumate-inducible expression of EYFP examined by flow cytometry. The corresponding cumate concentrations are indicated

feedback circuit responded to the input combinations similarly as the circuit without feedback (Additional file 4: Figure S3).

Discussion

We observed leaky expression in the 3-input circuit, specifically, the leakage of rtTAm and mCherry-TCP was observed (Fig. 5). The leaky expression usually causes undesired basal expression of regulated proteins in the absence of the respective inducers and results in compromise of the circuits [22, 23]. However, the leaky expression here might lead to the differential responses of the circuit to individual input combinations, which is not completely undesirable. It is possible that due to the leaky expression of rtTAm and mCherry-TCP, each inducer, i.e., light, cumate or Dox, exhibited different effects to the system. For instance, when cumate was added in the medium, the elevated rtTAm level could

enhance leaky expression of TRE3G promoter, in the absence of the other two inducers; when the cells were illuminated, increased mCherry-TCP interacted with leaky rtTAm, and resulted in higher rtTAm-dependent expression of luciferase. Previous study demonstrated the way to fine-tune basal and/or maximal expression of LightOn system [24], which provides insights for the future modification of the circuit, especially when stringent control of a specific promoter is needed. Similarly, lower leaky expression level would be achieved by modifying CMV5(CuO) promoter using another weaker enhancer element to replace the strong CMV enhancer or increase the level of CymR by using a stronger promoter.

Our circuits can be used as building blocks in a synthetic programmable system. For instance, it can be utilized in a synthetic bow-tie structure, which is designed to sense and convert multiple input signals into

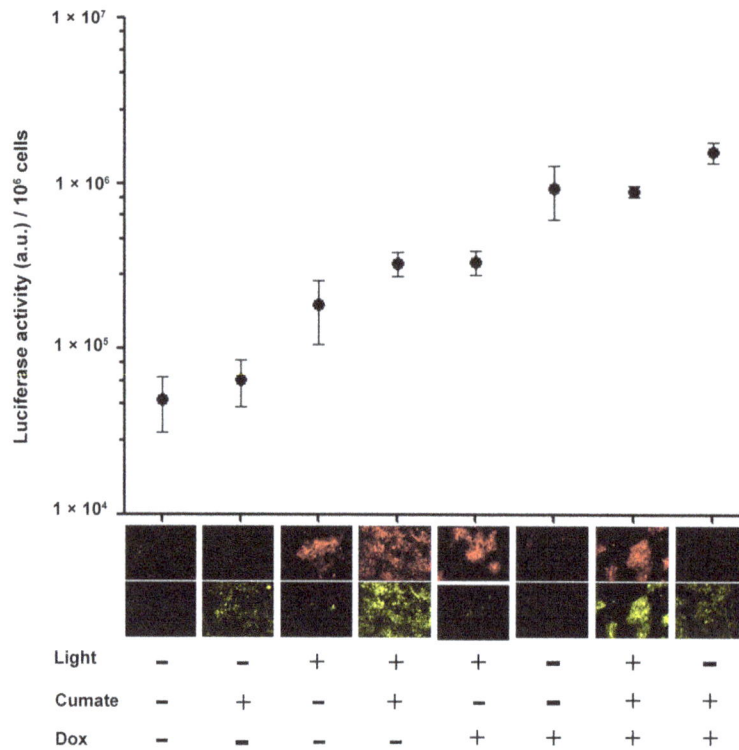

Fig. 5 Characterization of the complete 3-input circuit. Microscopy images, showing the expression of mCherry-TCP (Cy3 filter) and EYFP (YFP filter) of cells treated with all eight combinations of three inducers, are presented beneath the corresponding values of luciferase activity. Scale bar in the last image is 100 μl. All the images were taken at the same magnification. The x-axis indicates the specific treatments. "+" represents illumination with blue LED (1.25 W m^{-2}) for 24 h then incubated in the dark for another 24 h, treatment of 1 μg/ml of Dox, or treatment of 30 μg/ml of cumate, respectively. "-" represents no corresponding treatments. Expression of luciferase of the cells treated with various input signals combinations was examined. The y-axis shows luciferase activity. Data are presented as mean ± SD ($n = 6$)

the adjustment of master regulatory factors, for cell fate decision or cell cycle control. Recently, researchers from another group demonstrated optically programmed gene expression control [17, 18]. In their work, the light sequence generated by computer were used to control dynamics of synthetic circuits. Our circuit also can be modified to respond to a computer-generated sequence of light of different wavelengths. For instance, a red- or far-red light-responsive system [25], can be used to replace the cumate-inducible part in the current circuit. Thereby it may allow us to convert digital codes generated by an electronic-function-generator into gradual varying of cellular activities. It has been suggested that some master regulators displayed quantitative effects on cell fate decision, for instance, Oct/4 quantitatively influences differentiation, dedifferentiation or self-renewal of embryonic stem cells [26]; P53 quantitatively control cell fate decision between apoptosis and growth arrest [27]. Therefore, a synthetic circuit, which responds to various input programs and produced gradual output values, has the potential to be used as computer-aided cell fate controller. In addition, our 3-input circuit that responds to light and chemicals can be used to express a

therapeutic gene at a specific place and time, meanwhile, exhibits minimized undesirable expression at other places, which would improve the safety of the therapy.

In natural biology systems, analog behaviors are common, for instance, stimulation of stress-responsive gene could be operated in an analog regime [28] and neurons perform both digital and analog information processing [29]. Moreover, the analog computing system has been demonstrated in *E. coli* [30–32]. As mentioned above, DAC is widely used in electronic engineering. A biological DAC-like module that combines multiple signals, and process the digital combinations of stimuli into graded output values for reconstruction of analog signal is needed to achieve sophisticated bio-computation functions, for example, programmable logic controller and reliable environmental sensor. In this study, by constructing a circuit that converts discrete input signals into varying of transcriptional activity of the output promoter, we attempted to explore the possibility that confers DAC merit, to a certain degree, to mammalian cells.

However, the output steps of our circuit were nonmonotonic, therefore it may not act as a real DAC (Fig. 5 and Additional file 4: Figure S3). We hypothesize that modification of TCP sequence to increase its efficiency of rtTAm induction might be a potential way to increase the linearity of the output steps. Previous work has suggested the method to improve the function of TetR-inducing short peptides [20, 33]. Also, a further modification is required to expand the rationale to implement more inputs.

Conclusion

We presented mammalian circuits that processed multi-input of blue light and chemical molecules. The 2-input circuit displayed blue light illumination dose-dependent shifting of Dox response threshold. The results suggested that increased expression of the upstream repressor (TetR) resulted in an increased activation threshold with similar basal expression level and dynamic range to that of the downstream TetR-repressible promoter. The 3-input circuit converted the sequence of blue light and two chemical molecules into varying of promoter activities over two orders of magnitude.

Methods

Construction of DNA plasmid

The details of DNA cloning are described in Additional file 1: Supplementary note.

Cell culture, transient transfections, and generation of stable cell lines

Human Embryonic Kidney (HEK) 293 cells were obtained from the American Type Culture Collection (Manassas, VA), and were used in our previous study conducted by Dr. Zai Wang et al. [34]. The cells were grown in High Glucose Dulbecco's modified Eagle's medium (DMEM, Life Technologies, catalog number: 12800–017) supplemented with 10% FBS (Life Technologies, catalog number: 10270–106). The cells were sustained at 37 °C, in 5% CO_2 environment. In this study, all transfections were conducted by using Lipofectamine 2000 Transfection Reagent (Life Technologies, catalog number: 11668–019). 24 h before DNA transfection, 0.5×10^6 cells were seeded in each well of 6 well cell culture multiwell plate. On the day of DNA transfection, 5 µg DNA diluted in 500 µl Opti-MEM (Thermo Scientific, catalog number: 31985–070) mix with 10 µl Lipofectamine 2000 Transfection Reagent diluted in 500 µl Opti-MEM. The DNA-lipid complex was incubated at room temperature for 20 min, and then was added to the cells. To establish the desirable stable cell lines, 400 µg/ml Zeocin (InvivoGen, catalog number: ant-zn-1), 200 µg/ml Hygromycin B (Sigma-Aldrich, catalog number: H3274-50MG), and/or 2 µg/ml Puromycin (InvivoGen, catalog number: ant-pr-1) were correspondingly used to selected the cells for 10 to 14 days. Doxycycline (Sigma-Aldrich, catalog number 24390-14-

5) and/or cumate solution (System Biosciences, catalog number: QM100A-1) were correspondingly used to treat the cells. For light-inducible expression, cells were sustained in a dark environment for 7 days before illumination with blue LED. To induce the LightOn controlled gene expression, the cells were illuminated for 24 h and then were put back to the dark environment for further 48 h, 24 h or 0 h culture before the measurements.

Inducible gene expression

For light-inducible expression, cells were kept in darkness for 7 days before illumination with blue LED. Cells were exposed to blue light (1.25 W m^{-2}) for 5 min daily to maintain a minimal level of TetR::mCherry::NLS for the suppression of downstream GFP, thereby ensuring cells remained in the same condition (GFP negative) before Dox induction. In the tuning activation threshold experiment, the cells were exposed to blue light for different time durations and then kept in darkness before treatment with various concentrations of Dox. For cumate-inducible expression, the inducer was added into the culture medium 72 h before measurement of gene expression.

Fluorescence microscopy

The cells for microscopy images collection were grown in 60 mm dish, sustained in DMEM medium supplemented with 10% FBS, treated with the particular inducers, i.e., doxycycline, cumate and blue light. Images were taken at 72 h after induction started. Microscopy images were acquired on a Nikon TE2000-E inverted fluorescence microscope. The filter for fluorescent images of GFP was FITC (Ex 465–495 nm, Em 515–555 nm), exposure time was 3000 ms; for mCherry was Cy3 (Ex 530–560 nm, Em 573–648 nm), exposure time was 6000 ms; and for EYFP was YFP (Ex490-500, Em 520–560), exposure time was 3000 ms. Data processing was performed with software Image J.

Flow cytometry

The flow cytometry analysis was carried out on BD LSR Fortessa Analyzer. Before the analysis, the cells were trypsinized using 0.25% trypsin–EDTA and then were centrifuged. 100,000 to 150,000 cells were suspended in 700 µl PBS supplemented with 1% FBS before loading onto the analyzer. GFP was measured with a 488 nm blue laser and an FITC (530/30 nm) emission filter, whereas mCherry was measured with a 561 nm yellow-green laser and a PE-Texas Red (610/20 nm) emission filter. FlowJo software was used to perform the data collection and processing.

Luciferase assay

The Varioskan Flash Spectral Scanning Multimode Reader (Thermo Scientific) was used to measure the chemiluminescence catalyzed by luciferase. The Luciferase Assay System (Promega, catalog number E1500) was used for luciferase activity measurements. The measurements were carried out at 72 h after induction by following the kit instruments.

Additional files

Additional file 1: Supplementary note. This note describes the details of DNA cloning; information of plasmids used in this study; information of oligonucleotides used in this study; nucleic acid sequence of CMV(tetO2) promoter and CMV5(CuO) promoter; amino acid sequence of rtTAm, TCP, and mCherry-TCP.

Additional file 2: Figure S1. TCP fusion proteins induce rtTAm-dependent expression of GFP. We transfected HEK cells with mCherry, mCherry-TCP, or mCherry-NLS-TCP constitutive expression plasmids, respectively, also transfected TRE3G promoter controlled GFP expression plasmid to all the three groups of cells. We divided mCherry transfected cells into two dishes and treated one dish of the cells with 1 µg/ml Dox at 24 h after transfection. The images were collected at 72 h after transfection.

Additional file 3: Figure S2. Kinetics of the circuit. (A) Kinetics of the circuit responding to blue light illumination with constitutive rtTAm expression. The cells were illuminated by blue LED (1.25 W m^{-2}) for 0 h, 5 h, 10 h, 20 h and 40 h, respectively. The data are presented as mean ± SD ($n = 6$). (B) Kinetics of the circuit responding to Dox with constitutive rtTAm expression. The cells were treated with 1 µg/ml of Dox for 0 h, 5 h, 10 h and 20 h, respectively. The data are presented as mean ± SD ($n = 6$). (C) Kinetics of the circuit responding to cumate with constitutive mCherry-TCP expression. The cells were treated with 30 µg/ml of cumate for 0 h, 5 h, 10 h, 20 h and 40 h, respectively. The data are presented as mean ± SD ($n = 6$).

Additional file 4: Figure S3. (A) Scheme of the modified circuit with a conditional positive feedback loop in the third layer. We inserted a TetR and 3 × VP16 fusion (tTA) at downstream of the reporter luciferase. In the absence of Dox, the tTA binds to its own promoter and enhances the transcription from this promoter. (B) The output luciferase activities induced by different combinations of input signals. We used 10^5 cells for the luciferase activity measurement. We repeated this experiments for three times. Data acquired in one of the three experiments were presented in mean ± SEM ($n = 6$).

Abbreviations

CuO: p-cmt and p-cym operator site; CymR: p-cmt and p-cym operon repressor; Dox: Doxycycline; *E. coli*: *Escherichia coli*; EYFP: Enhanced yellow fluorescent protein; GAVPO: Gal4(65) and the smallest light-oxygen-voltage domain fusion protein; GFP: Green fluorescent protein; HEK 293 cells: Human embryonic kidney 293 cells; LED: Light-emitting diode; rtTA: Reverse tetracycline-controlled trans-activator; TCP: Tetracycline repressor co-repression peptides; tetO: Tetracycline operator sequence; TetR: Tetracycline repressor; TRE3G promoter: Third generation tetracycline-responsive element promoter; tTA: Tetracycline-controlled trans-activator; VP16: Herpes simplex virus protein VP16.

Acknowledgements

We would like to thank Dr. Yi Yang for providing LightOn plasmids.

Funding

This work was supported by National Basic Research Program of China (973 Program, 2014CB745202) from the Ministry of Science and Technology of PRC, Shenzhen Peacock project (201503313000502), Shenzhen Science and Technology Innovation Committee Basic Science Research Grant (JCYJ20150629151046896) to JDH, and Shenzhen Science and Technology Innovation Committee Basic Science Research Grant (JCYJ20160331115823245) to WH. The funding bodies have no contribution in the design of the study, collection, analysis, and interpretation of data.

Authors' contributions

LZL and JDH designed the research, WH provided essential comments. LZL performed experiments. LZL and JDH performed data analysis. The manuscript was written by LZL, edited by JDH and WH. All authors' read and approved the final manuscript.

Competing interests

The authors declare they have no competing interests.

Author details

School of Biomedical Sciences, Li Ka Shing Faculty of Medicine, University of Hong Kong, Pok Fu Lam, Hong Kong, People's Republic of China. [2]Shenzhen Institute of Research and Innovation, University of Hong Kong, Shenzhen 518057, People's Republic of China. [3]The Centre for Synthetic Biology Engineering Research, Shenzhen Institutes of Advanced Technology, Shenzhen 518055, People's Republic of China. [4]Department of Biology, Shenzhen Key Laboratory of Cell Microenvironment, South University of Science and Technology of China, Shenzhen 518055, People's Republic of China.

References

1. Elowitz MB, Leibler S. A synthetic oscillatory network of transcriptional regulators. Nature. 2000;403:335–8.
2. Gardner TS, Cantor CR, Collins JJ. Construction of a genetic toggle switch in Escherichia coli. Nature. 2000;403:339–42.
3. Friedland AE, Lu TK, Wang X, Shi D, Church G, Collins JJ. Synthetic gene networks that count. Science. 2009;324:1199–202.
4. Bonnet J, Yin P, Ortiz ME, Subsoontorn P, Endy D. Amplifying genetic logic gates. Science. 2013;340:599–603.
5. Guido NJ, Wang X, Adalsteinsson D, McMillen D, Hasty J, Cantor CR, Elston TC, Collins JJ. A bottom-up approach to gene regulation. Nature. 2006;439:856–60.
6. Chin JW. Modular approaches to expanding the functions of living matter. Nat Chem Biol. 2006;2:304–11.
7. Friedlander T, Mayo AE, Tlusty T, Alon U. Evolution of bow-tie architectures in biology. PLoS Comput Biol. 2015;11:e1004055.
8. Kitano H. Biological robustness. Nat Rev Genet. 2004;5:826–37.
9. Radulov G, P Q, Hegt H, van Roermund AHM. Smart and flexible digital-to-analog converters. 1st ed. Dordrecht: Springer; 2011.
10. Gossen M, Bujard H. Tight control of gene expression in mammalian cells by tetracycline-responsive promoters. Proc Natl Acad Sci USA. 1992;89:5547–51.
11. Gossen M, Freundlieb S, Bender G, Muller G, Hillen W, Bujard H. Transcriptional activation by tetracyclines in mammalian cells. Science. 1995;268:1766–9.
12. Shimizu-Sato S, Huq E, Tepperman JM, Quail PH. A light-switchable gene promoter system. Nat Biotechnol. 2002;20:1041–4.
13. Levskaya A, Chevalier AA, Tabor JJ, Simpson ZB, Lavery LA, Levy M, Davidson EA, Scouras A, Ellington AD, Marcotte EM, Voigt CA. Synthetic biology: engineering Escherichia coli to see light. Nature. 2005;438:441–2.
14. Wang X, Chen XJ, Yang Y. Spatiotemporal control of gene expression by a light-switchable transgene system. Nat Methods. 2012;9:266–9.
15. Tabor JJ, Salis HM, Simpson ZB, Chevalier AA, Levskaya A, Marcotte EM, Voigt CA, Ellington AD. A synthetic genetic edge detection program. Cell. 2009;137:1272–81.
16. Chen X, Li T, Wang X, Du Z, Liu R, Yang Y. Synthetic dual-input mammalian genetic circuits enable tunable and stringent transcription control by chemical and light. Nucleic Acids Res. 2015;44(6):2677–90.
17. Olson EJ, Hartsough LA, Landry BP, Shroff R, Tabor JJ. Characterizing bacterial gene circuit dynamics with optically programmed gene expression signals. Nat Methods. 2014;11:449–55.
18. Olson EJ, Tabor JJ. Optogenetic characterization methods overcome key challenges in synthetic and systems biology. Nat Chem Biol. 2014;10:502–11.

19. Mullick A, Xu Y, Warren R, Koutroumanis M, Guilbault C, Broussau S, Malenfant F, Bourget L, Lamoureux L, Lo R, Caron AW, Pilotte A, Massie B. The cumate gene-switch: a system for regulated expression in mammalian cells. BMC Biotechnol. 2006;6:1–18.

20. Goeke D, Kaspar D, Stoeckle C, Grubmüller S, Berens C, Klotzsche M, Hillen W. Short Peptides Act as Inducers, Anti-Inducers and Corepressors of Tet Repressor. J Mol Biol. 2012;416:33–45.

21. Wimmert C, Platzert S, Hillen W, Klotzsche M. A novel method to analyze nucleocytoplasmic transport in vivo by using short peptide tags. J Mol Biol. 2013;425:1839–45.

22. Deans TL, Cantor CR, Collins JJ. A tunable genetic switch based on RNAi and repressor proteins for regulating gene expression in mammalian cells. Cell. 2007;130:363–72.

23. Delerue F, White M, Ittner LM. Inducible, tightly regulated and non-leaky neuronal gene expression in mice. Transgenic Res. 2014;23:225–33.

24. Ma ZC, Du ZM, Chen XJ, Wang X, Yang Y. Fine tuning the LightOn light-switchable transgene expression system. Biochem Biophys Res Commun. 2013;440:419–23.

25. Muller K, Zurbriggen MD, Weber W. Control of gene expression using a red- and far-red light-responsive bi-stable toggle switch. Nat Protoc. 2014;9:622–32.

26. Niwa H, Miyazaki J, Smith AG. Quantitative expression of Oct-3/4 defines differentiation, dedifferentiation or self-renewal of ES cells. Nat Genet. 2000;24:372–6.

27. Kracikova M, Akiri G, George A, Sachidanandam R, Aaronson SA. A threshold mechanism mediates p53 cell fate decision between growth arrest and apoptosis. Cell Death Differ. 2013;20:576–88.

28. Stewart-Ornstein J, Nelson C, DeRisi J, Weissman JS, El-Samad H. Msn2 coordinates a stoichiometric gene expression program. Curr Biol. 2013;23:2336–45.

29. Li Z, Liu J, Zheng M, Xu XZ. Encoding of Both Analog- and Digital-like Behavioral Outputs by One C. elegans Interneuron. Cell. 2014;159:751–65.

30. Daniel R, Rubens JR, Sarpeshkar R, Lu TK. Synthetic analog computation in living cells. Nature. 2013;497:619–23.

31. Purcell O, Lu TK. Synthetic analog and digital circuits for cellular computation and memory. Curr Opin Biotechnol. 2014;29:146–55.

32. Rubens JR, Selvaggio G, Lu TK. Synthetic mixed-signal computation in living cells. Nat Commun. 2016;7:11658.

33. Urlinger S, Baron U, Thellmann M, Hasan MT, Bujard H, Hillen W. Exploring the sequence space for tetracycline-dependent transcriptional activators: novel mutations yield expanded range and sensitivity. Proc Nat Acad Sci USA. 2000;97:7963–8.

34. Wang Z, Zhou ZJ, Liu DP, Huang JD. Double-stranded break can be repaired by single-stranded oligonucleotides via the ATM/ATR pathway in mammalian cells. Oligonucleotides. 2008;18:21–32.

9

Quantitative reproducibility analysis for identifying reproducible targets from high-throughput experiments

Wenfei Zhang[2*], Ying Liu[1], Mindy Zhang[2], Cheng Zhu[2] and Yuefeng Lu[2]

Abstract

Background: High-throughput assays are widely used in biological research to select potential targets. One single high-throughput experiment can efficiently study a large number of candidates simultaneously, but is subject to substantial variability. Therefore it is scientifically important to performance quantitative reproducibility analysis to identify reproducible targets with consistent and significant signals across replicate experiments. A few methods exist, but all have limitations.

Methods: In this paper, we propose a new method for identifying reproducible targets. Considering a Bayesian hierarchical model, we show that the test statistics from replicate experiments follow a mixture of multivariate Gaussian distributions, with the one component with zero-mean representing the irreproducible targets.

Results: A target is thus classified as reproducible or irreproducible based on its posterior probability belonging to the reproducible components. We study the performance of our proposed method using simulations and a real data example.

Conclusion: The proposed method is shown to have favorable performance in identifying reproducible targets compared to other methods.

Keywords: Reproducibility, High-throughput experiment, Bayesian classification, Empirical Bayes, Gaussian mixture, EM algorithm

Background

In biological research, high-throughput assays, such as microarrays, are widely used to effectively select potential targets by studying a large number of candidates in a single experiment. However a high-throughput assay is often subject to substantial variability. Reproducibility of high-throughput assays, such as the level of agreement across replicate samples, test sites or data analytical platforms, is a concerned topic in scientific applications, and has been discussed in [1] for microarray and [2] for ChIP-seq technology. Therefore quantitative analysis for the reproducibility of high-throughput assays is an important exercise for evaluating the reliability and robustness of scientific discoveries across studies.

Reproducibility is nonstandard and unsettled across the sciences. Goodman et al. [3] provides a survey on the papers with the word reproducibility included in titles, abstracts and keywords, and concludes that the interpretation of reproducibility varies among different papers. Goodman et al. [3] further allies the word reproducibility in the papers and classifies them into three terms: methods reproducibility, results reproducibility and inferential reproducibility. In [3], methods reproducibility refers to the provision of enough detail about study procedures and data so the same procedures could, in theory or in actuality, be exactly repeated, such as [1] and [2]; results reproducibility refers to obtaining the same results from the conduct of an independent study whose procedures are as closely matched to the original experiment as possible, such as [4] and [5]; Inferential reproducibility refers to the drawing of qualitatively similar conclusions from

*Correspondence: wenfei.zhang@sanofi.com
[2]Sanofi, Framingham, MA, USA
Full list of author information is available at the end of the article

Table 1 The summary of misclassification rates for the four compared methods under different significant levels (α) and proportions of reproducible genes (γ)

	The proposed Method		The copula mixture method [10]		Benjamini & Heller method [9]		The rank product method [8]	
	$\alpha=0.1$	$\alpha=0.05$	$\alpha=0.1$	$\alpha=0.05$	$\alpha=0.1$	$\alpha=0.05$	$\alpha=0.1$	$\alpha=0.05$
$\gamma=80\%$	0.007(0.001)	0.008(0.0012)	0.24(0.0708)	0.271(0.0954)	0.025(0.0022)	0.032(0.0025)	0.197(0.0044)	0.25(0.0036)
$\gamma=60\%$	0.007(0.0013)	0.008(0.0013)	0.402(0.0022)	0.404(0.0028)	0.022(0.0017)	0.027(0.002)	0.073(0.0031)	0.099(0.0035)
$\gamma=40\%$	0.005(0.001)	0.006(0.001)	0.568(0.0059)	0.541(0.01)	0.016(0.0017)	0.02(0.0019)	0.02(0.0018)	0.028(0.0021)
$\gamma=20\%$	0.004(8e-04)	0.004(8e-04)	0.166(0.0026)	0.186(0.0015)	0.01(0.0014)	0.013(0.0015)	0.004(9e-04)	0.006(0.0011)
$\gamma=10\%$	0.002(6e-04)	0.002(6e-04)	0.058(0.0104)	0.077(0.0075)	0.007(0.001)	0.008(0.0011)	0.002(5e-04)	0.002(6e-04)
$\gamma=5\%$	0.001(5e-04)	0.001(5e-04)	0.011(0.0038)	0.025(0.0042)	0.004(9e-04)	0.005(0.001)	0.001(4e-04)	0.001(3e-04)
$\gamma=1\%$	0.001(4e-04)	0(4e-04)	0.001(6e-04)	0.002(9e-04)	0.001(7e-04)	0.002(7e-04)	0.001(4e-04)	0(3e-04)

either an independent replication of a study or a reanalysis of the original study, such as [1] and [2].

In this paper, our reproducibility analysis aims to identify reproducible targets with consistent and significant signals across replicate studies, which belongs to the category of inferential reproducibility as defined in [3]. Our reproducibility analysis is different from meta-analysis, such as [6] and [7]. Meta-analysis combines the data from multiple studies to gain extra power for identifying targets with signals. The identified targets may not necessarily be significant across all studies.

A few methods have been developed for our reproducibility analysis. Hong et al. [8] proposed a permutation based method through estimating the empirical distribution of the rank product. Benjamini & Heller [9] developed a framework for testing partial conjunction hypothesis that the discovery is true in at least u studies out of total n studies. Most recently, [10] proposed a copula mixture model for estimating the irreproducible discovery rate across studies.

However all existing methods potentially have limitations. The permutation based method [8] can be computationally expensive when dealing with a large number of candidates. Benjamini & Heller method [9] aims at identifying candidates with reproduced signals in a few but not all the studies, which is a related but generally weaker

goal than ours. The special case of Benjamini & Heller method testing whether signals are reproduced in all studies is identical to using the largest p-value. The copula mixture [10] method builds the copula mixture using the rank transformation of the original data, which might be less powerful than modeling the original data with a proper probabilistic model as in our proposed method. A major drawback of both Benjamini & Heller method [9] and the copula mixture [10] method is that they both use the significant score of signals, such as p-value, without taking into account the directionality of signals, hence is prune to selecting candidates with significant scores but different directions across studies. For example, in the context of two replicate microarray studies with a treatment and a control group, consider genes with significant p-values in both experiments, but are up-regulated in one study and down-regulated in the other. Although those genes have inconsistent signals across studies, both methods will likely classify them as reproducible based on p-values alone. In contrast, our proposed method models the test statistics directly and is expected to correctly classify those genes as irreproducible most of the time.

In this paper, we propose a Bayesian hierarchical model and show the test statistics from replicate studies can be approximated by a mixture of multivariate Gaussian distributions. The proposed Gaussian mixture model

Table 2 The summary of sensitivities for the four compared methods under different significant levels (α) and proportions of reproducible genes (γ)

	The proposed Method		The copula mixture method [10]		Benjamini & Heller method [9]		The rank product method [8]	
	$\alpha=0.1$	$\alpha=0.05$	$\alpha=0.1$	$\alpha=0.05$	$\alpha=0.1$	$\alpha=0.05$	$\alpha=0.1$	$\alpha=0.05$
$\gamma=80\%$	0.992(0.0014)	0.991(0.0016)	0.948(0.0881)	0.905(0.1184)	0.97(0.0027)	0.96(0.0031)	0.754(0.0055)	0.687(0.0045)
$\gamma=60\%$	0.99(0.002)	0.988(0.0021)	0.978(0.0071)	0.956(0.0119)	0.966(0.0028)	0.955(0.0033)	0.878(0.0052)	0.836(0.0058)
$\gamma=40\%$	0.989(0.0024)	0.987(0.0024)	0.975(0.0069)	0.937(0.0161)	0.962(0.0046)	0.951(0.005)	0.949(0.0045)	0.931(0.0051)
$\gamma=20\%$	0.985(0.0037)	0.983(0.004)	0.176(0.0149)	0.069(0.0081)	0.949(0.007)	0.937(0.0076)	0.978(0.0046)	0.972(0.0053)
$\gamma=10\%$	0.984(0.0048)	0.982(0.0051)	0.421(0.1033)	0.228(0.0746)	0.934(0.0098)	0.92(0.0108)	0.985(0.0053)	0.982(0.0055)
$\gamma=5\%$	0.984(0.0069)	0.983(0.0075)	0.773(0.0741)	0.509(0.0832)	0.925(0.0191)	0.909(0.0195)	0.99(0.0049)	0.988(0.0057)
$\gamma=1\%$	0.986(0.0176)	0.984(0.0177)	0.907(0.0592)	0.842(0.0882)	0.866(0.0673)	0.844(0.0706)	0.99(0.0163)	0.99(0.0163)

Table 3 The summary of specificities for the four compared methods under different significant levels (α) and proportions of reproducible genes (γ)

	The proposed Method		The copula mixture method [10]		Benjamini & Heller method [9]		The rank product method [8]	
	α=0.1	α=0.05	α=0.1	α=0.05	α=0.1	α=0.05	α=0.1	α=0.05
γ=80%	0.996(0.002)	0.997(0.0017)	0.009(0.0058)	0.025(0.0152)	0.994(0.0021)	0.999(0.001)	1(0)	1(0)
γ=60%	0.998(9e-04)	0.999(7e-04)	0.029(0.0075)	0.057(0.0144)	0.997(0.0015)	0.999(7e-04)	1(0)	1(0)
γ=40%	0.999(7e-04)	0.999(6e-04)	0.07(0.0136)	0.139(0.0268)	0.999(7e-04)	1(4e-04)	1(0)	1(0)
γ=20%	0.999(4e-04)	0.999(3e-04)	0.999(9e-04)	1(4e-04)	1(3e-04)	1(1e-04)	1(0)	1(0)
γ=10%	0.999(4e-04)	1(3e-04)	1(1e-04)	1(1e-04)	1(1e-04)	1(1e-04)	1(2e-04)	1(1e-04)
γ=5%	1(3e-04)	1(3e-04)	1(1e-04)	1(0)	1(1e-04)	1(0)	1(3e-04)	1(1e-04)
γ=1%	1(3e-04)	1(3e-04)	1(1e-04)	1(0)	1(0)	1(0)	1(4e-04)	1(2e-04)

classifies the signals into three components: one irreproducible component and two reproducible components for consistent up-regulated and down-regulated signals respectively. The posterior probability of belonging to the reproducible components is used as a measure for reproducibility.

Methods

For simplicity, we will introduce our method in the context of microarray studies but it can be generalized to studies of other high-throughput assays. We consider I replicate microarray studies for p genes. In this paper, we focus on the situation of two replicate studies $I = 2$, although our method can be readily extended to the case with more than two studies. We assume a study includes two groups, e.g., the treatment and control group, with sample size equal to n_{ik} for group k, k=1,2, in the i-th study. Let x_{gijk} be the normalized and transformed measurement of gene expression of the jth sample from group k for gene g in the i-th study. The test statistics of two-sample unpaired t-test for gene g in the i-th study is

$$d_{gi} = \frac{\bar{x}_{gi2} - \bar{x}_{gi1}}{s_{gi}}, \text{where}$$

$$\bar{x}_{gi1} = \sum_{j=1,\cdots,n_{i1}} x_{gij1}/n_{i1}, \bar{x}_{gi2} = \sum_{j=1,\cdots,n_{i2}} x_{gij2}/n_{i2}$$

$$s_{gi} = \left[(1/n_{i1} + 1/n_{i2}) \left\{ \sum_{j=1,\cdots,n_{i1}} (x_{gij1} - \bar{x}_{gi1})^2 \right.\right.$$

$$\left.\left. + \sum_{j=1,\cdots,n_{i2}} (x_{gij2} - \bar{x}_{gi2})^2 \right\} / (n_{i1} + n_{i2} - 2) \right]^{1/2}$$

We present an empirical Bayesian hierarchical model to account for various sources of variability. When the sample size n_{ik} is reasonably large, say $n_{i1} + n_{i2} \geq 30$, the test statistics d_{gi} is well approximated by a normal distribution:

$$d_{gi}|\mu_{gi} \sim \mathcal{N}(\delta_{S_i}\mu_{gi}, 1) \tag{1}$$

where μ_{gi} is the expected group mean difference for gene g in the i-th study, and $\delta_{S_i} = \tilde{\sigma}_i^{-1}(1/n_{i1} + 1/n_{i2})^{-1/2}$ with $\tilde{\sigma}_i$ being the common standard deviation for $\{x_{gij1}\}$, $j = 1, 2, \ldots, n_{i1}$ and $\{x_{gij2}\}$, $j = 1, 2, \ldots, n_{i2}$. When the sample size is small, the same procedure as in [11] can be applied to construct z-tests based on two-sample t-tests. For simplicity we assume the within-group between-sample standard deviation is the same for all the genes. The general case can be derived in a similar fashion but a bit more tedious.

For the expected group mean difference μ_{gi}, we assume it follows

$$\mu_{gi}|\mu_g \sim \mathcal{N}\left(\mu_g, \sigma_g^2\right) \tag{2}$$

where μ_g is the "true" group mean difference for gene g across all studies and σ_g^2 models the between-study variability due to various experiment conditions.

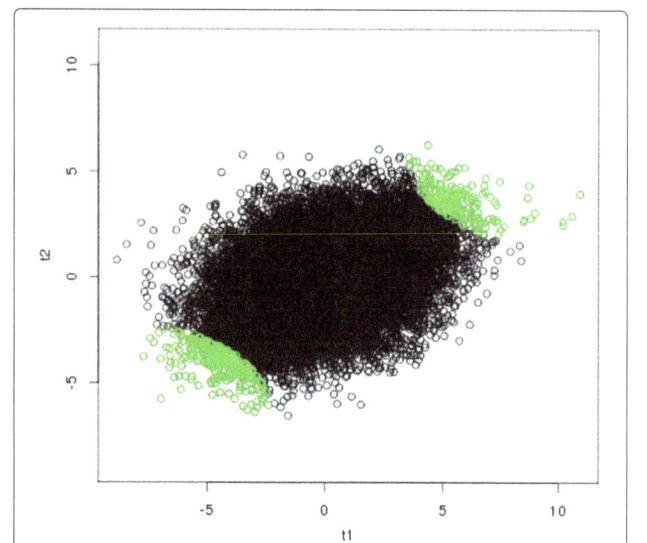

Fig. 1 Bivariate plot of test statistics from two studies. The x axis represents the test statistics from GSE 28042 study [18], and the y axis represents the test statistics from GSE 33566 [19]. The *green* points are the top 500 reproducible genes selected by the proposed method

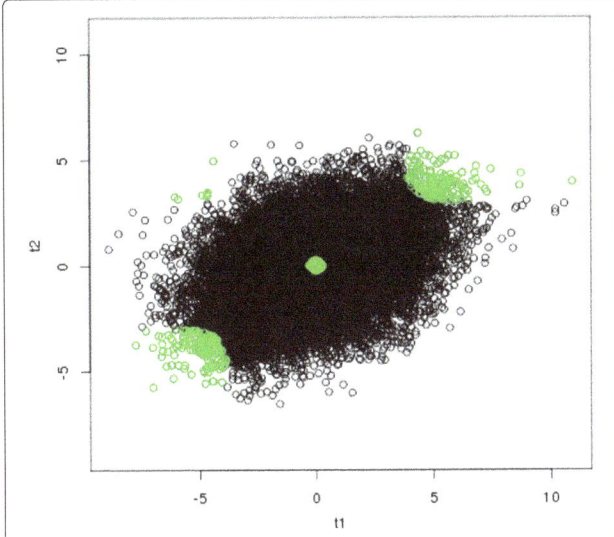

Fig. 2 Bivariate plot of test statistics from two studies. The x axis represents the test statistics from GSE 28042 study [18], and the y axis represents the test statistics from GSE 33566 [19]. The *green* points are the top 500 reproducible genes selected by the copula mixture model [10]

Furthermore we assume μ_g is from a mixture distribution

$$\mu_g \sim \pi_0 I_{\{0\}} + \pi_1 \mathcal{N}\left(\mu_{G_1}, \sigma_{G_1}^2\right) + \pi_2 \mathcal{N}\left(\mu_{G_2}, \sigma_{G_2}^2\right) \quad (3)$$

where $\pi_i \geq 0$, $i = 0, 1, 2$, with $\pi_0 + \pi_1 + \pi_2 = 1$, $\mu_{G_1} > 0$ and $\mu_{G_2} < 0$. The distribution has three components: the null case where there is no differentially expressed gene, the "up-regulated" case where the treatment stimulates the gene expression, and the "down-regulated" case where the treatment suppresses the gene expression. Generally for microarray studies $\pi_0 \simeq 1$. Similar mixture models have been considered in [11–16]. Particularly we choose to model the cluster of up-regulated (or down-regulated) genes with a Gaussian distribution for the computational convenience, same as in [12]. Alternative choices include the semiparametric mixture model in [11, 14], mixture of Gaussian distributions in [13, 15] and mixture of t-distributions in [16].

We can show that the test statistics (d_{g1}, d_{g2}) follow a Gaussian mixture model. The derivations are standard by repeatedly applying the law of total expectation and the law of total variance and thus omitted. The mixture model is

$$
\begin{aligned}
(d_{g1}, d_{g2}) \sim{} & \pi_0 \mathcal{N}(\mu_0, \Sigma_0) + \pi_1 \mathcal{N}(\mu_1, \Sigma_1) \\
& + \pi_2 \mathcal{N}(\mu_2, \Sigma_2),
\end{aligned}
\quad (4)
$$

where $\mathcal{N}(\mu_l, \Sigma_l)$ $(l = 0, 1, 2)$ is the biviariate normal distribution with mean vector μ_l and covariance matrix Σ_l. Let I_2 and J_2 be the identity matrix and the square matrix of ones respectively, both with order 2. This mixture model classify the candidates into three components:

$\mathcal{N}(\mu_0, \Sigma_0)$ is the irreproducible component with zero-mean $\mu_0 = (0, 0)^T$ and covariance structure $\Sigma_0 = \left(\sigma_g^2 + 1\right) I_2$; $\mathcal{N}(\mu_1, \Sigma_1)$ and $\mathcal{N}(\mu_2, \Sigma_2)$ are two reproducible components with $\mu_1 = (\delta_{S_1}\mu_{G_1}, \delta_{S_2}\mu_{G_1}) > 0$ and $\Sigma_1 = \left(\sigma_g^2 + 1\right) I_2 + \sigma_{G_1}^2 J_2$ representing the up-regulated genes, and $\mu_2 = (\delta_{S_1}\mu_{G_2}, \delta_{S_2}\mu_{G_2}) < 0$ and $\Sigma_2 = \left(\sigma_g^2 + 1\right) I_2 + \sigma_{G_2}^2 J_2$ representing the down-regulated genes, where the inequalities are meant to be interpreted component-wise.

Note with increased sample sizes or decreased within-group between-sample variability, the mean μ_1 and μ_2 of the reproducible components move further away from the origin, making the three components more separable. Also note the test statistics from replicate studies have zero correlations in the irreproducible components; in the reproducible components, the correlations become larger when the between-study variability becomes smaller; for all components, the variance is smaller with less between-study variability, resulting in more separable components.

Under the Gaussian mixture model, the posterior probability of (d_{g1}, d_{g2}) belonging to a component is

$$p_{gl} = \frac{\pi_l \phi(d_{g1}, d_{g2} | \mu_l, \Sigma_l)}{\sum_{\ell=0,1,2} \pi_\ell \phi(d_{g1}, d_{g2} | \mu_\ell, \Sigma_\ell)}, l = 0, 1, 2. \quad (5)$$

where $\phi(\cdot|\cdot)$ is the density function of bivariate normal distribution. According to [10], the posterior probability of being in the irreproducible/null component p_{i0} can be introduced as the individual significant score, namely local false discovery rate. When p_{g0} is less than a significant level α, gene g is classified as reproducible.

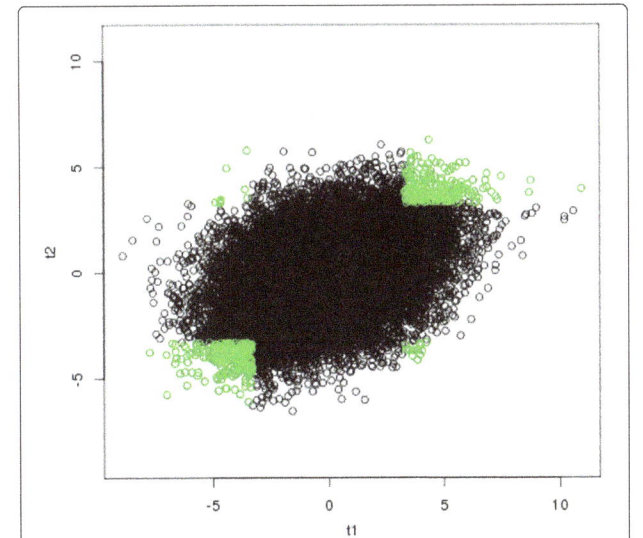

Fig. 3 Bivariate plot of test statistics from two studies. The x axis represents the t-statistics from GSE 28042 study [18], and the y axis represents t-statistics from GSE 33566 [19]. The *green* points are the top 500 reproducible genes selected by Benjamini & Heller method [9]

Table 4 The list of 23 selected genes, which are in the list of the top 500 reproducible genes selected by Benjamini & Heller method [9], but have opposite signs of signals in two studies

	Genes	t-statistics in GSE 28042 [18]	t-statistics in GSE 33566 [19]
1	A1BG	3.34	-3.63
2	ANKRD39	3.93	-3.35
3	CA4	-4.4	4.94
4	CDK14	-4.88	3.34
5	CHCHD2	3.5	-3.65
6	CXCR2	-4.67	3.38
7	HCG27	-4.68	3.29
8	KAT6A	-3.48	3.54
9	MFSD3	4.25	-3.29
10	MMP9	-3.51	5.77
11	MRPL14	4.06	-3.69
12	MRPL15	3.99	-3.38
13	MRPL55	3.63	-3.95
14	NDUFB7	3.79	-3.54
15	NDUFS3	3.98	-3.89
16	PRPS1	3.87	-4.13
17	RBBP6	3.66	-3.67
18	ROMO1	3.33	-3.41
19	SEPHS1	4	-3.44
20	TANC2	-3.59	3.95
21	TCN1	-4.69	3.36
22	TMEM141	3.45	-3.64
23	TRIM33	-4.64	3.47

Next, we consider estimation of the unknown parameters

$$\theta = (\mu_1, \mu_2, \Sigma_0, \Sigma_1, \Sigma_2, \pi_0, \pi_1, \pi_2) \qquad (6)$$

in the mixture model (4) to get the estimate of p_{g0} for individual genes. It is natural to use the expectation-maximization (EM) algorithm to estimate θ by maximizing the log-likelihood of the data [17], i.e.,

$$\ell(\theta) = \sum_{g=1}^{p} \log\{P(d_{g1}, d_{g1}|\theta)\}$$

$$= \sum_{g=1}^{p} \log \left\{ \sum_{l=0}^{2} \pi_l \phi(d_{g1}, d_{g2}|\mu_l, \Sigma_l) \right\} \qquad (7)$$

In our algorithm, we start with some initials value for the parameters θ^0, then iterate between two steps: (1) Evaluate the current posterior probabilities p_{gl} using the current parameters; (2) Maximize the likelihood estimator given current posterior probabilities. The details of the

EM procedures are provided in Appendix. Multiple random initial vaues are used to avoid being trapped at the local maximum.

Simulation studies

In this section, we present numerical simulations to illustrate the performance of our proposed method compared to three existing methods, the copula mixture model [10], Benjamini & Heller method [9], and the rank product method [8]. We use the following model to simulate data

$$x_{gijk} = \mu + \alpha_g + \beta_i + (\alpha\beta)_{gi} + \delta I(k = 2)$$
$$+ \gamma_g I(k = 2) + (\gamma\beta)_{gi} I(k = 2) + \epsilon_{gijk} \qquad (8)$$

From this model, the mean expression level of gene g for group 1 of study s is modeled as $\mu_{gs1} = \mu + \alpha_g + \beta_i + (\alpha\beta)_{gi}$, where μ is the overall mean; α_g is the main effect of gene g; β_i is the main effect of study i; $(\alpha\beta)_{gi}$ is the gene-study interaction. We set $\mu = 0$, $\alpha_g \sim \mathcal{N}(0,1)$, $\beta_i = 0.1$, and $(\alpha\beta)_{gi} \sim \mathcal{N}(0, 0.5^2)$. For non-differentially expressed genes, the mean expression level for both groups are the same, i.e., $\mu_{gs1} = \mu_{gs2}$. For differentially expressed genes, (8) models the difference between the two comparison groups as $\mu_{gi2} - \mu_{gi1} = \delta + \gamma_g + (\gamma\beta)_{gi}$, where δ is the fixed effect of group difference; γ_g is the effect of gene on the group difference; $(\gamma\beta)_{gi}$ is the gene-study interaction of the group difference. We set $\delta = 0$, generate γ_g from $\mathcal{N}(2, 0.5^2)$ or $\mathcal{N}(-2, 0.5^2)$ to mimic two possible directions of signals, $(\gamma\beta)_{gi} \sim \mathcal{N}(0, 0.5^2)$. ϵ_{gijk} is the random error term, and following the distribution $\mathcal{N}(0, 0.5^2)$.

For each simulation run, we generate 2 studies. Each study has two groups with 10 samples per group. We generate $G = 5000$ genes per sample and choose the proportions of reproducible genes (γ) from (80%, 60%, 40%, 20%, 10%, 5%, 1%). We apply the proposed method and the three existing methods to the simulated data, and classify the genes as reproducible based on two commonly used significant levels (α) 0.05 and 0.1. The performance of the four compared methods is evaluated by three criteria, i.e., sensitivity, specificity and misclassification rate. Results from 50 simulations are summarized in Tables 1, 2 and

Table 5 The list of 7 selected genes, which are in the list of the top 500 reproducbile genes selected by the copula mixture model [10], but have opposite signs of signals in two studies

	Gene	t-statistics in GSE 28042 [18]	t-statistics in GSE 33566 [19]
1	CA4	-4.4	4.94
2	CDK14	-4.88	3.34
3	CXCR2	-4.67	3.38
4	HCG27	-4.68	3.29
5	MME	-6.05	3.25
6	TCN1	-4.69	3.36
7	TRIM33	-4.64	3.47

3 respectively. The results shows our proposed method performs the best among the four methods with the smallest misclassification rates (Table 1), highest sensitivity (Table 2) and highest specificities (Table 3).

Results

In this section, we illustrate our proposed method using a real example. This example includes two microarray studies [18] and [19] comparing idiopathic pulmonary fibrosis (IPF) samples with healthy control samples. Data from both studies are obtained from Gene Expression Omnibus [20]. GSE 28042 [18] measures profiles of peripheral blood mononuclear cell (PBMC) for 75 IPF samples and 16 control samples through GeneChip Human 1.0 exon ST arrays, and GSE 33566 [19] measures profiles of peripheral blood RNA for 93 IPF patients and 30 control samples through Agilent Whole Human Genome Oligonucleotide Microarrays. We only consider the overlap 17708 common genes for reproducibility analysis.

We apply our proposed method, the copula mixture model [10] and Benjamini & Heller method [9]. The rank product method [8] is too computationally intensive to be applied to this example and thus excluded from this study. Figures 1, 2 and 3 show the results of selected reproducible genes from the three compared methods respectively (green). In all three figures, the x axis represents the test statistics from GSE 28042 [18], and the y axis represents the test statistics from GSE 33566 [19]. The top 500 reproducible genes selected by three methods are highlighted in green. As shown in Fig. 1, our proposed method only selects genes with consistently significant signals in both studies. Benjamini & Heller method [9] incorrectly identifies 23 genes (the upper left and bottom right corners of Fig. 2) as reproducible, which actually have opposite directions in two studies. The complete list of the 23 genes incorrectly selected by Benjamini & Heller method [9] is provided in Table 4. The copula mixture model [10] selects 7 genes (Table 5) with opposite directions of signals. It's also noted that the copula mixture model [10] appears to be less powerful in separating the irreproducible and reproducible genes and has incorrectly selected some insignificant genes (see the center of Fig. 3), likely resulting from the rank transformation. Overall, our method performs favorably in identifying reproducible genes.

Conclusion and discussion

This paper proposes a new method for identifying consistent and significant signals across replicate high-throughput experiments. Existing methods ignore the directionality of signals, and can incorrectly identify signals with opposite directions as reproducible ones. Our proposed method considers both the significant scores and directions of signals by modeling the test statistics

directly, leading to improved performance in selecting reproducible candidates. When the proposed method is applied to a real data example for identifying reproducible genes in studies of idiopathic pulmonary fibrosis samples, it is shown to have better performance in detecting significant and reproducible genes compared to other methods. Simulations also demonstrate that our method compares favorably to the existing methods.

Appendix

Expectation-maximization (EM) algorithm to estimate model parameters

The algorithm for estimating θ in (6) is an iterative algorithm between Expectation steps and maximization step. We use $\widehat{\theta}^v$ to denote the estimate at vth iteration. The algorithm includes the following steps:

Step 1: Initial Values Generate the initial values for θ and denote it as $\widehat{\theta}^0$

Step 2: Expectation-Step Continue from the vth iteration step with the estimate $\widehat{\theta}^v$. We can obtain the estimated posterior probability $\widehat{p_{gl}}^v$ of (d_{g1}, d_{g2}) from (5) by

$$\widehat{p_{gl}}^v = \frac{\widehat{\pi_l}^v \phi\left(d_{g1}, d_{g2} | \widehat{\mu_l}^v, \widehat{\Sigma_l}^v\right)}{\sum_{\ell=0,1,2} \widehat{\pi_\ell}^v \phi\left(d_{g1}, d_{g2} | \widehat{\mu_\ell}^v, \widehat{\Sigma_\ell}^v\right)}, l = 0, 1, 2. \quad (9)$$

Step 3: Maximization-Step Update the parameter $\widehat{\theta}^{v+1}$ by maximizing the log-likelihood function $\ell(\theta)$ in (7) given the current estimated posterior probability $\widehat{p_{gl}}^v$. The estimated parameters from the maximization are

$$\widehat{\pi_l}^{v+1} = \sum_{g=1}^{p} \widehat{p_{gl}}^v / p, l = 0, 1, 2.$$

$$\widehat{\mu_1}^{v+1} = \left(\widehat{\mu_{11}}^{v+1}, \widehat{\mu_{12}}^{v+1}\right) = \left(\frac{\sum_{g=1}^{p} \widehat{p_{g2}}^v d_{g1}}{\sum_{g=1}^{p} \widehat{p_{g2}}^v}, \frac{\sum_{g=1}^{p} \widehat{p_{g2}}^v d_{g2}}{\sum_{g=1}^{p} \widehat{p_{g2}}^v}\right)$$

$$\widehat{\mu_2}^{v+1} = \left(\widehat{\mu_{21}}^{v+1}, \widehat{\mu_{22}}^{v+1}\right) = \left(\frac{\sum_{g=1}^{p} \widehat{p_{g3}}^v d_{g1}}{\sum_{g=1}^{p} \widehat{p_{g3}}^v}, \frac{\sum_{g=1}^{p} \widehat{p_{g3}}^v d_{g2}}{\sum_{g=1}^{p} \widehat{p_{g3}}^v}\right)$$

$$\widehat{\sigma_g^2}^{v+1} = \frac{\sum_{g=1}^{p} \widehat{p_{g1}}^v \left(d_{g1}^2 + d_{g2}^2\right)}{2 \sum_{g=1}^{p} \widehat{p_{g1}}^v} - 1$$

$$\widehat{\sigma_{G_1}^2}^{v+1} = \frac{\sum_{g=1}^{p} \left[\widehat{p_{g2}}^v \left(d_{g1} - \widehat{\mu_{11}}^{v+1}\right)^2 + \widehat{p_{g2}}^v \left(d_{g2} - \widehat{\mu_{12}}^{v+1}\right)^2\right]}{2 \sum_{g=1}^{p} \widehat{p_{g2}}^v}$$

$$- \frac{\sum_{g=1}^{p} \widehat{p_{g1}}^v \left(d_{g1}^2 + d_{g2}^2\right)}{2 \sum_{g=1}^{p} \widehat{p_{g1}}^v}$$

$$\widehat{\sigma_{G_2}^2}^{v+1} = \frac{\sum_{g=1}^{p} \left[\widehat{p_{g3}}^v \left(d_{g1} - \widehat{\mu_{21}}^{v+1}\right)^2 + \widehat{p_{g3}}^v \left(d_{g2} - \widehat{\mu_{22}}^{v+1}\right)^2\right]}{2 \sum_{g=1}^{p} \widehat{p_{g3}}^v}$$

$$- \frac{\sum_{g=1}^{p} \widehat{p_{g1}}^v \left(d_{g1}^2 + d_{g2}^2\right)}{2 \sum_{g=1}^{p} \widehat{p_{g1}}^v}$$

Step 4: Solution The algorithm continues between Expectation-Step and Maximization-Step until the following two conditions are satisfied.

1. The difference between $\widehat{\theta}^{\nu}$ and $\widehat{\theta}^{\nu+1}$ is less than a small value δ_1 for all their elements;

2. The change in log-likelihood function $\ell(\theta)$ between two consecutive iterations does not exceed a small value δ_2.

Acknowledgements
We would like thank referees for their time on reviewing this manuscript.

Authors' contributions
All authors equally distributed. All authors read and approved the final manuscript.

Competing interests
The authors declare that they have no competing interests.

Author details
[1] Department of Biostatistics, Columbia University, New York, NY, USA. [2] Sanofi, Framingham, MA, USA.

References

1. Shi L, Reid LH, Jones WD, Shippy R, Warrington JA, Baker SC, Collins PJ, De Longueville F, Kawasaki ES, Lee KY, et al. The microarray quality control (maqc) project shows inter-and intraplatform reproducibility of gene expression measurements. Nat Biotechnol. 2006;24(9):1151–61.

2. Park PJ. Chip–seq: advantages and challenges of a maturing technology. Nat Rev Genet. 2009;10(10):669–80.

3. Goodman SN, Fanelli D, Ioannidis JP. What does research reproducibility mean?. Sci Transl Med. 2016;8(341):341–1234112.

4. Darbani B, Stewart CN. Reproducibility and reliability assays of the gene expression-measurements. J Biol Res (Thessaloniki). 2014;21(1):3.

5. Parmigiani G, Garrett-Mayer ES, Anbazhagan R, Gabrielson E. A cross-study comparison of gene expression studies for the molecular classification of lung cancer. Clin Cancer Res. 2004;10(9):2922–7.

6. Choi H, Shen R, Chinnaiyan AM, Ghosh D. A latent variable approach for meta-analysis of gene expression data from multiple microarray experiments. BMC Bioinforma. 2007;8(1):364.

7. Parmigiani G, Garrett ES, Anbazhagan R, Gabrielson E. A statistical framework for expression-based molecular classification in cancer. J R Stat Soc Ser B Stat Methodol. 2002;64(4):717–36.

8. Hong F, Breitling R, McEntee CW, Wittner BS, Nemhauser JL, Chory J. Rankprod: a bioconductor package for detecting differentially expressed genes in meta-analysis. Bioinformatics. 2006;22(22):2825–7.

9. Benjamini Y, Heller R, Yekutieli D. Selective inference in complex research. Philos Trans R Soc Lond A Math Phys Eng Sci. 2009;367(1906):4255–71.

10. Li Q, Brown JB, Huang H, Bickel PJ. Measuring reproducibility of high-throughput experiments. Annals Appl Stat. 2011;5:1752–79.

11. Efron B. Microarrays, empirical bayes and the two-groups model. Stat Sci. 2008;23(1):1–22.

12. Chen MH, Ibrahim JG, Chi YY. A new class of mixture models for differential gene expression in dna microarray data. J Stat Plan Infer. 2008;138(2):387–404.

13. Najarian K, Zaheri M, Rad AA, Najarian S, Dargahi J. A novel mixture model method for identification of differentially expressed genes from dna microarray data. BMC Bioinforma. 2004;5(201):201–10.

14. Newton MA. Detecting differential gene expression with a semiparametric hierarchical mixture method. Biostatistics. 2004;5(2):155–76.

15. Wei Pan JL, Le CT. A mixture model approach to detecting differentially expressed genes with microarray data. Funct Integr Genomics. 2003;3: 117–24.

16. G.J. McLachlan RWB, Peel D. A mixture model-based approach to the clustering of microarray expression data. Bininformatics. 2002;18(3): 413–22.

17. Dempster AP, Laird NM, Rubin DB. Maximum likelihood from incomplete data via the em algorithm. J R Stat Soc Ser B Methodol. 1977;38:1–38.

18. Herazo-Maya JD, Noth I, Duncan SR, Kim S, Ma SF, Tseng GC, Feingold E, Juan-Guardela BM, Richards TJ, Lussier Y, et al. Peripheral blood mononuclear cell gene expression profiles predict poor outcome in idiopathic pulmonary fibrosis. Sci Transl Med. 2013;5(205):205–136205136.

19. Yang IV, Luna LG, Cotter J, Talbert J, Leach SM, Kidd R, Turner J, Kummer N, Kervitsky D, Brown KK, et al. The peripheral blood transcriptome identifies the presence and extent of disease in idiopathic pulmonary fibrosis. PLoS One. 2012;7(6):37708.

20. Gene Expression Omnibus. http://www.ncbi.nlm.nih.gov/geo/.

10

Superpixel-based segmentation of muscle fibers in multi-channel microscopy

Binh P. Nguyen[1,2], Hans Heemskerk[2,1], Peter T. C. So[3,2] and Lisa Tucker-Kellogg[1,2]*

Abstract

Background: Confetti fluorescence and other multi-color genetic labelling strategies are useful for observing stem cell regeneration and for other problems of cell lineage tracing. One difficulty of such strategies is segmenting the cell boundaries, which is a very different problem from segmenting color images from the real world. This paper addresses the difficulties and presents a superpixel-based framework for segmentation of regenerated muscle fibers in mice.

Results: We propose to integrate an edge detector into a superpixel algorithm and customize the method for multi-channel images. The enhanced superpixel method outperforms the original and another advanced superpixel algorithm in terms of both boundary recall and under-segmentation error. Our framework was applied to cross-section and lateral section images of regenerated muscle fibers from confetti-fluorescent mice. Compared with "ground-truth" segmentations, our framework yielded median Dice similarity coefficients of 0.92 and higher.

Conclusion: Our segmentation framework is flexible and provides very good segmentations of multi-color muscle fibers. We anticipate our methods will be useful for segmenting a variety of tissues in confetti fluorecent mice and in mice with similar multi-color labels.

Keywords: Superpixel, Segmentation, Muscle fibers, Confetti fluorescence, Multi-channel microscopy

Background

Cells can be genetically engineered with fluorescence genes that are inherited when the cell divides, meaning the descendents use the genes to continuously synthesize fluorescent proteins. This allows a form of biology experiment called "lineage tracing" to see the long-term impact of specific populations of labelled cells. The labelled cells can be bred into an animal from birth, can be injected, transplanted, etc.

Snippert et al. [1] developed a 4-color "confetti" transgene for labelling stem cells. The confetti transgene exploits genetic recombination to achieve a random choice of color (red, yellow, cyan, green) in each stem cell.

When a stem cell is induced to divide (whether by natural turnover or by injury), each daughter cell expresses the same color as its ancestor. This creates a patch of homogeneous color in the regenerated tissue. Regenerated cells of different color must have originated from different stem cells. Confetti fluorescence and other multi-color cell labelling strategies are useful for tracking regeneration in adult mice, for evaluating the potency of stem cells in vivo, or for judging the effectiveness of stem cell therapies.

In this project we address the analysis of multi-color stem cells after muscle regeneration. Skeletal muscle is a highly regenerative tissue in which each mature muscle cell is a long thin fiber with many nuclei. This muscle fiber is surrounded by a basal lamina, which gives the muscle fiber its firmness during contraction. Muscle-resident stem cells, called satellite cells, are located between the muscle fiber and the basal lamina. If the muscle fiber is severely damaged, it will become necrotic and induce an immune reaction. This activates the satellite cells, which migrate to the injured area and divide into a set

*Correspondence: lisa.tucker-kellogg@duke-nus.edu.sg
[1]Centre for Computational Biology, and Program in Cancer and Stem Cell Biology, Duke-NUS Medical School, 169857 Singapore, Singapore
[2]BioSystems and Micromechanics (BioSyM) Singapore – MIT Alliance for Research and Technology, 138602 Singapore, Singapore
Full list of author information is available at the end of the article

of myoblasts. The myoblasts each have one nucleus, but upon differentiation they fuse together in a linear configuration to generate a multi-nucleated myotube or myofiber. The myoblasts can also fuse to pre-existing or partially-damaged fibers [2]. Figure 1 shows a cross-section and a lateral-section of regenerated muscle fibers from confetti-fluorescent mice.

Cells with multiple fluorescent proteins are imaged using multi-channel microscopy (such as confocal or two-photon imaging). For example, the image datasets used in our project include four channels: cyan, green, yellow and red (Fig. 2). Each of the fluorescent proteins emitting those colors was excited by a laser at its respective excitation wavelength. The light emitted from the sample contains autofluorescence, so the light is passed through a band-pass filter specific for each fluorescent protein, before detection with a camera. The resulting images show which muscle fibers are positive for which fluorescent proteins. Because muscle fibers are multinucleated cells, an overlay of the four colors can show muscle fibers positive for more than one color.

To analyze the regeneration results, the images must be segmented. A cursory glance at a composite color image in Fig. 2e may lead to the conclusion that this segmentation is similar to the segmentation of real-world color images. However, our problem is different or more difficult in some ways as follows. In our images, there is extensive contact or overlap between objects (squeezed together), meaning that contour-closing (e.g., in snake- or level sets-based methods) does not work very well for segmentation. We also cannot re-use methods that combine object recognition with segmentation unless we develop domain-specific object models. In addition, many

of the boundaries are blurry; some of the objects are in the process of fusing; and there is tremendous variation in the fiber brightness. Using a conventional color difference measure may not be appropriate in our problem since color similarity in multi-channel imaging is different from color similarity in a normal spectrum of visible light. Furthermore, the images have random noise and non-random artifacts including optical aberration from the imaging device; damaged tissue or fracture planes during sample preparation; and ice crystals which cause small empty holes in the image. The ice crystals often have clear boundaries but they should be omitted from the segmentation results. Finally, the four colors of the confetti construct are in different locations within the cell. Green is located in the nucleus, yellow and red are in the cytosol and cyan is on the membrane. In mature muscle fibers, the membrane becomes a sarcolemma with many invaginations. As a result, cyan fluorescence can be seen inside the muscle fibers as well as along the cell-cell edges.

We propose a novel method called SLIC-MMED (simple linear iterative clustering on multi-channel microscopy with edge detection) which uses superpixels for the segmentation of muscle fibers in muli-channel microscopy. A superpixel is a perceptual grouping of neighboring pixels that aligns better with image edges than a rectangular patch [3]. Superpixels are widely used in numerous applications in computer vision including image segmentation. Among existing superpixel generation methods [4], simple linear iterative clustering (SLIC) [5] was chosen for our project because of its effectiveness, scalability and speed. However, SLIC needs to be modified to adapt to our problem. Nuclei are orders of magnitude

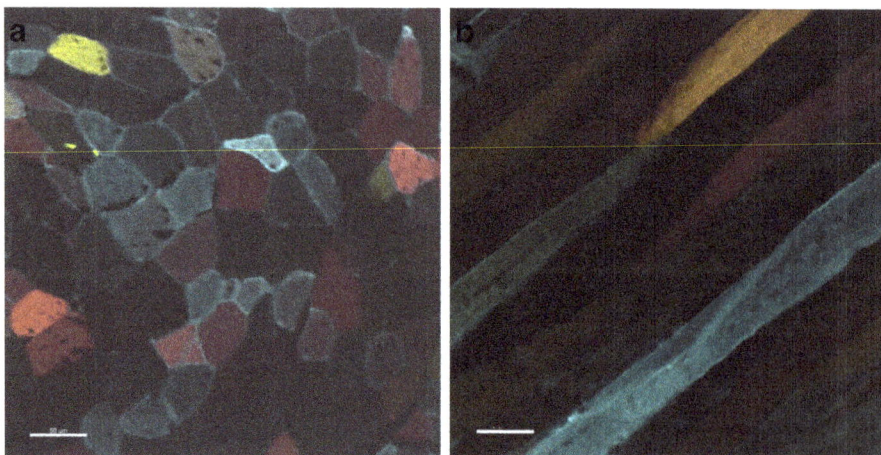

Fig. 1 A cross-section (**a**) and a lateral-section (**b**) of regenerated muscle fibers from confetti-fluorescent mice. A five-month-old male (**a**) and female (**b**) mice were injected with 100 $\mu g/g$ tamoxifen on 5 consecutive days to achieve transgene recombination. Ten days after the last tamoxifen injection, muscle injury was induced by injecting the tibialis anterior with 50 μl 10 μM cardiotoxin. Sixteen days after injury mice were sacrificed, the tibialis anterior was fixed and frozen, and 10 μm sections were cut. Scalebars are 50 μm

Fig. 2 A multi-channel microscopy image of muscle fibers. (**a**) *Cyan* channel; (**b**) *Green* channel; (**c**) *Yellow* channel; (**d**) *Red* channel; (**e**) Composite color image from all the four channels; (**f**) Composite color image from *cyan, yellow* and *red* channels after preprocessing. This section is from the same muscle of the cross-section in Fig. 1a. Scalebars are 50 μm. Each channel is a 12-bit image of 1024 × 1024 pixels

smaller than mature muscle cells, but when colored green they have a very strong color boundary. This difference in the scale of color change (due to scale difference in underlying objects) could confound the superpixel generation. So we first remove the nuclei. A simple method

for segmenting the nuclei turns out to be extremely accurate by using domain-specific information (rounded or ellipsoidal morphology, green color).

To make existing image processing algorithms useful for our domain, we had to perform several modifications: (1)

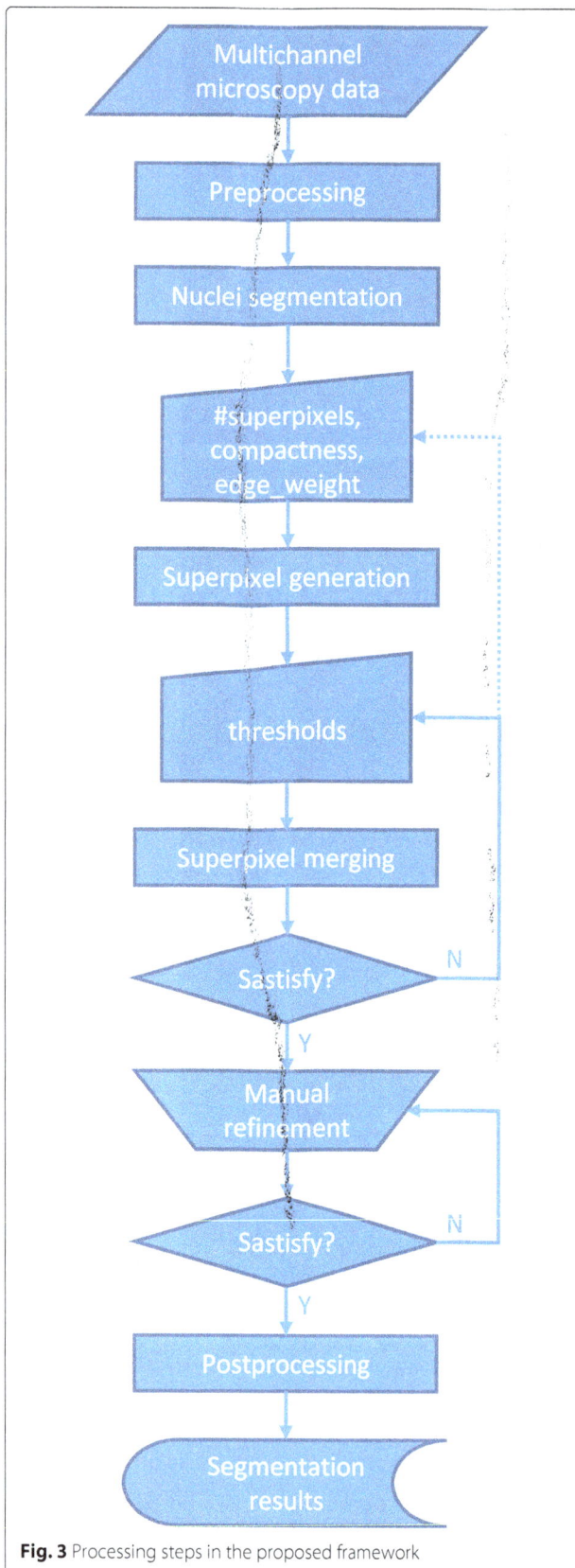

Fig. 3 Processing steps in the proposed framework

forking different channels to different methods, based on an object model for nuclei, (2) introducing an enhanced superpixel method named *simple linear iterative clustering on multi-channel microscopy with edge detection* (SLIC-MMED), and (3) developing a semi-automatic segmentation framework based on superpixels that can produce very good results for our problem. We believe that all three of these modifications will be useful for other forms of multi-channel cell microscopy, for non-neuronal eukaryotic cell types.

Methods

Overview of the framework

An overview of our framework is shown in Fig. 3. First, each channel undergoes intensity normalization and noise reduction filtering. Then the green channel is processed to extract the nuclei. The remaining channels (cyan, yellow and red) are used to generate superpixels according to our SLIC-MMED algorithm (Fig. 4). Next, an automatic superpixel merging algorithm (Fig. 5) is executed to merge a subset of the generated superpixels to form the muscle fibers. The superpixel generation and/or this superpixel merging step can be performed repeatedly with different user-defined parameters until users are satisfied or until termination criteria are reached. After that, the resulting superpixels and merged regions can be further revised through a user-friendly graphical user interface (GUI). Lastly, this segmentation is combined with the nuclear segmentation to form the final result. In this section, we use the dataset shown in Fig. 2 to describe our proposed framework. The primary steps in our framework are detailed below.

Preprocessing and nuclear segmentation

The first preprocessing is to normalize each channel to the same range, e.g., [0, 255]. Another optional preprocessing step is to apply a noise reduction filter, e.g., median filter, to each channel in cases of noisy images. Figure 2f shows an example of a composite color image from cyan, yellow and red channels after preprocessing.

In microscopy, cell nuclei are often the easiest morphological features to identify (whether by eye or by algorithm), and many microscopy protocols include nuclear staining, or more recently, fluorescent proteins genetically engineered for nuclear localization. Because each fluorescent tag corresponds to one (or at least one) channel, we can analyze channels individually, based on this knowledge of the underlying signal sources. In other words, we can analyze the green channel for nucleus-like objects. By using image thresholding and morphological techniques, the nuclei in the dataset shown in Fig. 2 can be segmented using only the green channel as illustrated in Fig. 6.

Require: K, m, α, and $[c_i, y_i, r_i, x_i, y_i]^T$ with p_i for each pixel i in the multichannel image I

1: **function** SLIC_MMED(I)

2:     $S \leftarrow \sqrt{N/K}$

3:     **for** k = 1 to K **do**

4:        $(\mathrm{x}_k, \mathrm{y}_k) \leftarrow$ center of grid cell k^{th} in a regular grid with the interval of S

5:        $C_k \leftarrow$ the feature vector at the lowest gradient position of $[c, y, r]$ in a 3×3 neighborhood around
      $(\mathrm{x}_k, \mathrm{y}_k)$

6:     **end for**

7:     **for all** pixel i **do**

8:        $label(i) \leftarrow -1$

9:        $distance(i) \leftarrow +\infty$

10:    **end for**

11:    $count \leftarrow 0$

12:    **while** $count < 10$ **do**

13:       **for all** C_k **do**

14:          **for all** pixel i in the $2S \times 2S$ region around $(\mathrm{x}_k, \mathrm{y}_k)$ **do**

15:             $D \leftarrow$ distance between feature vector at pixel i and C_k

16:             **if** $D < distance(i)$ **then**

17:                $label(i) \leftarrow k$

18:                $distance(i) \leftarrow D$

19:             **end if**

20:          **end for**

21:       **end for**

22:       **for all** C_k **do**

23:          $C_k \leftarrow$ mean feature vectors of all the pixels having $label = k$

24:       **end for**

25:       $count \leftarrow count + 1$

26:    **end while**
         return $label$

27: **end function**

Fig. 4 Simple Linear Iterative Clustering on Multichannel Microscopy with Edge Detection (SLIC-MMED) Algorithm

---

SLIC-MMED for superpixel generation

The original SLIC algorithm operates on color images in the CIELAB color space with one input parameter is K, the desired number of superpixels. Each pixel is represented by a 5-dimensional feature vector, $[l, a, b, \mathrm{x}, \mathrm{y}]^T$, containing 3 color components and 2 pixel coordinates. At the initialization step, K initial cluster centroids $C_k = [l_k, a_k, b_k, \mathrm{x}_k, \mathrm{y}_k]^T$ are sampled on a regular grid with the interval of $S = \sqrt{N/K}$, where N is the number of pixels. To reduce the chance of centering a superpixel on an edge or on a noisy pixel, each centroid is moved to the lowest gradient position in a 3×3 neighborhood. Next, in the iteration step, each pixel is associated with its nearest centroid. In order to speed up the algorithm (compared with k-means clusturing), the size of the search space is reduced to a region proportional to the superpixel size. Here, for each cluster centroid C_k, only the pixels in the $2S \times 2S$ region around C_k are evaluated, meaning that if the distance from a pixel to C_k is less than the distance

from that pixel to its current associated centroid, then the associated centroid will be changed to C_k. Once all the pixels have been assigned to their nearest centroids, an update process adjusts each centroid to be the mean feature vector of all the pixels belonging to the corresponding cluster. In practice, repeating this iteration step 10 times is sufficient for most images. Finally, a postprocessing step assigns all disjoint pixels (if any) to nearby superpixels.

In SLIC, the distance D between two pixels i and j is a combination of two distances, d_c and d_s, representing color proximity and spatial proximity, respectively, as below:

$$
\begin{aligned}
d_c &= \sqrt{(l_i - l_j)^2 + (a_i - a_j)^2 + (b_i - b_j)^2}, \\
d_s &= \sqrt{(\mathrm{x}_i - \mathrm{x}_j)^2 + (y_i - y_j)^2}, \\
D &= \sqrt{(d_c/m)^2 + (d_s/S)^2},
\end{aligned}
\tag{1}
$$

Require: image I, superpixel labels *label*, edge probability matrix *edge_prob*, thresholds *thres_dB, thres_edge, thres_size, thres_bg*

1: **function** SUPERPIXELMERGING(I, *label*, *edge_prob*)
2:   **repeat**
3:     $C\_dist \leftarrow$ ComputeChiSquaredDistances(I, *label*)
4:     $E\_strength \leftarrow$ ComputeEdgeStrengths(I, *label*, *edge_prob*)
5:     $size \leftarrow$ ComputeSuperpixelSizes(*label*)
6:     **for all** superpixel i **do**
7:       $nearest[i] \leftarrow$ the neighbor having smallest C_dist to superpixel i
8:       $C\_dist\_min[i] \leftarrow C\_dist[i, nearest[i]]$
9:       $visited[i] \leftarrow$ FALSE
10:     **end for**
11:     sort(C_dist_min, *org_index*)
12:     **for all** i such that $C\_dist\_min \leq thres\_dC$ **do**
13:       $idx \leftarrow org\_index[i]$
14:       **if** not $visited[idx]$ **then**
15:         **for all** neighbor j of superpixel idx **do**
16:           **if** (not $visited[j]$) and ($E\_strength[idx, j] \leq thres\_edge$) and (($size[idx] \leq thres\_size$) or (MeanColor($idx$) $\leq thres\_bg$)) **then**
17:             **for all** k such that $label[k] = j$ **do**
18:               $label[k] \leftarrow idx$
19:             **end for**
20:             $visited[j] \leftarrow$ TRUE
21:           **end if**
22:         **end for**
23:       **end if**
24:       $visited[idx] \leftarrow$ TRUE
25:     **end for**
26:   **until** (no more superpixels are merged)
      return *label*
27: **end function**

Fig. 5 Superpixel Merging Algorithm

where the *compactness m* is used to to weight the relative importance of color similarity versus spatial proximity. When *m* is large, spatial proximity is more important, and the resulting superpixels are more compact. In contrast, a small value of *m* leads to superpixels that are less regular in size and shape; however, since in this case color proximity is more important, the resulting superpixels follow the image boundaries more closely.

Since the input data in our project are multichannel microscopy images, not real-world color images, the feature vector needs to be modified. Instead of using 3 CIELAB color components (l, a, b), we use each image channel as a component of the feature vector. Since the green channel represents only nuclei, it is discarded from the feature vector and processed separately as described in the previous subsection. In short, the feature vector representing each pixel in our data is $[c, y, r, x, y]^T$ which consists of the pixel intensities in the 3 channels cyan, yellow, red, and the pixel coordinates (x, y), respectively.

In the superpixel method benchmarking, boundary recall is used to measure the fraction of the ground truth edges falling within at least two pixels of a superpixel boundary. A good superpixel segmentation should adhere to object boundaries, meaning that it should produce a high boundary recall. Although SLIC demonstrates very good boundary recall performance for real-world color images [4, 5], this is not the case for our datasets as shown in Fig. 7c. To overcome these problems, we propose to integrate an additional score d_e into the pixel distance measure. d_e represents the presence of edges between two pixels, suggesting the likelihood that an object boundary falls between the two points. Before starting superpixel generation, an edge detection algorithm is executed to compute a value p_i for each pixel i, indicating its

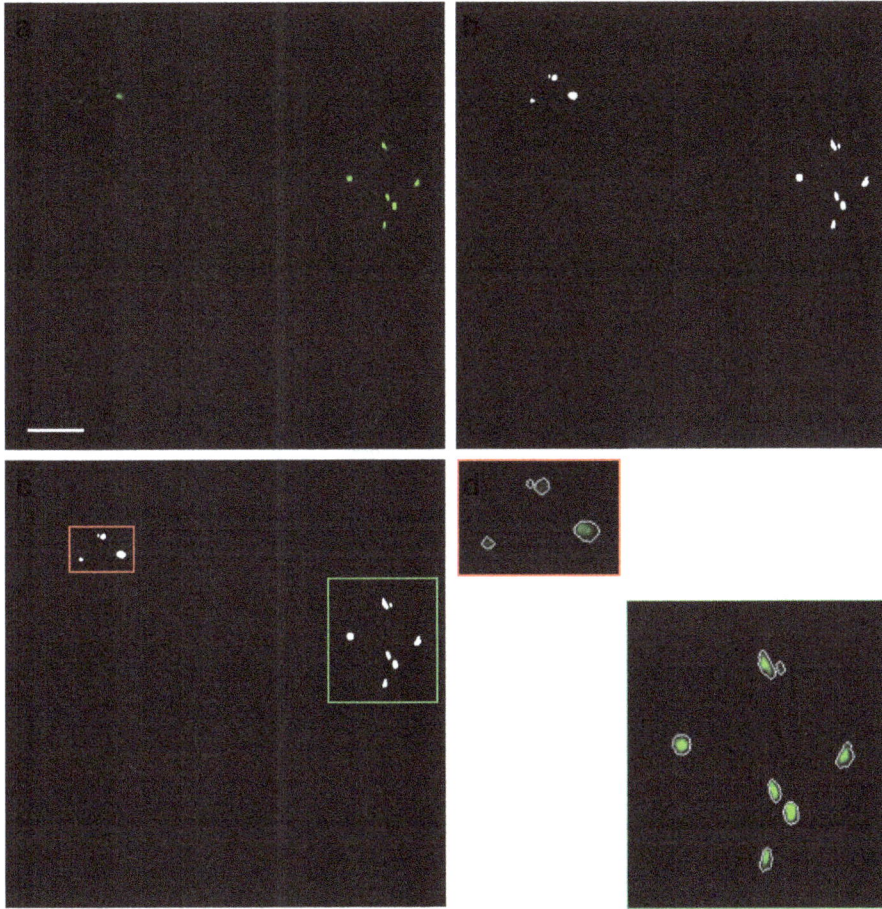

Fig. 6 Nuclear segmentation. (**a**) *Green* channel after preprocessing (scalebar is 50 μm); (**b**) After image thresholding; (**c**) Using morphological techniques to remove noise; (**d**) Segmented nuclei (enlarged for detail)

probability of being on an edge (boundary). Then the distance d_e between two pixels i and j is calculated as the maximum edge probability over all the pixels lying on the line connecting pixel i and pixel j. The new distance is calculated as below.

$$d'_c = \sqrt{(c_i - c_j)^2 + (y_i - y_j)^2 + (r_i - r_j)^2},$$
$$d_s = \sqrt{(x_i - x_j)^2 + (y_i - y_j)^2},$$
$$d_e = \max_{\forall t \in line(i,j)} p_t,$$
$$D' = \sqrt{(d'_c/m)^2 + (d_s/S)^2 + \alpha \times (d_e)^2},$$

(2)

A wide range of edge detection algorithms would be appropriate for computing the edge probabilities p_i, and we chose a detector based on the photometric invariance theory and tensor-based features [6]. Figure 7a shows an edge map generated using that edge detector. Note that if no edge detection were involved (i.e., if $p_i = 0 \forall i$) then the revised distance measure D' (Eq. 2) would reduce to

the original SLIC distance measure D (Eq. 1). The new algorithm is presented in Fig. 4.

Automatic superpixel merging

After superpixel generation, all the superpixels within each muscle fiber need to be merged together to form the muscle fiber boundary. The similarity measure used in our method, to determine if two neighboring superpixels should be merged, is the Chi-squared (χ^2) histogram distance [7]. The χ^2 distance between two histograms P and Q is defined as

$$\chi^2(P,Q) = \frac{1}{2} \sum_k \frac{(P_k - Q_k)^2}{P_k + Q_k}.$$

(3)

We represent the intensity distribution in each channel of each superpixel as a histogram and use the following

Fig. 7 Superpixel generation using different algorithms. (**a**) Computed edge map. The segmentation results from (**b**) our method, SLIC-MMED; (**c**) SLIC; (**d**) VCells. *White arrows* indicate poorly segmented superpixels

formula to measure the similarity distance $D_C(i,j)$ between superpixels i and j,

$$D_C(i,j) =$$
$$\sqrt{\left(\chi^2(H_i^c, H_j^c)\right)^2 + \left(\chi^2(H_i^y, H_j^y)\right)^2 + \left(\chi^2(H_i^r, H_j^r)\right)^2},$$
(4)

where H^c, H^y, H^r are the superpixel histograms of channels cyan, yellow, and red, respectively.

In addition to $D_C(i,j)$, another measure called *edge strength* is used for the superpixel merging decision. The edge strength E_{ij} between superpixels i and j is defined as the average edge probability over all the pixels in superpixel i having at least one neighbor belonging to superpixel j. The D_C and edge strength are then used in a series of thresholds. If superpixels i and j have similar colors, it might be because they are part of the same fiber, or it might be because they come from different fibers that happen to have similar colors. Therefore, whenever two superpixels have similar D_C (the χ^2 distance

is not greater than a threshold *thres_dC*), they can only be merged if they have low edge strengths (their edge strengths are not greater than a threshold *thres_edge*). In addition, we use another predefined threshold *thres_size* to avoid muscle fibers having unrealistic sizes which are formed from over-merging. However, this size limitation is not applied to superpixels representing "background" (namely, black-colored superpixels with mean color intensity below *thres_bg*).

Our iterative superpixel merging algorithm starts with a calculation of the χ^2 distances and the edge strengths between each superpixel and its neighbors. Then the method for merging superpixels is a series of thresholded criteria as described in Fig. 5. The algorithm stops when there are no more superpixels merged. Figure 8a shows a result after this processing step.

Manual refinement

For challenging datasets, it is impossible to produce an error-free segmentation. Errors from the automatic superpixel merging process include two types:

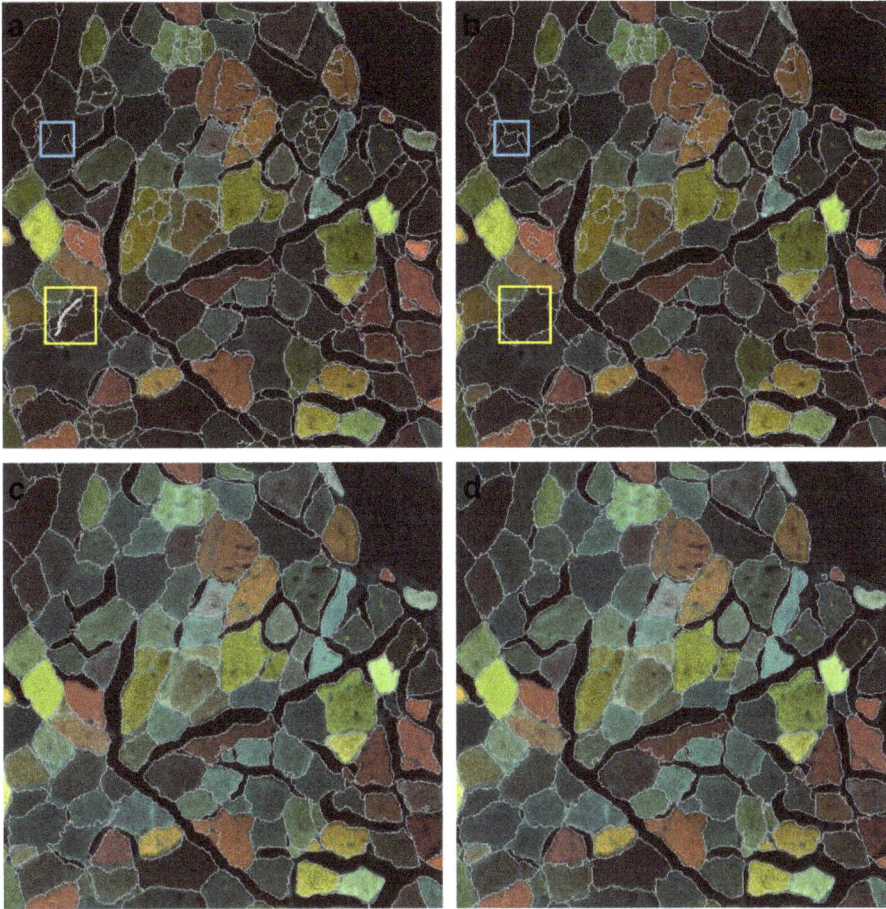

Fig. 8 Segmentation using superpixels. (**a**) Automatic superpixel grouping; (**b**) Example of manual refinement: *yellow box* - draw a curve to merge all the regions along the curve, *blue box* - restore a superpixel to merge it with a neighboring region in another way; (**c**) Final result; (**d**) Ground truth

1. Over-merging: merging of superpixels from different muscle fibers, or from a muscle fiber and neighboring background/artifacts.
2. Under-merging: some neighboring superpixels from the same muscle fiber have not been merged yet.

A manual refinement step is introduced to our framework through a user-friendly GUI in order to fix the superpixel merging errors (Fig. 8b). The main supported operations include

- Drawing a freehand curve to merge all the superpixels or regions along the curve.
- Restoring the original superpixels surrounding a selected position to allow manually merging them in another way.

If the superpixel generation produces a high boundary recall, using these two operations can guarantee a very good segmentation result (Fig. 8c).

After this manual refinement, the superpixel segmentation is combined with the nuclear segmentation in a postprocessing step to form the final segmentation result.

Results

Superpixel evaluation

We use two error metrics, *boundary recall* and *under-segmentation error*, to evaluate our SLIC-MMED algorithm and compare it with the original SLIC and another advanced superpixel algorithm named VCells [8]. Boundary recall is the fraction of ground truth edges that fall within a certain distance d ($d = 2$ in our experiments) of at least one superpixel boundary. A good superpixel segmentation should produce a high boundary recall. Under-segmentation error compares segment areas to measure to what extend superpixels flood over the ground truth segment borders. Details about the calculation of these two measures can be found in [4].

We used the dataset in Fig. 2 for this evaluation. The corresponding ground truth was created by one computer expert under the supervison of our muscle biology expert (Fig. 8d). The compactness was chosen as 20 for all the three algorithms. The edge weight α in Eq. 2 was 20 for SLIC-MMED. The input image for the original SLIC and VCells algorithms was the composite color image from the cyan, yellow and red channels after preprocessing (Fig. 2f).

Figure 9 shows the performance of the three super-pixel algorithms, scored using boundary recall, under-segmentation error and processing time (in seconds). As can be seen, our proposed SLIC-MMED outperforms the other two algorithms in term of superpixel quality. In term of processing time, our algorithm is slower than the original SLIC due to the extras processes, but is still very fast (about 2–5 seconds for a 1024×1024-pixel three-channel image) and much faster than VCells. Algorithmic efficiency is not a limiting factor in our context because the slowest step of our pipeline, preparing tissue sections for microscopy, is much slower than any of the image analysis algorithms.

Figure 7 shows the superpixels generated when each of the candidate algorithms was run on the same image (using a moderate parameter value of 800 for the number of superpixels). The superpixel shapes generated by SLIC-MMED look less regular than those of the other two algorithms due to the introduction of an edge map. However, the superpixels created by SLIC-MMED adhere more closely to the image boundaries, so they have virtually no significant errors. Significant errors are annotated by small white arrows on the images.

Segmentation evaluation

The final stage of evaluation considers the total sementation accuracy. The methods SLIC and VCells provide only superpixels, not segmentation, so they are not covered in this section. Our method, SLIC-MMED, merges superpixels to create segmented regions, so the quality of its segmentation is inherently related to the quality of its superpixels. We assess the segmentation using an absolute score, defined with respect to ground truth, called the Dice similarity coefficient (DSC) [9]. The DSC measures the spatial overlap between two segmentations, X and Y, and is defined as

$$DSC = \frac{2\,|X \cap Y|}{|X| + |Y|}, \tag{5}$$

where $|X|$ and $|Y|$ are the number of pixels in X and Y, respectively. It should be noted that Eq. 5 is for the evaluation of one resulting segment. Our segmentation problem is a multiple-object segmentation with multiple fibers and other regions. We propose the median DSC (*medDSC*) which is computed as in Eq. 6 to measure

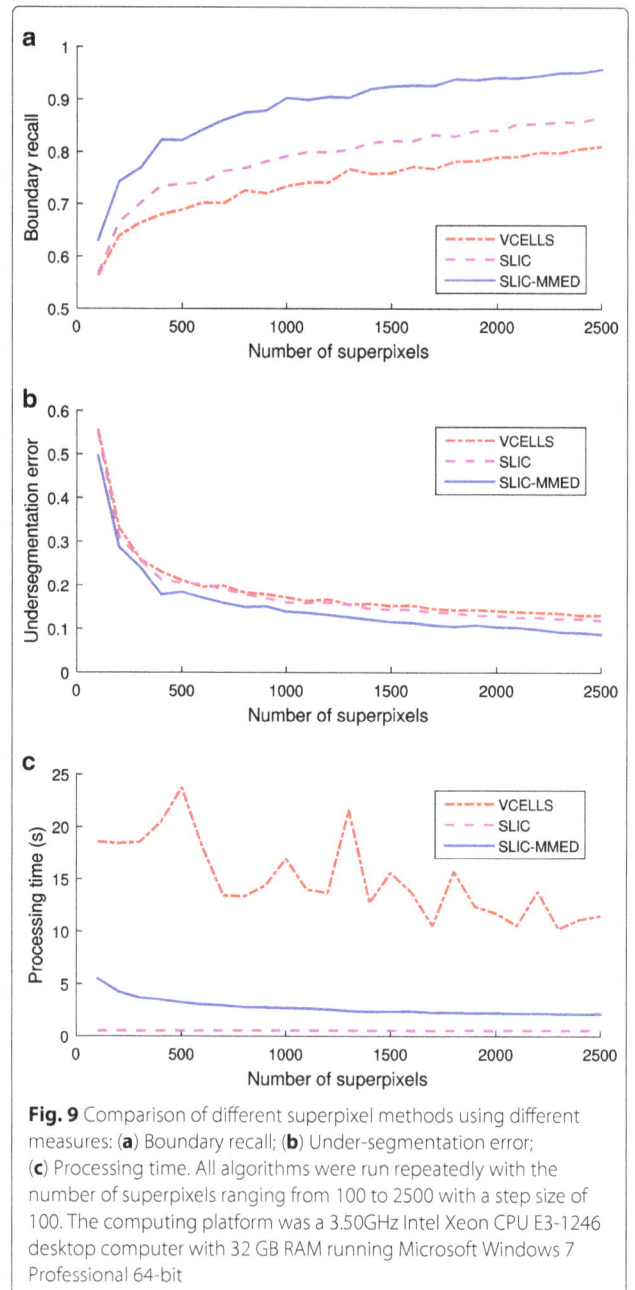

Fig. 9 Comparison of different superpixel methods using different measures: (**a**) Boundary recall; (**b**) Under-segmentation error; (**c**) Processing time. All algorithms were run repeatedly with the number of superpixels ranging from 100 to 2500 with a step size of 100. The computing platform was a 3.50GHz Intel Xeon CPU E3-1246 desktop computer with 32 GB RAM running Microsoft Windows 7 Professional 64-bit

the similarity between the segmentation result S and the ground truth G.

$$medDSC = \underset{i=1,..,N_G}{median}\,(DSC_i),$$
$$DSC_i = \frac{2\left|g_i \cap s_{f(i)}\right|}{\left|g_i\right| + \left|s_{f(i)}\right|}, \tag{6}$$

where N_G and N_S are the respective total number of segments in G and S, g_i is one segment in G, and $s_{f(i)}$ with $f(i) \in [1, N_S]$ is the corresponding segment in S having the largest overlap with g_i.

For the example data set (Fig. 2), we used SLIC-MMED to generate 1500 superpixels (with compactness = 20 and

$\alpha = 20$). After the automatic superpixel merging step (Fig. 8a), we manually refined the segmentation (Fig. 8b) and got the final result as in Fig. 8c. The resulting *medDSC* was 0.92 for this dataset.

Figure 10 presents the final SLIC-MMED segmentation results for the images from Fig. 1. After the superpixel merging phase and prior to the refinement phase, the scores were *medDSC* 0.72 and 0.67 for the cross-section dataset and the lateral-section dataset, respectively. The lower score for lateral sections reflects the extreme aspect ratio (non-compactness) of the underlying objects. Fortunately, the under-segmentation of a fiber into fiber segments is nearly instantaneous for manual refinement to merge correctly. After the refinement phase, the *medDSCs* were 0.93 and 0.95 for the cross-section dataset and the lateral-section dataset, respectively. As recommended by Zijdenbos et al. [10], a segmentation is considered good if $DSC > 0.70$.

We also propose DSC_X, where X is a number between 0 and 100, to measure the fraction of segments in the ground truth are segmented with a $DSC \geq X/100$. The final segmentation in Fig. 8c has DSC_{75} of 0.78, meaning that 78% of the segments were segmented with a $DSC \geq 0.75$. For the cross-section dataset and the lateral-section dataset in Fig. 10, the scores were DSC_{75} 0.93 and 0.91, respectively.

Discussion and conclusion

In our experiments, the proposed method can correctly segment muscle fibers in very heterogeneous sections having both bright and dark regions, a wide range of fiber sizes, homogeneous red or yellow but more irregular cyan segments in cytoplasm. The fact that the method can handle a variety of cell sizes and morphologies in these confetti-fluorescent images suggests that it may be useful for analyzing confetti-fluorescent images in other tissues.

With an accurate segmentation, we can count the number of muscle fibers that contain each of the confetti colors. The method works well for even weakly fluorescent areas, as in Figs. 1a and 10a. Using the same segmentation, we also can measure the diameter and cross sectional area for each fiber. In most labs, measuring the diameter and cross-sectional area would require cutting a set of adjacent tissue sections, which would then undergo labor-intensive staining, followed by imaging and registration of the adjacent sections, to provide a superpositioning of the stained section and the confetti-labelled section. Staining requires doubling the number of sections because staining eliminates the endogenous fluorescence. In this work we show it is possible to obtain segmentation from the endogenous fluorescence, allowing us to skip the costly process of staining.

In fluorescence microscopy, each fluorophore emits light over a range of wavelengths (its emission spectrum), causing nearby colors to overlap. For example, the emission spectrum of green overlaps with the emission spectrum of yellow. It is for this reason that multi-color labelling strategies have engineered the similar fluorescent proteins to have different sub-cellular localizations (e.g., nuclear localization of the green and cytosolic localization of the yellow). Localization allows the identity of the label to be disambiguated. In our images, the green signal bled into the yellow channel (see Fig. 2c). SLIC-MMED includes explicit management of different sub-cellular localizations, and this may be why we had no segmentation errors due to green-yellow spectral overlap.

The observed colors arise from fluorescent protein molecules that are diffusible in their compartment (cytosol, membrane, or nucleus). The Brownian nature of

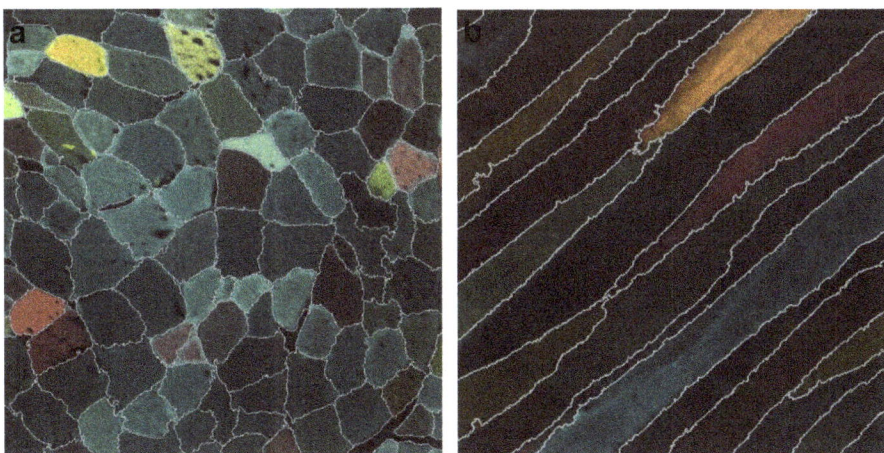

Fig. 10 Segmentation results applying SLIC-MMED to the images in Fig. 1. The formats of each image channel of the data are (**a**) 8-bit 1024 × 1024 pixels and (**b**) 12-bit 512 × 512 pixels. Each segmentation is displayed on the composite color image from the *cyan, yellow* and *red* channels after preprocessing

diffusion suggests that color distribution might be nearly uniform across the space of the relevant compartment. In other words, the limit of molecular diffusion becomes the boundary of the color, which we identify as the boundary of the segmentation. Cell membranes are 2-dimensional surfaces which can appear as 1-dimensional curves when imaged from a cross-section. The cytosol and nucleus of a cell are 3-dimensional compartments, which appear as 2-dimensional continuous regions. Edge detection is a natural approach for analyzing 2-dimensional membrane-targeted fluorescence, such as the cyan channel in our images. Meanwhile, superpixel-based region detection methods are a natural approach for analyzing 3-dimensional compartments. If the spatial distribution of the fluorescent proteins were punctate (0-dimensional) or fibrillar (1-dimensional), then our method would be less appropriate.

Our segmentation framework, SLIC-MMED, is a "hybrid" method that combines the advantages of a region-based clustering algorithm (SLIC) and an edge detector through the integrated edge map. The introduction of an user-friendly superpixel refinement module provide flexibility for the framework. As long as the superpixel generation provides a high performance in boundary recall, the framework provides very good segmentations. Our experimental results show a high degree of agreement with experts. In the final scoring, the differences between different trials are also heavily dependent on the specific dataset, the number of superpixels to be revised, and the user's expertise at performing the manual refinement step. In future, we will intensively analyze the contribution of automation to the effectiveness of the framework.

The algorithm is potentially applicable to other multi-channel microscopy applications besides muscle. Mouse transgenes with confetti, brainbow or other multi-color stochastic labels have become extremely popular [1] and the scientific community is rapidly generating multi-channel images that require analysis. Our image analysis method is particularly valuable for such applications.

Acknowledgments
We thank Paul Matsudaira for hosting the mouse experiments.

Declarations
This article has been published as part of BMC Systems Biology Volume 10 Supplement 5, 2016. 15th International Conference On Bioinformatics (INCOB 2016): systems biology. The full contents of the supplement are available online http://bmcsystbiol.biomedcentral.com/articles/supplements/volume-10-supplement-5.

Funding
Funding for this research was provided by Duke-NUS SRP Phase 2 Research Block Grant and by the National Research Foundation (NRF), Prime Minister's Office, Singapore, under its CREATE programme, Singapore-MIT Alliance for Research and Technology (SMART) BioSystems and Micromechanics (BioSyM) IRG.

Authors' contributions
Method was designed and implemented by BPN with advice from LTK. Mouse experiments and microscopy were performed by HH. PTCS and LTK supervised the imaging studies. All authors edited and approved the manuscript.

Competing interests
The authors declare that they have no competing interests.

Author details
[1]Centre for Computational Biology, and Program in Cancer and Stem Cell Biology, Duke-NUS Medical School, 169857 Singapore, Singapore. [2]BioSystems and Micromechanics (BioSyM) Singapore – MIT Alliance for Research and Technology, 138602 Singapore, Singapore. [3]Department of Mechanical Engineering, Massachusetts Institute of Technology, 02139 Cambridge, MA, USA.

References
1. Snippert HJ, van der Flier LG, Sato T, van Es JH, van den Born M, Kroon-Veenboer C, Barker N, Klein AM, van Rheenen J, Simons BD, Clevers H. Intestinal crypt homeostasis results from neutral competition between symmetrically dividing lgr5 stem cells. Cell. 2010;143(1):134–44.
2. Schmalbruch H. The morphology of regeneration of skeletal muscles in the rat. Tissue Cell. 1976;8(4):673–92.
3. Veksler O, Boykov Y, Mehrani P. Superpixels and supervoxels in an energy optimization framework In: Daniilidis K, Maragos P, Paragios N, editors. Proceedings of the 11th European Conference on Computer Vision (ECCV 2010). LNCS. Berlin, Germany: Springer; 2010. p. 211–24.
4. Neubert P, Protzel P. Proceedings of Forum Bildverarbeitung In: Puente León F, Heizmann M, editors. Karlsruhe, Germany: KIT Scientific Publishing; 2012. p. 205–18.
5. Achanta R, Shaji A, Smith K, Lucchi A, Fua P, Süsstrunk S. SLIC superpixels compared to state-of-the-art superpixel methods. IEEE Trans Pattern Anal Mach Intell. 2012;34(11):2274–82.
6. van de Weijer J, Gevers T, Smeulders AWM. Robust photometric invariant features from the color tensor. IEEE Trans Image Process. 2006;15(1): 118–27.
7. Pele O, Werman M. The quadratic-chi histogram distance family In: Daniilidis K, Maragos P, Paragios N, editors. Proceedings of the 11th European Conference on Computer Vision (ECCV 2010). LNCS. Berlin, Germany: Springer; 2010. p. 749–62.
8. Wang J, Wang X. VCells: Simple and efficient superpixels using edge-weighted centroidal voronoi tessellations. IEEE Trans Pattern Anal Mach Intell. 2012;34(6):1241–7.
9. Dice LR. Measures of the amount of ecologic association between species. Ecology. 1945;26(3):297–302.
10. Zijdenbos AP, Dawant BM, Margolin, Palmer AC. Morphometric analysis of white matter lesions in MR images: method and validation. IEEE Trans Med Imaging. 1994;13(4):716–24.

11

Dynamic optimization of biological networks under parametric uncertainty

Philippe Nimmegeers, Dries Telen, Filip Logist and Jan Van Impe[*]

Abstract

Background: Micro-organisms play an important role in various industrial sectors (including biochemical, food and pharmaceutical industries). A profound insight in the biochemical reactions inside micro-organisms enables an improved biochemical process control. Biological networks are an important tool in systems biology for incorporating microscopic level knowledge. Biochemical processes are typically dynamic and the cells have often more than one objective which are typically conflicting, e.g., minimizing the energy consumption while maximizing the production of a specific metabolite. Therefore multi-objective optimization is needed to compute trade-offs between those conflicting objectives. In model-based optimization, one of the inherent problems is the presence of uncertainty. In biological processes, this uncertainty can be present due to, e.g., inherent biological variability. Not taking this uncertainty into account, possibly leads to the violation of constraints and erroneous estimates of the actual objective function(s). To account for the variance in model predictions and compute a prediction interval, this uncertainty should be taken into account during process optimization. This leads to a challenging optimization problem under uncertainty, which requires a robustified solution.

Results: Three techniques for uncertainty propagation: linearization, sigma points and polynomial chaos expansion, are compared for the dynamic optimization of biological networks under parametric uncertainty. These approaches are compared in two case studies: *(i)* a three-step linear pathway model in which the accumulation of intermediate metabolites has to be minimized and *(ii)* a glycolysis inspired network model in which a multi-objective optimization problem is considered, being the minimization of the enzymatic cost and the minimization of the end time before reaching a minimum extracellular metabolite concentration. A Monte Carlo simulation procedure has been applied for the assessment of the constraint violations. For the multi-objective case study one Pareto point has been considered for the assessment of the constraint violations. However, this analysis can be performed for any Pareto point.

Conclusions: The different uncertainty propagation strategies each offer a robustified solution under parametric uncertainty. When making the trade-off between computation time and the robustness of the obtained profiles, the sigma points and polynomial chaos expansion strategies score better in reducing the percentage of constraint violations. This has been investigated for a normal and a uniform parametric uncertainty distribution. The polynomial chaos expansion approach allows to directly take prior knowledge of the parametric uncertainty distribution into account.

Keywords: Dynamic optimization, Optimization under uncertainty, Biological networks, Multi-objective

Abbreviations: CPU, Central processing unit; KKT, Karush-Kuhn-Tucker; LIN, Linearization; MC, Monte Carlo; NBI, Normal boundary intersection; NLP, Nonlinear programming problem; PCE, Polynomial chaos expansion; PCE1, First order polynomial chaos expansion; PCE2, Second order polynomial chaos expansion; SP, Sigma points

*Correspondence: jan.vanimpe@kuleuven.be
KU Leuven, Department of Chemical Engineering, BioTeC+ & OPTEC,
Gebroeders De Smetstraat 1, 9000 Ghent, Belgium

Background

The application of micro-organisms in chemical industry and life sciences is paramount. In industrial biotechnology, on the one hand, microbial growth is stimulated in order to enhance the production of (high added value) chemical and pharmaceutical products. On the other hand, in food industry the aim is to avoid the growth of pathogens and food spoilage to ensure food safety.

Therefore, a profound biochemical insight in microbial dynamics and the reactions inside micro-organisms is important. Integrating insights obtained at systems biology (microscopic) level contributes to an improved (macroscopic level) biochemical process control (i.e., enabling advanced model based monitoring, control and optimization of bioprocesses) [1].

A basic tool in systems biology for incorporating microscopic level information are biological networks, e.g., metabolic reaction networks in which the knots represent the metabolites (chemical substances produced/consumed in the micro-organisms) and the connections indicate the mass fluxes between those metabolites. A biological network is a systematic representation of the cellular processes and the interactions between the molecules in the cells: e.g., proteins and metabolites. Such a network comprises (a subset of) all reactions which occur inside a cell and the knots represent the metabolites (i.e., products consumed/produced by the cells) and the links represent the intracellular reactions or reactions between the cell and its environment. A cell can be seen on microscopic scale as a combination of interactions between different layers: fluxome, metabolome, proteome, transcriptome and genome. In terms of network complexity (i.e., the number of metabolites and fluxes), fluxome level biological networks have the lowest level of complexity, while genome-scale biological networks have the highest level of complexity [2–4].

Insight in the dynamic behavior of micro-organisms can be obtained by simulation of metabolic networks. Optimization of biological networks can be used to analyze and also influence the regulation of pathways, e.g., to stimulate the production of high added value products. In practice, cells often have more than one objective, which are conflicting, e.g., minimizing the energy consumption while maximizing the production of a certain metabolite. Therefore, dynamic (multi-objective) optimization, which provides optimal (possibly time-varying) control profiles, is an important tool. Multi-objective optimization of biological networks has been investigated in [5–7]. The multi-objective design of bioprocesses and solution strategies have for instance been presented in [8] with application to a well-stirred, aerobic fermentor in which *Saccharomyces cerevisiae* grows in a medium of sugar cane molasses.

However, in practice, uncertainty on the model parameters and external process disturbances are inherently present. Uncertainty can originate from unmodeled process variables (*process noise*), e.g., inherent biological variability between cells which are genetically identical [9] or from a parameter estimation procedure based on noisy measurements (*measurement noise*), such that the true parameter values (which are different from the model parameters) are unknown. Not taking this uncertainty into account, possibly leads to the violation of constraints and erroneous estimates of the actual objective function(s). Therefore, the information about the uncertainty has to be taken into account to obtain *robustified controls* (i.e., variables that can be manipulated throughout the process) that ensure that constraints are met and an overall better objective function estimate is guaranteed. In this work the nature of uncertainty is assumed to be *stochastic*, i.e., following a probability distribution, and the uncertainty is modeled in the model parameters, i.e., *parametric uncertainty* [10].

Including robustness in an optimization problem is often tedious, since this typically leads to semi-infinite optimization problems that are challenging to solve in practice [10]. Three methods are compared in this work to approximately solve the *(multi-objective) dynamic optimization problem under parametric uncertainty* for biological networks: linearization [11], sigma points [12] and polynomial chaos expansion [13, 14]. Each of these methods requires increasing levels of information on the parametric uncertainty distribution to propagate the parametric uncertainty towards the states, constraints or objectives of interest.

The authors want to highlight that enzyme activation in biological networks has been studied in terms of dynamic optimization, single objective as well as multi-objective. In this work, for the first time, parametric uncertainty is taken into account for prediction and control of biological networks. Another novelty is the critical comparison of the linearization, sigma points and polynomial chaos expansion approaches for dynamic optimization of biological networks under uncertainty. Single objective as well as multi-objective optimization case studies have been investigated in this work. Therefore the general formulations in this work have been presented for multi-objective optimization problems.

The paper is structured as follows. In the 'Methods' section the multi-objective dynamic optimization problem formulation under parametric uncertainty is first presented. Then the concept of uncertainty propagation is introduced, together with the three applied approximation techniques for uncertainty propagation. Subsequently, multi-objective optimization methods are briefly discussed. To conclude this section the software and case studies are presented. A validation and assessment of the approximation techniques for uncertainty propagation based on the case studies is presented in the 'Results and discussion' section, together with a physical/biological interpretation. Finally, the 'Conclusions' section summarizes the main results of this work.

Methods

In this section the robustified multi-objective dynamic optimization formulation is presented. Subsequently, the different approximation techniques for uncertainty propagation that enable a robustified dynamic optimization under parametric uncertainty are discussed. Next, the approach for the Monte Carlo simulations is presented. In addition, multi-objective optimization methods are introduced, followed by a brief discussion on the software used in this work. To conclude the case studies are presented.

Multi-objective dynamic optimization under parametric uncertainty

Consider the system $\dot{\mathbf{x}} = \mathbf{f}(\mathbf{x}, \mathbf{u}, \boldsymbol{\theta}, t)$, with $\mathbf{x} \in \mathbb{R}^{n_x}$ the state vector (e.g., metabolite concentrations), $\mathbf{u} \in \mathbb{R}^{n_u}$, the control vector (e.g., enzyme expression rates), $\boldsymbol{\theta} \in \mathbb{R}^{n_\theta}$ the vector containing the uncertain parameters (e.g., kinetic constants such as the maximum reaction rate) and t the time. The aim of a multi-objective dynamic optimization problem is to design a control, which minimizes several objective functions $\{J_1, \ldots, J_{n_J}\}$, subject to the constraints (i.e., model as dynamic constraint and other constraints). The multi-objective dynamic optimization problem in the time interval $t \in [0, t_f]$ and constraints $\mathbf{c}(\mathbf{x}, \mathbf{u}, \boldsymbol{\theta}, t) \in \mathbb{R}^{n_c}$ (e.g., bounds on the metabolite concentrations or fluxes for cell viability) is formulated as in Eq. (1).

$$\min_{\mathbf{u}, \mathbf{x}, t_f} \quad \{J_1, \ldots, J_{n_J}\}$$

$$\text{s.t.} \begin{cases} \dot{\mathbf{x}} &= \mathbf{f}(\mathbf{x}, \mathbf{u}, \boldsymbol{\theta}, t) \\ \mathbf{x}(0) &= \mathbf{x}_0 \\ 0 &\geq \mathbf{c}(\mathbf{x}, \mathbf{u}, \boldsymbol{\theta}, t) \end{cases} \quad (1)$$

An inherent problem in the modeling of biological processes is uncertainty. This uncertainty can originate from model uncertainty and external disturbances [10]. The emphasis in this work is on parametric uncertainty, i.e., the uncertainty is present in several model parameters, which for instance can originate from biological variability. Not taking this uncertainty into account can possibly lead to constraint violations or erroneous estimates of the actual objective function of the process. In the field of robust optimization these uncertainties are taken into account to guarantee that critical constraints are not violated [10].

If knowledge about the parametric uncertainty distribution is present, expected values for the states and chance constraints can be formulated [15]. Chance constraints express that the probability of a constraint to be valid must be larger than a specific value [16, 17].

Consider that the constraints $0 \geq \mathbf{c}(\mathbf{x}, \mathbf{u}, \boldsymbol{\theta}, t)$ can be replaced by $n_{c_{\text{prob}}}$ chance constraints $c_{\text{prob},i}$, expressing that the probability that a constraint is satisfied is larger than a preset probability β_i, with $i = 1, \ldots, n_{c_{\text{prob}}}$. In this work only single chance constraints are considered.

$$\beta_i \leq \mathbf{Pr}\left[0 \geq c_{\text{prob},i}(\mathbf{x}, \mathbf{u}, \boldsymbol{\theta}, t)\right] \quad (2)$$

If the uncertainty is fully known within a specific bounded set, the optimization problem is solved for the *worst-case scenario* in which all constraints have to be satisfied [15]. This approach typically leads to minmax problems which are hard to solve [18]. Since the worst-case scenario is often highly unlikely to occur, this approach can lead to poor results [11]. In order to solve this, a trade-off between the nominal case (i.e., the non-robustified case in which uncertainty is not taken into account and the nominal parameter values are used) and worst-case scenario can be made [15].

The main limitation of the dynamic optimization problem with chance constraints is solving the problem in a computationally efficient way. The propagation of the parameter uncertainties through the nonlinear model and obtaining computationally tractable expressions for the dynamic optimization problem with chance constraints remains challenging [19]. Therefore, the chance constraints can be approximated by deterministic constraints as in Eq. (3).

$$0 \geq \mathbf{E}\left[c_{\text{prob,i}}\right] + \alpha_{c_{\text{prob},i}} \sqrt{\mathbf{Var}\left[c_{\text{prob,i}}\right]} \quad (3)$$

In Eq. (3), $\mathbf{E}\left[c_{\text{prob,i}}\right]$ and $\mathbf{Var}\left[c_{\text{prob,i}}\right]$ express the expected value and variance of the chance constraint function $c_{\text{prob,i}}$, respectively. The coefficient $\alpha_{c_{\text{prob},i}}$ is introduced as a *backoff parameter* (e.g., [11, 20]) to take the uncertainty on the chance constraints into account. The choice of the backoff can for instance correspond to a probability that the specific constraint is violated, i.e., so-called *single chance constraints*.

A first way to choose the backoff parameter $\alpha_{c_{\text{prob},i}}$ is with Cantelli-Chebyshev's inequality. In [17] it is shown that an upper bound for the expected value on an individual chance constraint can be calculated. This equation holds for any underlying distribution of the chance constraint. Computing the backoff parameter via Cantelli-Chebyshev's inequality for a probability of 95 % for the chance constraint to be satisfied, results in a backoff parameter $\alpha_{c_{\text{prob},i}} = 4.36$, while for a normal probability distribution the backoff parameter would be 1.96.

From this, it is clear that Cantelli-Chebyshev's inequality generally leads to a very conservative bound with too high backoff parameter values for use in practice, leading potentially to infeasibilities.

If a probability distribution is assumed for the considered constraint(s) or objective function(s), a second way is to choose the backoff parameter based on the quantiles. For this a procedure as in [10] can be followed to obtain a desired confidence level for the constraint to be satisfied or to cover the objective function in a prediction interval. In this work the choice of the backoff parameter is based on the quantiles, assuming that the states follow a normal distribution, as shown in Table 1.

Similarly to the reformulation of constraints, the objective function J_i can be reformulated by adding the term $\alpha_{J_i} \sqrt{\mathbf{Var}\,[J_i]}$. Since an objective function has to be minimized, an increase in variance is penalized by this reformulation. The reformulated robustified multi-objective dynamic optimization problem with deterministic constraints is formulated in Eq. (4).

$$
\min_{\mathbf{u},\mathbf{x},t_f} \left\{ \mathbf{E}\,[J_1] + \alpha_{J_1}\sqrt{\mathbf{Var}\,[J_1]}, \ldots, \mathbf{E}\,[J_{n_J}] + \alpha_{J_{n_J}}\sqrt{\mathbf{Var}\,[J_{n_J}]} \right\}
$$

$$
\text{s.t.} \begin{cases} \dot{\mathbf{x}} &= \mathbf{f}(\mathbf{x},\mathbf{u},\boldsymbol{\theta},t) \\ \mathbf{x}(0) &= \mathbf{x}_0 \\ 0 &\geq \mathbf{E}\left[c_{\text{prob},i}\right] + \alpha_{c_{\text{prob},i}}\sqrt{\mathbf{Var}\left[c_{\text{prob},i}\right]} \\ & i = 1,\ldots,n_{c_{\text{prob}}} \end{cases}
$$

$$(4)$$

In practice, not all constraints have to be replaced by probabilistic constraints (e.g., bounds on the controls and states) and constraints of the form $0 \geq \mathbf{c}(\mathbf{x},\mathbf{u},\boldsymbol{\theta},t)$ can still be present in the optimization problem formulation.

Approximation techniques for uncertainty propagation

In this paper, the parametric uncertainty is propagated to the states, constraints and objectives of interest. This can be illustrated with an example shown in Fig. 1. Consider the simple nonlinear model $y = g(x)$ (blue curve), with the parameter x that is uncertain with a known parametric uncertainty distribution (green curve). The principle of uncertainty propagation will propagate the parametric uncertainty distribution (green curve) through the nonlinear model (blue curve) in order to obtain the uncertainty distribution of the output y (purple curve).

Table 1 Backoff parameter α_i with corresponding quantiles and confidence levels

α_i	0.84	1.28	1.65	1.96
Quantile	0.20	0.10	0.05	0.025
Confidence level	0.80	0.90	0.95	0.975

The parametric uncertainty can be propagated via a numerical integration over the parameter distribution [21]. However, this integration is typically computationally expensive for realistic models and more efficient approximative uncertainty propagation techniques exist.

An alternative to this numerical integration is the use of Monte Carlo simulations in optimization. A large number of N realizations is drawn from the assumed parametric uncertainty distribution with variance-covariance matrix $\boldsymbol{\Sigma}$ and the empirical confidence regions can be determined by using the appropriate quantiles. However, this is in practice computationally extremely expensive. Due to the large amount of simulations, no computationally tractable procedure for gradient based optimization schemes is available. In addition there is no clear rule on how many noise realizations have to be taken in order to obtain an accurate estimate [10]. For these reasons, Monte Carlo simulations are not pursued in the dynamic optimization procedure.

Approaches that exploit the availability of measurements as described in [11, 22, 23] are also used in the field of robust optimization. However, these are not considered in this paper, since in an industrial setting intracellular measurements are typically not available on a routine basis.

The first type of the employed techniques is a so-called *linearization approach*, which is based on first-order Taylor series approximations of the model functions with respect to the uncertainty [10, 11]. This approximation can be used if higher order terms can be neglected. This is the case when the uncertainty is small compared to the model curvature [15]. In this linearization approach a linear approximation of the variance-covariance matrix [11] of the states is made. On the other hand efficient sampling-based uncertainty propagation techniques exist as, e.g., using Hammersley sequences [24], the unscented transformation or *sigma points approach* [12] and the *polynomial chaos expansion approach* [19, 25].

In addtition to the *linearization approach* [11], two other techniques are considered: the *sigma points approach* [12] and the *polynomial chaos expansion approach* [19].

In practice, one is often interested in the violation of a path constraint (e.g., fluxes or concentrations that should not exceed their bounds for cell viability), a terminal constraint (e.g., a minimum amount of a specific metabolite to be produced) or the robustness of the objective (e.g., a minimum enzymatic cost), i.e., minimizing the uncertainty on the objective by taking into account the variance on the objective. The constraint or objective function to which the uncertainty is propagated, is denoted by $R_k(\mathbf{x},\mathbf{u},\boldsymbol{\theta},t)$ in the following. The three techniques consist of propagating the parametric uncertainty by approximating the expected value $\mathbf{E}\,[R_k]$ and variance $\mathbf{Var}\,[R_k]$ of

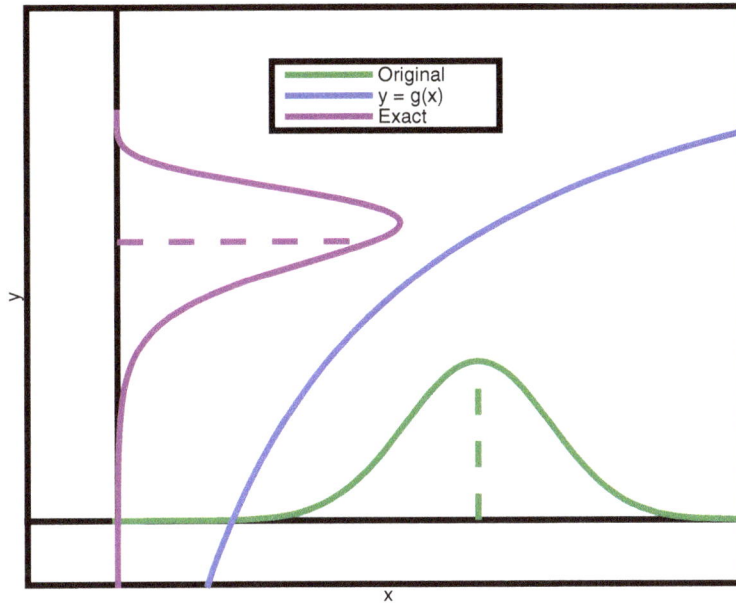

Fig. 1 Principle of uncertainty propagation. Uncertainty propagation of x towards y via nonlinear transformation g(x) in an exact way

$R_k(\mathbf{x}, \mathbf{u}, \boldsymbol{\theta}, t)$. The approximated expected value and variance are denoted by $\bar{R}_{k,\mathrm{LIN}}$, $\bar{R}_{k,\mathrm{SP}}$, $\bar{R}_{k,\mathrm{PCE}}$ and $\mathbf{P}_{\mathrm{LIN}}$, $\mathbf{P}_{\mathrm{SP}}$, $\mathbf{P}_{\mathrm{PCE}}$, respectively.

An overview of these techniques with the robustified multi-objective dynamic optimization problem formulation is provided in Table 2. A graphical representation of the approximation techniques for uncertainty propagation is shown in Fig. 2. A more detailed review of these techniques is presented in Additional file 1. Note that these approaches are also applicable to uncertainty present in the model right hand side. However, in this work only the application to parametric uncertainty propagation is considered.

Note: The approximation techniques for uncertainty propagation can also be used to quantify the effect of parametric uncertainty on model predictions. If the controls are fixed, then an additional benefit could be that the uncertainty on the states can be displayed. This enhances the insight in whereto the system can evolve with an a priori known or assumed parametric uncertainty. In fact, in literature it has been pointed out that the polynomial chaos expansion could be considered as an alternative to Monte Carlo simulations, but with less computational cost [25].

Multi-objective optimization methods

In practice, multiple objectives, which are very often conflicting with each other, have to be considered simultaneously, e.g., minimizing the enzyme consumption while maximizing the production of a certain metabolite. Therefore, a single optimal solution will not exist, but a set

of *trade-off solutions*, called the Pareto front, is obtained when solving a multi-objective problem [16].

Two categories of methods can be distinguished for the calculation of Pareto fronts: *scalarization* methods ([26–30]) that convert the multi-objective optimization problem into a series of single objective optimization problems by using scalar variables, and *vectorization* methods [16, 32–34] that start from a population of candidate solutions that gradually evolve to the Pareto front. Scalarization methods can take advantage of fast and efficient gradient based methods to find an optimum for the series of single objectives, while vectorization methods often use derivative-free optimization methods as evolutionary or stochastic optimization approaches. In this work the Normal Boundary Intersection (NBI) method [30, 31], i.e., a scalarization method, is used. For a more detailed description of the frame of multi-objective dynamic optimization and the NBI method see, e.g., [35] and Additional file 2.

Implementation and software

The dynamic optimization problems in this work are solved using a direct approach, in which first the optimal control problem is discretized into a nonlinear optimization problem that can be solved afterwards with NLP solvers. It is chosen to discretize the problems using an orthogonal collocation discretization scheme. The rationale of orthogonal collocation is that the states and controls are fully discretized with respect to time in finite elements. Per finite element there are four *collocation points* of which the first one is fixed and the three other

Table 2 Overview of the approximation techniques for uncertainty propagation with R_k the variable to which the parametric uncertainty is propagated

	Linearization	Sigma points	Polynomial chaos
Rationale	Linearization of state equations around $\bar{\theta}$	Approximate distribution by a fixed number of parameters (sigma points)	Approximate response of the model (at sampling points) as a pth order polynomial function of θ
Uncertainty distribution	Normal	Any symmetric, unimodal distribution	Any
Equations	State equations + Sensitivity equations: $\begin{cases}\dot{\mathbf{S}}_{LIN}(t) = \frac{\partial \mathbf{f}(\mathbf{x},\mathbf{u},\theta_{nom},t)}{\partial \mathbf{x}}\mathbf{S}_{LIN} + \frac{\partial \mathbf{f}(\mathbf{x},\mathbf{u},\theta_{nom},t)}{\partial \theta},\\ \mathbf{S}_{LIN}(0) = \mathbf{0}\end{cases}$	State equations for SPs: $\dot{\mathbf{x}}_i = \mathbf{f}(\mathbf{x}_i,\mathbf{u},\boldsymbol{\pi}_i,t)$ with $i=0,\ldots,2n_\theta$	State equations for sampling points: $\dot{\mathbf{x}}_i = \mathbf{f}(\mathbf{x}_i,\mathbf{u},\boldsymbol{\pi}_i,t)$ with $i=0,\ldots,n_S-1$
Total n_{states}	$(n_\theta+1)n_x$	$(2n_\theta+1)n_x$	$\frac{(n_\theta+p)!}{n_\theta!p!}n_x$
Sampling points	–	Sigma points: $2n_\theta+1$	Collocation points: $\frac{(n_\theta+p)!}{n_\theta!p!}$
Expected value of R_k	R_k	$\frac{1}{n_\theta+\kappa}\left(\kappa R_k(\boldsymbol{\pi}_0) + \frac{1}{2}\sum_{i=1}^{2n_\theta}R_k(\boldsymbol{\pi}_i)\right)$	$a_{R_k,0}^{(p)}$
Variance on R_k	$\mathbf{P}_{R_kR_k,LIN} = \frac{\partial R_k}{\partial \mathbf{x}}\mathbf{P}_{LIN}\left(\frac{\partial R_k}{\partial \mathbf{x}}\right)^T$ with: $\mathbf{P}_{LIN}=\mathbf{S}_{LIN}(t)\boldsymbol{\Sigma}\mathbf{S}_{LIN}(t)^T$	$\mathbf{P}_{R_kR_k,SP} = \frac{1}{n_\theta+\kappa}\left(\kappa(R_k(\boldsymbol{\pi}_0)-\bar{R}_k)(R_k(\boldsymbol{\pi}_0)-\bar{R}_k)^T\right)$ $+\frac{1}{n_\theta+\kappa}\left(\frac{1}{2}\sum_{i=1}^{2n_\theta}(R_k(\boldsymbol{\pi}_i)-\bar{R}_k)(R_k(\boldsymbol{\pi}_i)-\bar{R}_k)^T\right)$ with: $\begin{cases}\boldsymbol{\pi}_0 = \theta_{nom},\\ \boldsymbol{\pi}_i = \theta_{nom} + \sqrt{(n_\theta+\kappa)\boldsymbol{\Sigma}_i}\text{ with }i=1,\ldots,n_\theta,\\ \boldsymbol{\pi}_i = \theta_{nom} - \sqrt{(n_\theta+\kappa)\boldsymbol{\Sigma}_{i-n_\theta}}\text{ with }i=n_\theta+1,\ldots,2n_\theta.\\ \kappa = 3-n_\theta\end{cases}$	$\mathbf{P}_{R_kR_k,PCE}^{(p)} = \sum_{j=1}^{L-1}\left(a_{R_k,j}^{(p)}\right)^2\mathbf{E}\left[\Phi_j^2(\theta)\right]$ with: $\mathbf{a}=(\Lambda\Lambda^T)^{-1}\Lambda R_{k,S}$
Optimization problem	$\min_{\mathbf{u},\mathbf{x}_{t_f}}\ \{J_1,\ldots,J_{n_J}\}$ $\bar{J}_{LIN}+\alpha_J\sqrt{\mathbf{P}_{J,J,SP}}$ s.t. $\begin{cases}\dot{\mathbf{x}} = \mathbf{f}(\mathbf{x},\mathbf{u},\theta,t)\\ \dot{\mathbf{S}}_{LIN}(t)=\frac{\partial\mathbf{f}(\mathbf{x},\mathbf{u},\theta_{nom},t)}{\partial\mathbf{x}}\mathbf{S}_{LIN}+\frac{\partial\mathbf{f}(\mathbf{x},\mathbf{u},\theta_{nom},t)}{\partial\theta},\\ \mathbf{S}_{LIN}(0)=\mathbf{0}\\ \mathbf{x}(0)=\mathbf{x}_0\\ 0\geq\bar{c}_{prob,i,LIN}\\ \quad+\alpha_{c_{prob,i}}\sqrt{\mathbf{P}_{c_{prob,i},c_{prob,i},LIN}}\end{cases}$	$\min_{\mathbf{u},\mathbf{x}_{t_f}}\ \{J_1,\ldots,J_{n_J}\}$ $\bar{J}_{SP}+\alpha_J\sqrt{\mathbf{P}_{J,J,SP}}$ s.t. $\begin{cases}\dot{\mathbf{x}}_i = \mathbf{f}(\mathbf{x}_i,\mathbf{u},\boldsymbol{\pi}_i,t)\text{ with }i=0,\ldots,2n_\theta\\ \mathbf{x}(0)=\mathbf{x}_0\\ 0\geq\bar{c}_{prob,i,SP}\\ \quad+\alpha_{c_{prob,i}}\sqrt{\mathbf{P}_{c_{prob,i},c_{prob,i},SP}}\end{cases}$	$\min_{\mathbf{u},\mathbf{x}_{t_f}}\ \{J_1,\ldots,J_{n_J}\}$ $\bar{J}_{PCE}^{(p)}+\alpha_J\sqrt{\mathbf{P}_{J,J,PCE}^{(p)}}$ s.t. $\begin{cases}\dot{\mathbf{x}}_i = \mathbf{f}(\mathbf{x}_i,\mathbf{u},\boldsymbol{\pi}_i,t)\\ \quad\text{with }i=0,\ldots,n_S-1\\ \mathbf{x}(0)=\mathbf{x}_0\\ \mathbf{R}_{k,S}^{(p)}=(\Lambda^{(p)})^T\mathbf{a}_{R_k}^{(p)}\\ \quad\text{with }k=1,\ldots,n_R\\ 0\geq\bar{c}_{prob,i,PCE}^{(p)}\\ \quad+\alpha_{c_{prob,i}}\sqrt{\mathbf{P}_{c_{prob,i},c_{prob,i},PCE}^{(p)}}\\ \quad\text{with }i=1,\ldots,n_{c_{prob}}\end{cases}$

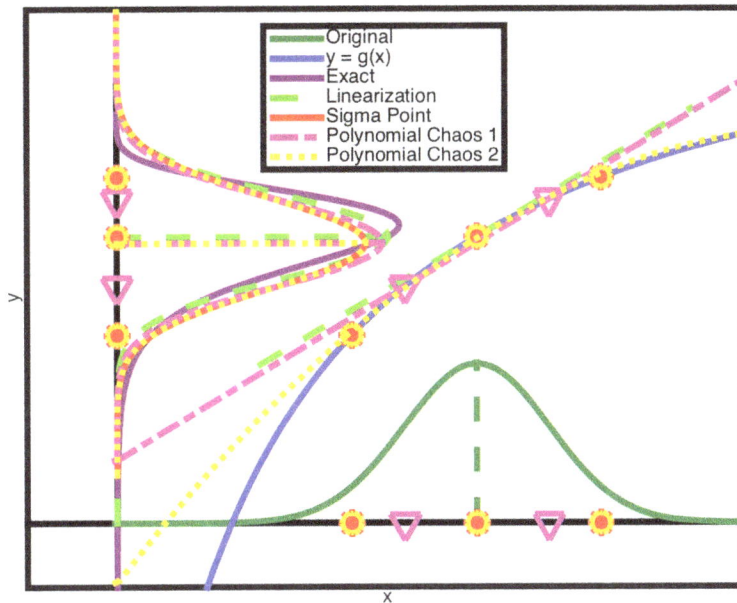

Fig. 2 Approximation techniques for uncertainty propagation. Overview of the linearization, sigma points and polynomial chaos expansion methods for uncertainty propagation of x towards $y = g(x)$

ones should obey the model equations and are seen as equality constraints (i.e., so-called *collocation constraints*). Between each finite element there is also a constraint that ensures continuity (i.e., so-called *continuity constraints*). As interpolation between the *collocation points* a cubic Lagrange polynomial is used, with four collocation points situated at the Radau roots on each interval. State bounds are easily added in this technique. The fact that orthogonal collocation has hardly any problem with stiff systems is advantageous in case of numerically unstable systems.

An inhouse developed software package, called Pomodoro, is used for the implementation of both case studies. The Pomodoro software contains a collection of algorithms and tools for dynamic optimization and is implemented in Python. Pomodoro uses CasADi [36] as a backbone for the dynamic optimization problem formulation. CasADi is a software package for rapid prototyping of large-scale optimization problems with automatic differentiation using a symbolic/numeric approach. For solving the NLP, an interior point algorithm, IPOPT [37], has been used. The Pomodoro software can be downloaded from https://perswww. kuleuven.be/~u0093798/software.php. For review purposes the work describing Pomodoro (Bhonsale SS, Telen D, Vercammen D, Vallerio M, Hufkens J, Nimmegeers P, Logist F, Van Impe J. Pomodoro - A novel toolkit for (multiobjective) dynamic optimization, model based control and estimation, submitted), can be found on http://www.student.kuleuven.be/~s0212066/pomodoro/. For more information on the optimization methods and

implementation, the reader is referred to Additional file 2 and [35].

Case studies

Two case studies are considered: *(a)* a three-step linear pathway with mass-action kinetics [7, 38] and *(b)* a glycolysis based network with 1 output [7, 39]. It should be noted that the models for the case study are partially taken from [7]. The networks for the two case studies are presented in Fig. 3.

Case 1: three-step linear pathway

The first case study is a three-step linear pathway producing one product S_4 from a buffered substrate S_1 [7, 38]. This pathway consists of three enzymatic reactions (with reaction fluxes $\mathbf{v} = [v_1\ v_2\ v_3]^\top$.) following mass-action kinetics, between 4 metabolites $\mathbf{S} = [S_1\ S_2\ S_3\ S_4]^\top$. Each reaction is catalyzed by a specific enzyme e_i. The first metabolite S_1 is the substrate, intermediate metabolites are S_2 and S_3, while S_4 is the product, produced by this three step linear pathway. The substrate is considered to be buffered, which means that the substrate concentration remains constant.

The model contains four differential states and an additional state x_{extra} for the objective function corresponding to the integral of the intermediate accumulation. The model together with its constraints is presented in Eq. (5). The first constraint is to ensure that a minimum amount of product is obtained at t_f. The second constraint expresses that the sum of all enzyme concentrations

(a)

Network for Case 1: Three step linear pathway with four metabolites and three fluxes.

(b)

Network for Case 2: Glycolysis based network with five metabolites and four fluxes.

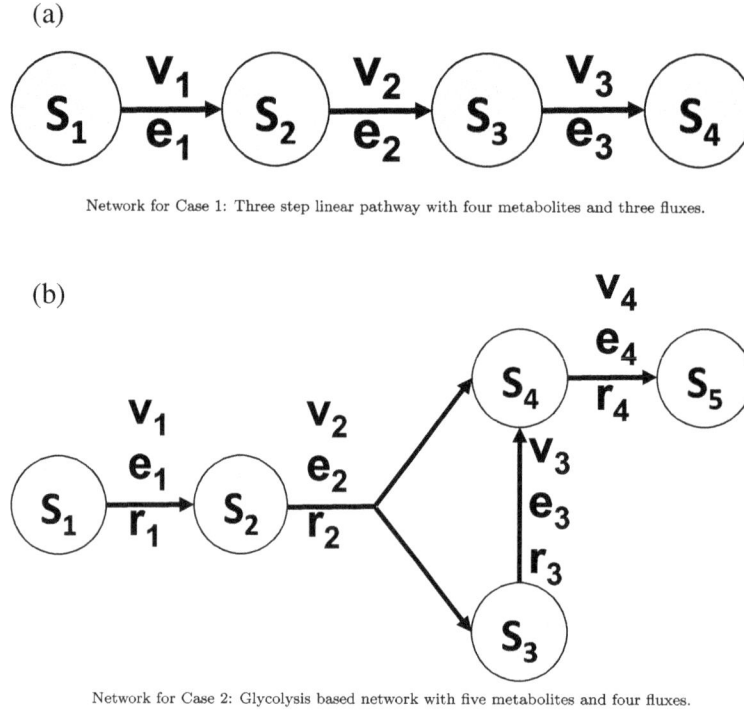

Fig. 3 Networks for Case 1 and Case 2. Biological networks based on [7]: **a** Three step linear pathway with four metabolites and three fluxes and **b** Glycolysis based network with five metabolites and four fluxes

cannot exceed the total enzymatic concentration E_T of 1 mM. The enzyme concentrations $\mathbf{e} = [e_1\ e_2\ e_3]^\top$ are the controls in this case study.

$$\begin{cases} \frac{d\mathbf{S}}{dt} & = \mathbf{N}\mathbf{v} \\ \frac{dx_{\text{extra}}}{dt} & = S_2 + S_3 \\ S_4(t_f) & \geq 0.90\text{mM} \\ \sum_{i_e=1}^{n_e=3} e_i & \leq E_T \end{cases} \tag{5}$$

with:

$$v_1 = k_1 \cdot S_1 \cdot e_1 \tag{6}$$
$$v_2 = k_2 \cdot S_2 \cdot e_2 \tag{7}$$
$$v_3 = k_3 \cdot S_3 \cdot e_3 \tag{8}$$

$$\mathbf{N} = \begin{bmatrix} 0 & 0 & 0 \\ 1 & -1 & 0 \\ 0 & 1 & -1 \\ 0 & 0 & 1 \end{bmatrix} \tag{9}$$

with $\mathbf{N} \in \mathbb{R}^{4\times3}$ the stoichiometric matrix, containing the stoichiometric coefficients N_{ij} of metabolite i in the j-th reaction and k_j, the maximum reaction rate of reaction j.

The three uncertain parameters are the three reaction rate constants k_1, k_2 and k_3, for which the nominal value equals 1 $(\text{mMs})^{-1}$. Since high concentrations of inter-

mediate metabolites S_2 and S_3 can be harmful for cell viability, the intermediate accumulation is minimized for this case study.

$$J = x_{\text{extra}}(t_f) = \int_{t_0=0s}^{t_f} S_2 + S_3 dt \tag{10}$$

Case 2: glycolysis based network with 1 output

The second case study is a glycolysis based network with the production of one product S_5, starting from one substrate S_1 from [7]. In this pathway four enzymatic reactions are taking place, each catalyzed by a specific enzyme. The fluxes are modeled with Michaelis-Menten kinetics. The intracellular metabolites in this network are S_2, S_3 and S_4. It is assumed that the substrate S_1 is buffered. This case study is particularly interesting due to the branch that is present. Such branches often occur in biological networks and the presented problem formulation can be modified/extended to many scenarios.

The expressions for this model are presented in Eq. (11) where $\mathbf{N}$ is the stoichiometric matrix, with $\mathbf{v} = [v_1\ v_2\ v_3\ v_4]^\top$ the flux vector, $\mathbf{S} = [S_1\ S_2\ S_3\ S_4\ S_5]^\top$ the vector containing the metabolite concentrations , $\mathbf{e} = [e_1\ e_2\ e_3\ e_4]^\top$ the enzyme concentration vector, the vector of manipulated variables $\mathbf{r} = [r_1\ r_2\ r_3\ r_4]^\top$ containing the expression rates, $k_{\text{cat},j}$ the maximum reaction rate for reaction j, dependent on the enzyme that is catalyzing the reaction j which is assumed to be the same for each reaction and therefore considered as the model parameter k_{cat} and K_M the Michaelis constant.

$$
\begin{cases}
\frac{d\mathbf{S}}{dt} & = \mathbf{Nv} \\
\frac{d\mathbf{e}}{dt} & = \mathbf{r} - \lambda\mathbf{e} \\
\frac{dx_{extra}}{dt} & = e_1 + e_2 + e_3 + e_4 \\
S_5(t_f) & \geq 0.675 \text{ mM} \\
\sum_{i_e=1}^{n_e=4} e_i & \leq E_T \\
r_1(t_0) & = 0.5 \\
r_{i_r}(t_0) & = 0, \sum_{i_r=1}^{n_r=4} r_{i_r} \leq 0.5 \quad i_r = 2, \ldots, n_r = 4
\end{cases}
\tag{11}
$$

with:

$$
v_1 = \frac{k_{cat} \cdot S_1}{K_M + S_1} \cdot e_1
\tag{12}
$$

$$
v_2 = \frac{k_{cat} \cdot S_2}{K_M + S_2} \cdot e_2
\tag{13}
$$

$$
v_3 = \frac{k_{cat} \cdot S_3}{K_M + S_3} \cdot e_3
\tag{14}
$$

$$
v_4 = \frac{k_{cat} \cdot S_4}{K_M + S_4} \cdot e_4
\tag{15}
$$

$$
\mathbf{N} = \begin{bmatrix}
0 & 0 & 0 & 0 \\
1 & -1 & 0 & 0 \\
0 & 1 & -1 & 0 \\
0 & 1 & 1 & -1 \\
0 & 0 & 0 & 1
\end{bmatrix}
\tag{16}
$$

The following values for the parameters are assumed [7]: $k_{cat} = 1 \text{ s}^{-1}$, $K_M = 1 \text{ mM}$ and $\lambda = 0.5 \text{ s}^{-1}$.

Two objectives are considered for which the enzyme expression rates in $\mathbf{r}$ are optimized: the minimization of the time to reach a given steady state and the enzyme consumption (or enzymatic cost) as shown in Eqs. (17)-(18):

$$
J_1 = t_f
\tag{17}
$$

$$
J_2 = \int_{t_0=0s}^{t_f} \left(\sum_{i_e=1}^{n_e=4} e_{i_e} dt \right).
\tag{18}
$$

Results and discussion

This section discusses the obtained results in this work. In the first subsection the approach followed to obtain these results is clarified. Next, the results for the three-step linear pathway are presented. In the third subsection the results for the glycolysis inspired network are described.

Approach

The approach consists of the four steps in Fig. 4. This approach is formulated for the generic case of the multi-objective dynamic optimization of biological networks under uncertainty. First, the (multi-objective) dynamic optimization problem is solved for the nominal (non-robustified) case. Then, desired confidence levels for the robustified constraint are set (*Step 1*). Robustified terminal constraints of the form $c_{min} \leq \mathbf{E}\left[c(t_f)\right] - \alpha_c \sqrt{\mathbf{Var}\left[c(t_f)\right]}$ are considered. Table 1 presents the backoff parameter values that are used for the computation of the robustified controls together with the corresponding quantiles and preset confidence levels for a normal distribution of the constraint.

In *Step 2* robustified Pareto fronts are calculated with the linearization, sigma points and polynomial chaos expansion approaches (PCE1 and PCE2) to include parametric uncertainty in the multi-objective dynamic optimization problem. These robustified Pareto fronts are computed for the different backoff parameter values.

A first comparison of the approximation techniques for uncertainty propagation is based on the Pareto fronts: comparison of the CPU time for a single Pareto point, comparison of the objective function vectors J_{LIN}, J_{SP}, J_{PCE1} and J_{PCE2}, calculated with the linearization, sigma points, PCE1 and PCE2 respectively. Also the expected values and variance approximations for the robustified model outputs with the approximation techniques for uncertainty propagation are compared.

Subsequently, a Pareto point is selected from the Pareto front for further analysis (*Step 3*). This analysis can be performed for any Pareto point or confidence level set (i.e., corresponding to different backoff parameter values α_i), without a loss of generality. In this work, the considered point corresponds to one of the objectives, i.e., the minimization of the intermediate accumulation for Case 1 (single objective case study) and the minimization of the enzymatic cost for Case 2 (multi-objective case study).

In *Step 4*, Monte Carlo simulations are done for the considered Pareto point by sampling 1000 randomly generated parameter sets from the parametric uncertainty distribution. The robustified controls, determined with linearization, sigma points and polynomial chaos expansion approaches (PCE1 and PCE2), are fixed in the Monte Carlo simulations.

The further analysis of the Pareto point consists of assessing the robustness of the optimal control profiles obtained with the different approximation techniques for uncertainty propagation: i.e., (i) checking the reduction of constraint violations by applying a robustified control in comparison with applying a nominal (non-robustified) control profile, (ii) evaluating the backoff taken in objective function when uncertainty is taken into account and (iii) a comparison between the predicted expected value

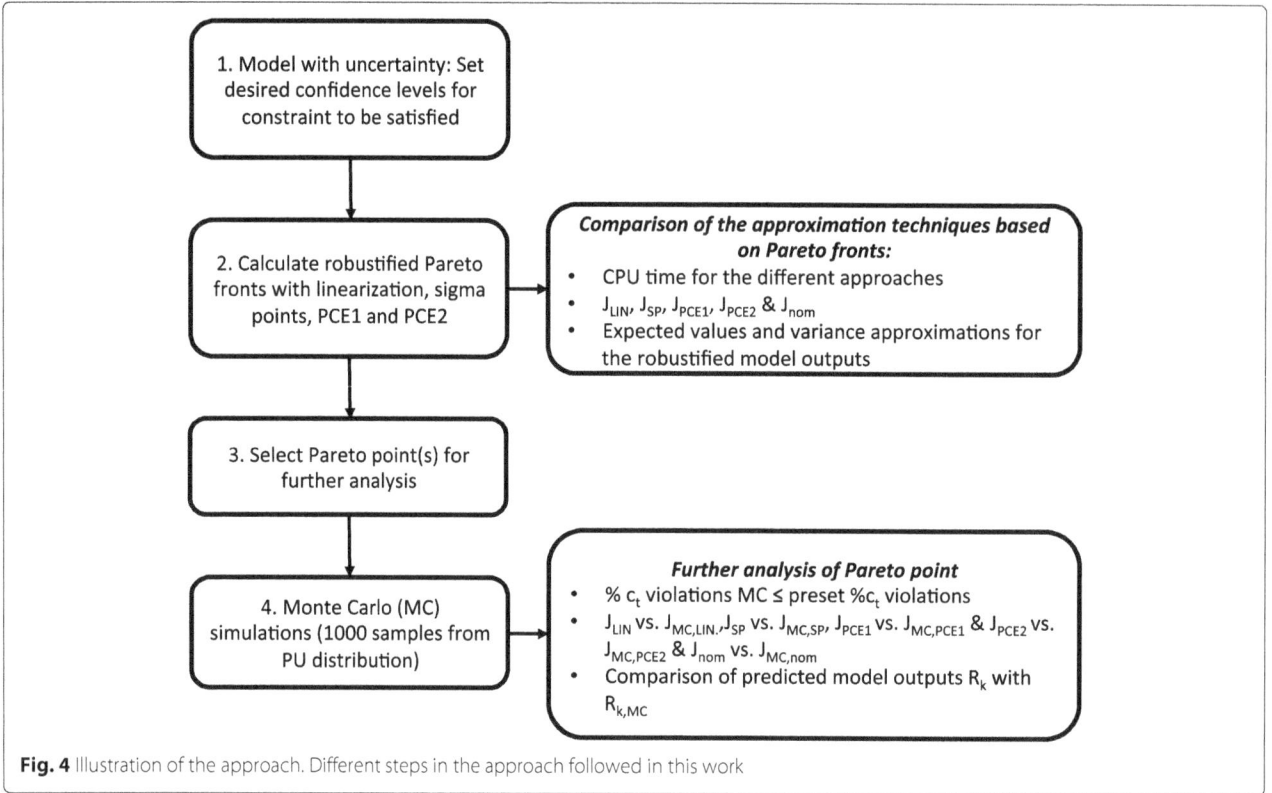

Fig. 4 Illustration of the approach. Different steps in the approach followed in this work

and variance for the robustified model output with the approximation techniques for uncertainty propagation and the calculated mean and variance with the Monte Carlo simulations. Furthermore, for the robustified single chance constraint it is investigated whether the preset confidence is reached for different backoff parameter values. This is done by checking whether the percentage of constraint violations in Monte Carlo simulations does not exceed the preset percentage of constraint violations. For instance, a confidence level of 0.95 is associated with a backoff parameter value of 1.65, meaning that in case the robustified control is applied, the percentage of constraint violations is not allowed to exceed 5 %. Alternatively, if the confidence level is not sufficient, the preset confidence can be increased by increasing the backoff parameter. An iterative procedure to determine the quantiles and backoff parameter can be followed as presented in [10].

Both case studies have been implemented in Pomodoro. The KKT tolerance is set to 10^{-5} and an orthogonal collocation discretization scheme is used for the dynamic optimization problems. Since the polynomial chaos expansion allows to take a priori information on the parametric uncertainty distribution directly into account via the orthogonal polynomials, two parametric uncertainty distributions have been studied: a priori normal and a priori uniform parametric uncertainty distribution with as mean the nominal parameter values and 20 %

relative standard deviation in Case 1 and 10 % relative standard deviation in Case 2. For the normal parametric uncertainty distribution, Hermite polynomials are used, while for the uniform distribution another type is used. The reader is referred to Additional files 3 and 4.

Case 1: three step linear pathway

In this case study the terminal constraint expressing that the concentration of S_4 at time t_f should exceed or equal 0.90 mM is robustified.

$$0.90\text{mM} \leq \mathbf{E}\left[S_4(t_f)\right] - \alpha_{S_4}\sqrt{\mathbf{Var}\left[S_4(t_f)\right]} \qquad (19)$$

Since this constraint only looks at one bound that has to be exceeded, the 95 % confidence region should be covered when a backoff parameter of $\alpha_{S_4} = 1.65$ is chosen. The intermediate accumulation objective function is not robustified for this case study (i.e., α_{J_2}). The three uncertain parameters are k_1, k_2 and k_3.

In this section, the single objective optimization (i.e., the minimization) of the intermediate accumulation is considered. Intermediate accumulation can be harmful for the cell and should therefore be minimized. It is assumed that the final time is fixed at 10 seconds. First the computational aspects are discussed. Subsequently a physical/biological interpretation of the results is given. To conclude Case 1 the approximation techniques for uncertainty propagation are compared based on the results from the single objective optimization and Monte Carlo simulations for a normal and uniform parametric uncertainty distribution.

Computational aspects
A first aspect is the computational cost of including uncertainty in dynamic optimization. The number of required states and variables, together with the CPU time for the different approximation techniques for uncertainty propagation are presented in Table 3. The CPU times are presented for the largest backoff parameter used in this work, i.e., $\alpha_{S_4} = 1.96$, to give an upper bound on the computation times that are required. The results in Table 3 confirm that taking uncertainty into account, leads to an increased computational time. An inherent property of the considered uncertainty propagation techniques, is the increase in number of states when the number of uncertain parameters increases. The increase in computational time for this case study is thus related to the increase in the size of the optimization problem. From this it is clear that the linearization and PCE1 approaches have a similar computation time and are the fastest of the considered approximation techniques for uncertainty propagation, followed by the sigma points approach and PCE2 approach.

Physical/biological interpretation
The enzyme concentration profiles are (from a computational point of view) seen as the optimal controls. The enzyme concentration profiles are illustrated in Fig. 5(a)-(c) for $\alpha = 1.65$. Different phases can be distinguished in the process (i.e., a sequential activation of the controls): *(i)* a first phase in which enzyme e_1 is activated to produce S_2 as fast as possible, *(ii)* a second phase in which both S_2 and S_3 are consumed and produced (by activation of e_2 and e_3) *(iii)* a third phase in which a novel activation of e_1 takes place, followed by *(iv)* a phase of activity of e_2 and e_3 and a final phase in which the third enzyme is fully activated for the production of S_4.

Table 3 Case 1 - Overview of the number of states, CPU time, expected values of the objective function, expected values of the terminal constraint and standard deviations for the different approximation techniques for uncertainty propagation when the enzymatic cost is minimized for $\alpha = 1.96$ for 3 uncertain parameters k_1, k_2 and k_3

	Nominal	Linearization	Sigma points	PCE1	PCE2
States	5	20	29	17	41
CPU time [s]	0.156	1.332	4.507	2.087	7.780
E [J]	3.65	6.61	7.06	7.74	7.04
E [S_4]	0.90	1.50	1.52	1.53	1.52
$\sqrt{\text{Var}}[S_4]$	0	0.30	0.32	0.33	0.32

From the enzyme concentration profiles in Fig. 5(a)-(c), it is clear that, the robustified enzyme concentrations will increase earlier than the nominal enzyme concentrations. The sequential activation of the controls (i.e., the increasing and decreasing enzyme concentrations e_i) will be early enough and sufficient in the robustified case to ensure that the robustified constraint is satisfied, i.e., a sufficient amount of S_4 is produced. In Fig. 5(d) it is shown that for the PCE1 approach the minimum treshold of 0.90 mM is reached after 5.3 seconds, for PCE2 after 6.52, for the sigma points after 6.58 and for linearization after 6.7 seconds, while in the nominal case the minimum treshold of 0.90 mM is reached at the end time of 10 seconds. The enzyme concentrations, calculated with the approximative uncertainty propagation techniques, ensure more robustness towards satisfying the minimum end concentration of 0.90 mM for S_4. However, including robustness towards satisfying the minimum end concentration leads to a higher intermediate accumulation as shown in Table 3. This is acceptable, as long as cell viability is not compromised. Comparing this with experimental results for amino-acid biosynthetic pathways in *Escherichia coli* [38, 40, 41] a sequential activation of the enzymes can be observed in the enzyme concentration profiles.

Comparison of the approximation techniques for uncertainty propagation
The number of constraint violations, expected value of S_4 and the variance on S_4 for the different approaches and α_{S_4} are shown in Tables 4 and 5 for a normal and uniform parametric uncertainty distribution, respectively. More extensive results are given in Additional file 3.

Expected value and 95 % confidence bound First the expected value and 95 % confidence bound of S_4 (based on $\alpha_{S_4} = 1.65$) are compared. This is done in Fig. 5(d) and it can be seen that the expected state and 95 % confidence bounds for S_4 are very similar when computed with the linearization, sigma points and PCE2 approaches. The expected value for the PCE1 approach differs slightly from the others: initially it is taking more distance from the nominal profile, indicating that this approach is more conservative than the other approaches and will lead to less constraint violations. In general, linear approximation techniques (as the PCE1 approach) tend to be more conservative, but it cannot be predicted upfront whether this is in a positive sense or not.

Constraint violations In order to investigate the performance of the different approximation techniques for uncertainty propagation with respect to the constraint violations, a Monte Carlo simulation with 1000 realizations (i.e., randomly generated parameter samples from

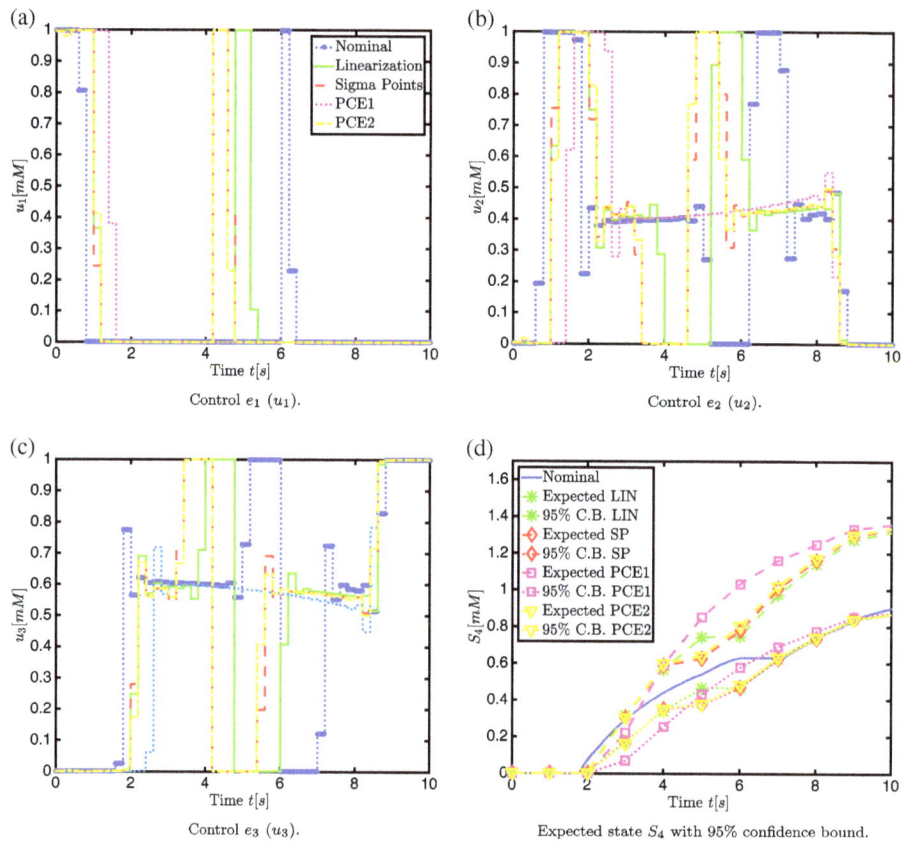

Fig. 5 Results for Case 1. Comparison of the control profiles e_1 (**a**), e_2 (**b**) and e_3 (**c**) calculated with linearization, sigma points approach, PCE1 and PCE2 for $\alpha = 1.65$ with the nominal control profile and (**d**) comparison of the expected state S_4 and its 95 % confidence bound calculated with linearization, sigma points, PCE1 and PCE2 with the nominal case ($\alpha = 1.65$) in case of 3 uncertain parameters k_1, k_2 and k_3

Table 4 Case 1 - Results Monte Carlo simulations ($N = 1000$) in case of three normally distributed uncertain parameters k_1, k_2 and k_3 for robustified terminal constraint with the number of constraint violations, mean terminal constraint values and variance on the terminal constraint

	Nominal case	Linearization	Sigma points	PCE 1	PCE 2
$\alpha = 1.96$					
$\bar{J}$	11.903	13.4600	13.600	13.829	13.583
σ_J	0.5578	0.9737	1.0087	1.016	0.9981
$\bar{S}_4$	0.88734	1.4739	1.5242	1.5500	1.5316
σ_{S_4}	0.1733	0.2878	0.2970	0.2993	0.2993
c_t violations	509 (50.9 %)	30 (3.0 %)	21 (2.1 %)	20 (2.0 %)	20 (2.0 %)
$\alpha = 1.65$					
$\bar{J}$	11.903	13.051	13.127	13.331	13.139
σ_J	0.5578	0.8624	0.8869	0.9049	0.8892
$\bar{S}_4$	0.8712	1.3357	1.3765	1.3910	1.3781
σ_{S_4}	0.1733	0.2614	0.2691	0.2683	0.2690
c_t violations	509 (50.9 %)	51 (5.1 %)	44 (4.4 %)	39 (3.9 %)	43 (4.3 %)

Table 5 Case 1 - Results Monte Carlo simulations ($N = 1000$) in case of three uniformly distributed uncertain parameters k_1, k_2 and k_3 for robustified terminal constraint with the number of constraint violations, mean values and variances on the objective function and terminal constraint, respectively

	Nominal case	Linearization	Sigma points	PCE1	PCE2	PCE2 Uniform
$\alpha = 1.96$						
$\bar{J}$	11.913	13.477	13.618	13.847	13.601	13.591
σ_J	0.5470	0.9601	0.9942	1.0000	0.9835	0.9790
$\bar{c}_t$	0.88917	1.4768	1.5273	1.5533	1.5347	1.5261
σ_{c_t}	0.1818	0.3015	0.3110	0.3132	0.3137	0.3118
c_t violation	516 (51.6 %)	4 (0.4 %)	1 (0.1 %)	0 (0.0 %)	1 (0.1 %)	2 (0.2 %)
$\alpha = 1.65$						
$\bar{J}$	11.913	13.066	13.143	13.347	13.155	13.330
σ_J	0.5470	0.8489	0.8730	0.6813	0.8750	0.8881
$\bar{c}_t$	0.88917	1.3384	1.3792	1.3940	1.3809	1.3660
σ_{c_t}	0.1818	0.2740	0.2820	0.2808	0.2819	0.2754
c_t violation	516 (51.6 %)	35 (3.5 %)	22 (2.2 %)	11 (1.1 %)	22 (2.2 %)	22 (2.2 %)

the parametric uncertainty distribution) has been performed for the four approaches and is compared with the nominal case.

From these simulations, it is observed in case of a normal parametric uncertainty distribution that all four methods reduce the amount of constraint violations significantly: from 50.9 % in the nominal case to even 2.0 % for PCE1 and PCE2, when a backoff parameter value of $\alpha = 1.96$ is chosen. The same holds for a uniform parametric uncertainty distribution.

In practice, the most interesting backoff parameter values are 1.65 and 1.96, corresponding to 5 % and 2.5 % violations in case of a normal distribution. From the results in Tables 4 and 5 it is clear that the PCE1 method scores the best with respect to constraint violations, followed by PCE2 and sigma points. The performance of the sigma points method and the second order polynomial chaos expansion are, as shown throughout this case study, very similar.

Parametric uncertainty distribution If an uncertainty distribution is assumed for the constraint and the back-off parameters are chosen in accordance with the quantiles (which is the case for the normal parametric uncertainty distribution), the level of constraint violations should correspond exactly with the confidence level. A too low degree of violations is also not wished, since the uncertainty is not propagated correctly in that case.

Furthermore, the expected values and variances of S_4 for a backoff parameter of $\alpha = 1.96$ from Table 3 which are predicted with the approximation techniques, are very close to the empirically calculated expected values and variances by Monte Carlo simulation in Tables 4 and 5. For the sigma points, PCE1 and PCE2 approaches these

predicted expected values and variances are an accurate estimation of the ones obtained by Monte Carlo simulation. For the linearization approach, this is not the case. However, the expected value of the intermediate accumulation objective function is not accurately predicted. This objective function is not robustified in this case study, as the variance on the objective function is not taken into account. Therefore, the predictions of the expected value of the objective functions are not accurate, when compared with a Monte Carlo setting.

There is less backoff from the objective function when a uniform parametric uncertainty distribution is assumed, also the variance is reduced for the second order polynomial chaos expansion with respect to the assumption of a normal parametric uncertainty distribution. The percentage of constraint violations on the other hand is slightly higher when a uniform distribution is considered. However, for this case study, the difference in performance between a uniform parametric uncertainty distribution and a normal parametric uncertainty distribution for the polynomial chaos expansion, is small. Therefore, it should be stressed that including the additional information on the parametric uncertainty distribution can be useful. However, gathering information on the parametric uncertainty distribution is quite intensive and does not lead to a drastic improvement in performance for this case study.

Case 2: glycolysis based network with 1 output
The terminal constraint and enzymatic cost objective function are robustified in this case study as shown in following equations. This is done in order to reduce the variance on the objective function. In contrast to Case 1, this should allow to have a better prediction

of the expected value and variance on the objective function.

$$0.675\text{mM} \leq \mathbf{E}\left[S_5(t_\mathrm{f})\right] - \alpha_{S_5}\sqrt{\mathbf{Var}\left[S_5(t_\mathrm{f})\right]} \qquad (20)$$

$$J_2 = \mathbf{E}\left[x_{\text{extra}}(t_\mathrm{f})\right] + \alpha_{x_{\text{extra}}}\mathbf{Var}\left[x_{\text{extra}}(t_\mathrm{f}))\right] \qquad (21)$$

For simplicity, the backoff parameters α_{S_5} and $\alpha_{x_{\text{extra}}}$ are assumed to be the same and are called α in the remainder of the text. The objective function is robustified by adding the term $\alpha_{x_{\text{extra}}}\mathbf{Var}\left[x_{\text{extra}}(t_\mathrm{f}))\right]$, since the objective function has to be minimized and an increase in variance is penalized.

In this case study three parameters (k_{cat}, K_M and λ) are considered uncertain. First, the multi-objective optimization results are discussed, followed by a more in depth analysis of the minimization of the enzymatic cost. Note that this in depth analysis can be performed for any Pareto point. Subsequently the computational aspects, a physical/biological interpretation of the results and comparison of the approaches are presented for this case study, based on the dynamic optimization results and Monte Carlo simulations for a normal and uniform parametric uncertainty distribution. For more extensive results, the reader is referred to Additional file 4.

Multi-objective optimization results

The multi-objective optimization problem consists of minimizing the time needed to reach at least 0.675 mM of product S_5 and minimizing the enzymatic cost, i.e., the total enzyme consumption over the whole time span.

In the robustified problem formulation, both the objective function for the enzymatic cost and the terminal constraint are robustified with backoff parameter α. It is assumed that the final time t_f cannot exceed 30 seconds. The two objectives, final time and enzymatic cost, are clearly conflicting: reducing the time needed to reach a level of 0.675 mM of S_5 leads to an increase in the enzymatic cost and vice versa.

Receeding Pareto fronts from the nominal optimal solution with increasing backoff parameter values can be observed in Fig. 6: both anchor points shift away from the nominal optimal solution. This is the price in performance (i.e., minimum time and enzymatic cost) that has to be paid to ensure a minimum concentration of 0.675 mM for S_5. However, it is also observed that the Pareto fronts change shape and range, when the backoff parameter increases. This is related to the feasibility of the Pareto points.

A comparison of the different approximation techniques for uncertainty propagation based on the Pareto fronts is presented in Additional file 4. From this comparison it is firstly seen that the linearization and sigma points approach take more backoff than the polynomial chaos approaches. Secondly, when the backoff parameters

decrease, it is observed that there is a similarity between the Pareto fronts for the PCE2 and sigma points approach. This can be explained from the variance that is taken less into account when the backoff parameter value decreases. Since the difference between the PCE2 and sigma points approach lies in the variance calculation, this explains the increasing difference in Pareto fronts with an increasing backoff parameter value. The expected value calculation is the same for the sigma points and PCE2 approach in case of a normal parametric uncertainty distribution. For this case study, the sigma points are a subset of the PCE2 sampling points as shown in Additional file 4.

Computational aspects

For the discussion of the computational aspects, the single objective optimization of the enzymatic cost is considered. It is assumed that the final time is fixed at 30 seconds.

In Table 6 an overview is presented of the number of states, the CPU time, objective function values, terminal constraint and their expected values and standard deviations for the different approximation techniques for uncertainty propagation for 3 uncertain parameters K_M, λ and k_{cat}. This is done for the largest backoff parameter value $\alpha = 1.96$ for the same reasons as mentioned in the first case study.

In this case study the linearization approach is computationally the most expensive. While in Case 1, the increase in computational time is related to the increase in the size of the optimization problem, this cannot be the explanation for why the linearization approach takes the most CPU time. One explanation for the long CPU time of the linearization approach, can be the nonlinearity of the model in Case 2 and solving the sensitivity equations. The interconnection of the states in the sensitivity equations, makes the linearization approach computationally more challenging.

Physical/biological interpretation

In Fig. 7 the enzyme expression rates are shown together with the corresponding enzyme concentration profiles, which are computed by minimizing the enzymatic cost for the nominal case (a) and with the different approximation techniques for uncertainty propagation: linearization (b), sigma points (c), PCE1 (d) and PCE2 (e) ($\alpha = 1.65$). It is observed that the expression rate profiles show an on/off behavior that leads to the sequential activation of the different enzymes in the network. The physical interpretation of this on/off behavior of the enzyme expression rates is that the previous enzyme first has to be degraded, before the other enzyme can be synthesized. This makes sense from a biological point of view, since the cells only have a limited amount of proteins available. This also corresponds to the satisfaction of the constraint on the

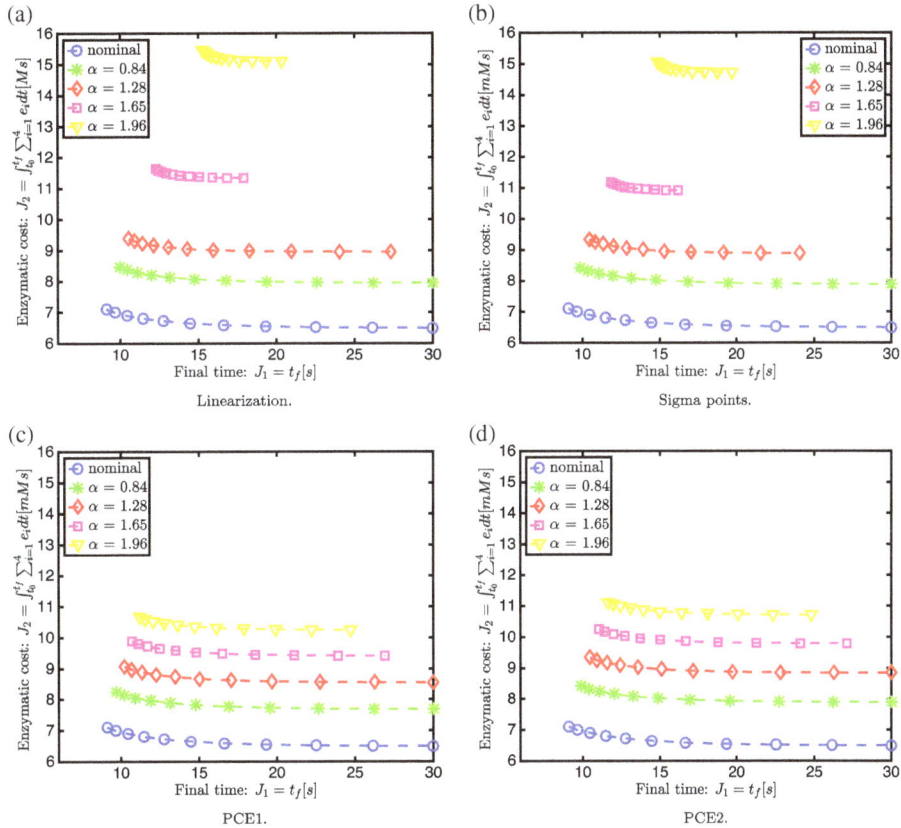

Fig. 6 Receeding Pareto fronts Case 2. Receeding Pareto fronts with increasing backoff parameter α for linearization (**a**), sigma points (**b**), first (**c**) and second order polynomial chaos expansion (**d**) approach in case of 3 uncertain parameters K_M, λ and k_{cat}

total enzymatic content. When minimizing the end time, there is a high accumulation of metabolites. This accumulation can affect cell viability negatively. The optimal control profiles in Fig. 8(a)-(d) clearly show a switching pattern, corresponding to the sequential activation of the pathways. According to [42] there is a mechanism leading to more pronounced transcriptional control of costly enzymes which can be explained by the trade-off between enzymatic cost minimization and time. Similarly to the

Case 1, it can be seen in Fig. 8(e)-(f) that the minimum threshold of 0.675 mM for S_5 is reached sooner when applying the approximation techniques for uncertainty propagation. However, this robustness with respect to the terminal constraint on S_5 comes together with an increase in enzymatic cost as shown in Table 6. To avoid a too high enzymatic cost that is harmful for cell viability, an upper bound on the enzymatic cost can be introduced and robustified.

Table 6 Case 2 - Overview of the number of states, CPU time, objective function values, terminal constraint values and their expected values and standard deviations for the different approximation techniques for uncertainty propagation when the enzymatic cost is minimized for $\alpha = 1.96$ for 3 uncertain parameters (K_m, λ and k_{cat})

	Nominal	Linearization	Sigma points	PCE1	PCE2
States	9	36	63	36	90
CPU time [s]	0.547	117.474	25.65	4.574	25.847
$\mathbf{E}[J]$	6.500	12.945	8.338	8.799	9.168
$\sqrt{\mathbf{Var}[J]}$	0	1.113	0.697	0.739	0.791
c_t	0.675	0.675	0.675	0.675	0.675
$\mathbf{E}[c_t]$	0.675	1.000	1.067	1.163	1.210
$\sqrt{\mathbf{Var}[c_t]}$	0	0.166	0.200	0.249	0.273

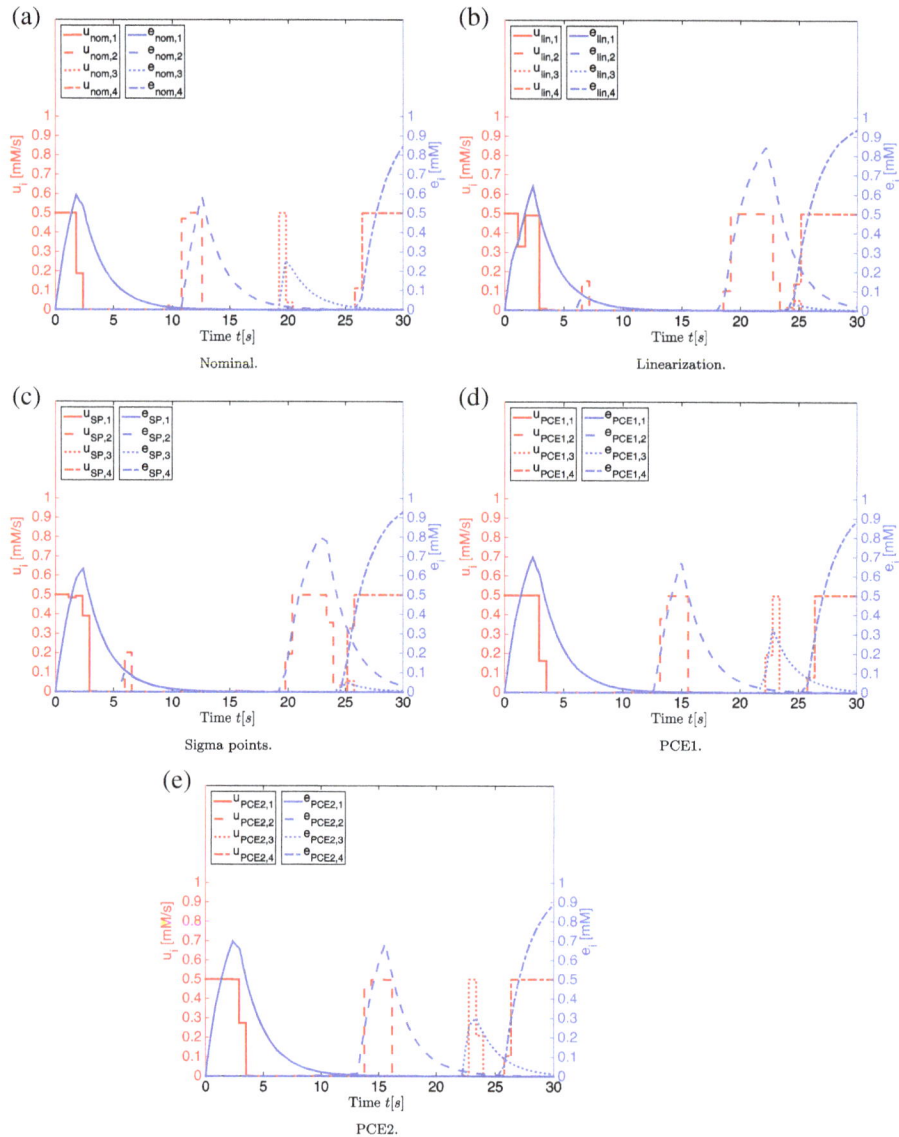

Fig. 7 Enzyme expression rates and enzyme concentration profiles Case 2. Enzyme expression rates together with the enzyme concentration profiles following from the minimization of the enzymatic cost for the nominal case (**a**) and with the different approximation techniques for uncertainty propagation: linearization (**b**), sigma points (**c**), PCE1 (**d**) and PCE2 (**e**) ($\alpha = 1.65$)

In Fig. 7(b) and (c) some remarkable observations are made for the linearization and sigma points approaches. After approximately 6 seconds a small expression rate u_2 is observed, leading to a short activation of e_2, producing intermediates S_3 and S_4. Furthermore after approximately 18 seconds in the linearization approach and 20 seconds in the sigma points approach, e_2 is expressed for a longer time. The accumulated S_2 is intensively consumed in this period for the production of intermediates S_3 and S_4. Eventually this will lead to a very low concentration of S_2 in comparison with the other approaches. Furthermore, a very small expression rate u_3 is observed in Fig. 7(b) and (c) for the optimization with the linearization and sigma points approaches, leading to a weak activation of e_3. This means that the branch in the glycolysis inspired network involving the conversion of S_3 to S_4 is practically inactive, leading to accumulation of S_3 for the linearization and sigma points approaches. This behavior is not observed in the profiles obtained with the other techniques and explains the higher objective function values for enzymatic cost in case of the linearization and sigma points approaches.

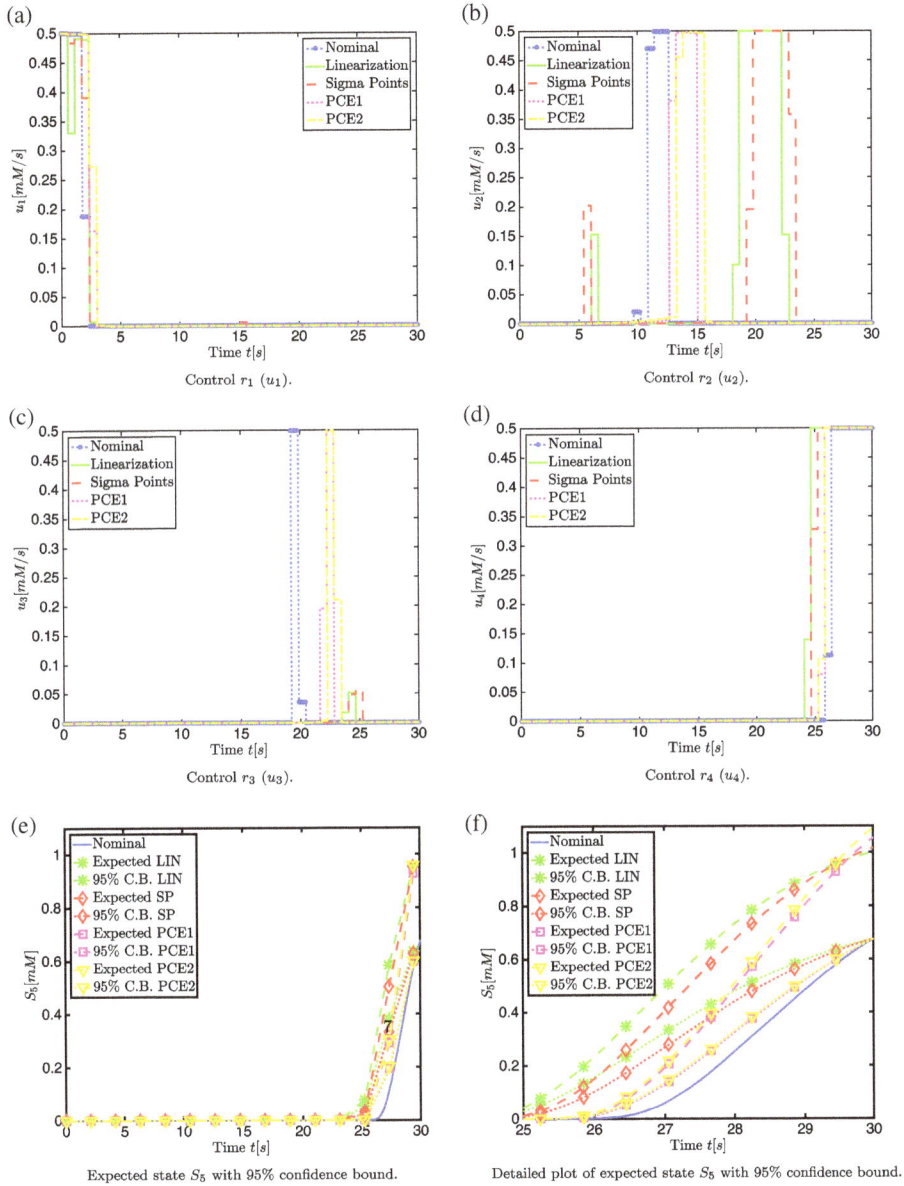

Fig. 8 Results for minimization of enzymatic cost in Case 2. Comparison of the control profiles r_1 (**a**), r_2 (**b**), r_3 (**c**) and r_4 (**d**) calculated with linearization, sigma points approach, PCE1 and PCE2 for $\alpha = 1.65$ with the nominal control profile and comparison of expected state S_5 and its 95 % confidence bound calculated with linearization, sigma points, PCE1 and PCE2 with the nominal case ($\alpha = 1.65$) ((**e**) and (**f**)) in case of 3 uncertain parameters K_M, λ and k_{cat}

Comparison of the approximation techniques for uncertainty propagation

The performance of the different approximation techniques for uncertainty propagation with respect to the constraint violations is investigated by performing a Monte Carlo simulation with 1000 noise realizations from a normal distribution. The number of constraint violations on S_5, mean values, and the variances for the objective function and terminal constraint, respectively are shown in Table 7 for the different approaches and

backoff parameter values α_i. More extensive results are presented in Additional file 4.

Expected value and 95 % confidence bound First the expected value and 95 % confidence bound of S_5 (based on $\alpha = 1.65$) are compared. This is done in Figs. 8(e)-(f). From Fig. 8(e)-(f) it can be seen that the expected state and 95 % confidence bounds for S_5 are similar for the PCE1 and PCE2 approaches. However, the linearization approach and sigma points approach take initially

Table 7 Case 2 - Results Monte Carlo simulations ($N = 1000$) in case of three normally distributed uncertain parameters (K_m, λ and k_{cat}) for robustified terminal constraint and objective function with the number of constraint violations, mean values and variances on the objective function and terminal constraint, respectively

	Nominal case	Linearization	Sigma points	PCE1	PCE2
$\alpha = 1.96$					
$\bar{J}$	6.5268	12.996	12.549	8.7702	9.1358
σ_J	0.57543	1.1893	1.15230	0.78540	0.82107
$\bar{c}_t$	0.69319	1.0106	1.0106	1.1406	1.2099
σ_{c_t}	0.18291	0.17462	0.17752	0.27117	0.28089
c_t violation	477 (47.7 %)	15 (1.5 %)	18 (1.8 %)	23 (2.3 %)	13 (1.3 %)
$\alpha = 1.65$					
$\bar{J}$	6.5268	10.04	9.6011	8.2477	8.5514
σ_J	0.57543	0.88430	0.83741	0.73621	0.76632
$\bar{c}_t$	0.69319	1.0143	1.0157	1.0339	1.0933
σ_{c_t}	0.18291	0.20620	0.21414	0.25176	0.26120
c_t violation	477 (47.7 %)	31 (3.1 %)	39 (3.9 %)	53 (5.3 %)	34 (3.4 %)

more distance from the nominal profile. In this case, similarly to Case1, a linear approximation technique (i.e., the linearization approach) is more conservative, implying less constraint violations. All 95 % confidence bounds are at 0.675 mM at the end as required by the imposed constraint in the implementation of the approximation techniques for uncertainty propagation.

Both PCE1 and PCE2 have a more accurate prediction of the expected value and variance of the objective function J and the terminal constraint function value c_t than the linearization and sigma points approach (in Table 6), when comparing with the empirically calculated expected values and variances with Monte Carlo simulations (in Table 7).

Constraint violations The percentage of constraint violations is investigated by performing a Monte Carlo simulation with 1000 noise realizations from the parametric uncertainty distribution. From these simulations, it is observed that all four methods reduce the amount of constraint violations significantly: from 47.7 % in the nominal case to even 1.3 % for PCE2, when a backoff parameter value of $\alpha = 1.96$ is chosen in case of a normal parametric uncertainty distribution. For this case study, the PCE2 method is superior in performance, when considering number of constraint violations.

Parametric uncertainty distribution For this case study, the integration of prior information on the parametric uncertainty distribution in the polynomial chaos expansion approaches is studied. The orthogonal polynomials for K_m, λ and k_{cat} are derived via the definition of orthogonal polynomials. Details can be found in Additional file 4.

For a normal and uniform parametric uncertainty distribution, a Monte Carlo simulation procedure with 1000 noise realizations has been followed and the results are summarized in Tables 7 and 8.

As in the first case study, the percentage of constraint violations is slightly higher when a uniform distribution is considered. For this case study, the difference in performance between a uniform parametric uncertainty distribution and a normal parametric uncertainty distribution for the polynomial chaos expansion, is small. Gathering information on the parametric uncertainty distribution from a parameter identification procedure is intensive and does not lead to a drastic improvement in performance for this case study.

Conclusions

In this work parametric uncertainty has been taken into account for prediction and control of biological networks. A critical comparison of three approximation techniques for uncertainty propagation, i.e., the linearization, sigma points and polynomial chaos expansion approaches has been made for dynamic optimization of biological networks under parametric uncertainty. The main advantage of the polynomial chaos expansion is its ability to tackle more easily non-normal parametric uncertainty distributions. Two case studies are investigated: *(i)* the minimization of intemediate metabolite accumulation in a basic three-step linear pathway model (with 3 metabolites, 3 fluxes, 4 differential states and 3 controls) and *(ii)* the multi-objective optimization (i.e., the minimization) of the final time and enzymatic cost a glycolysis inspired network model (with 4 metabolites, 4 fluxes, 8 differential states and 4 controls). For further analysis of

Table 8 Case 2 - Results Monte Carlo simulations ($N = 1000$) in case of three uniformly distributed uncertain parameters (K_m, λ and k_{cat}) for robustified terminal constraint and objective function with the number of constraint violations, mean values and variances on the objective function and terminal constraint, respectively

	Nominal case	Linearization	Sigma points	PCE1	PCE2	PCE2 uniform
$\alpha = 1.96$						
$\bar{J}$	6.5517	13.048	12.599	8.8045	9.1717	9.1498
σ_J	0.5323	1.1015	1.0665	0.7270	0.7600	0.7600
$\bar{c}_t$	0.7023	1.0205	1.0208	1.1555	1.2254	1.2240
σ_{c_t}	0.1744	0.1659	0.1687	0.2590	0.2684	0.2690
c_t violation	470 (47.0 %)	10 (1.0 %)	13 (1.3 %)	21 (2.1 %)	7 (0.7 %)	9 (0.9 %)
$\alpha = 1.65$						
$\bar{J}$	6.5517	10.079	9.6385	8.2798	8.5848	8.5683
σ_J	0.5323	0.8195	0.7762	0.6813	0.7092	0.7086
$\bar{c}_t$	0.7030	1.0259	1.0277	1.0476	1.1077	1.1056
σ_{c_t}	0.1744	0.1966	0.2042	0.2404	0.2495	0.2496
c_t violation	470 (47.0 %)	28 (2.8 %)	36 (3.6 %)	55 (5.5 %)	29 (2.9 %)	29 (2.9 %)

the robustness, emphasis was put on a single objective: in Case 1 the minimization of the intermediate accumulation and in Case 2 the minimization of the enzymatic cost. In a next step, the robustness of the optimal control profiles obtained with the different approximation techniques for uncertainty propagation is investigated. Monte Carlo simulations are used for the assessment of these control profiles. From the results for both case studies, the different uncertainty propagation strategies each offer a robust solution under parametric uncertainty. When making the trade-off between computation time and the robustness of the obtained profiles, the sigma points and polynomial chaos expansion strategies score the best. In both case studies the effect of taking a uniform probability distribution for the parametric uncertainty has been taken into account. However, the gain by taking a uniform probability distribution into account instead of a normal probability distribution is low. There is a reduction on the considered backoff and the variance of the objective functions and constraint. On the other hand, an upfront identification procedure for the parametric uncertainty distribution is time-consuming, expensive and does not offer a substantial advantage for the considered case studies in comparison with assuming a normal parametric uncertainty distribution. The linearization, sigma points and polynomial chaos expansion approaches offer a great potential for optimization and modeling under uncertainty in systems biology. The application of these approximation techniques for uncertainty propagation to large scale biological network models is the subject of future work. The integration of these approximation techniques for uncertainty propagation in an interactive tool for multi-objective dynamic optimization [43] is also part of the future work.

Additional files

Additional file 1: Detailed review on approximation techniques for uncertainty propagation. A detailed review of the approximation techniques for uncertainty propagation: linearization, sigma points and polynomial chaos expansion approach.

Additional file 2: Optimization methods and software. A more in depth description of the used optimization methods and software.

Additional file 3: Numerical results Case 1. A detailed overview of the numerical results for Case 1.

Additional file 4: Numerical results Case 2. A detailed overview of the numerical results for Case 2.

Acknowledgements
This work was supported by KU Leuven [PFV/10/002 Center-of-Excellence Optimization in Engineering (OPTEC), DT is supported by PDM grant 2015/134], Fonds Wetenschappelijk Onderzoek Vlaanderen [G.0930.13 and KAN2013 1.5.189.13] and the Belgian Science Policy Office (DYSCO) [IAP VII/19].

Funding
This research was supported by KU Leuven [PFV/10/002 Center-of-Excellence Optimization in Engineering (OPTEC), DT is supported by PDM grant 2015/134], Fonds Wetenschappelijk Onderzoek Vlaanderen [G.0930.13 and KAN2013 1.5.189.13] and the Belgian Science Policy Office (DYSCO) [IAP VII/19].

Authors' contributions
Authors jointly designed the study, PN and DT implemented the algorithms, performed the computations and wrote the manuscript. FL and JVI supervised the study and participated in writing the manuscript. All authors have read and approved the final manuscript.

Competing interests
The authors declare that they have no competing interests.

References

1. Vercammen D, Logist F, Van Impe J. Dynamic estimation of specific fluxes in metabolic networks using non-linear dynamic optimization. BMC Syst Biol. 2014;8(132):1–22.
2. Llaneras F, Picó J. Stoichiometric modelling of cell metabolism. J Biosci Bioeng. 2008;105(1):1–11.
3. Systems Metabolic Engineering In: Wittmann C, Lee SY, editors. Springer Science+Business Media. 1st ed. Netherlands: Springer; 2012. p. XII, 388.
4. Monk J, Nogales J, Palsson BO. Optimizing genome-scale network reconstructions. Nat Biotechnol. 2014;32(5):447–52.
5. Sendin JOH, Exler O, Banga JR. Multi-objective mixed integer strategy for the optimisation of biological networks. Syst Biol IET. 2010;4(3):236–48.
6. Higuera C, Villaverde AF, Banga JR, Ross J, Morán F. Multi-criteria optimization of regulation in metabolic networks. PLoS ONE. 2012;7(7): 41122.
7. de Hijas-Liste G, Klipp E, Balsa-Canto E, Banga J. Global dynamic optimization approach to predict activation in metabolic pathways. BMC Syst Biol. 2014;8(1):1.
8. Sendiín J-OH, Otero-Muras I, Alonso AA, Banga JR. Improved optimization methods for the multiobjective design of bioprocesses. Ind Eng Chem Res. 2006;45(25):8594–603.
9. Kaern M, Elston TC, Blake WJ, Collins JJ. Stochasticity in gene expression: from theories to phenotypes. Nat Rev Genet. 2005;6(6):451–64.
10. Telen D, Vallerio M, Cabianca L, Houska B, Van Impe J, Logist F. Approximate robust optimization of nonlinear systems under parametric uncertainty and process noise. J Process Control. 2015;33:140–54.
11. Srinivasan B, Bonvin D, Visser E, Palanki S. Dynamic optimization of batch processes: II, role of measurements in handling uncertainty. Comput Chem Eng. 2003;27(1):27–44.
12. Julier S, Uhlmann JK. A general method for approximating nonlinear transformations of probability distributions. Oxford, OX1 3PJ United Kingdom: Robotics Research Group, Department of Engineering Science, University of Oxford; 1996.
13. Wiener N. The homogeneous chaos. Am J Math. 1938;60:897–936.
14. Xiu D, Karniadakis GE. The wiener-askey polynomial chaos for stochastic differential equations. SIAM J Sci Comput. 2002;24:619–44.
15. Nagy ZK, Braatz RD. Open-loop and closed-loop robust optimal control of batch processes using distributional and worst-case analysis. J Process Control. 2004;14(4):411–22.
16. Logist F, Houska B, Diehl M, Van Impe J. Robust multi-objective optimal control of uncertain (bio)chemical processes. Chem Eng Sci. 2011;66: 4670–82.
17. Mesbah A, Streif S. A probabilistic approach to robust optimal experiment design with chance constraints. In: 9th IFAC Symposium on Advanced Control of Chemical Processes ADCHEM. Whistler, Canada: Elsevier; 2015. p. 100–5.
18. Houska B, Logist F, Diehl M, Van Impe J. Robust optimization of nonlinear dynamic systems with application to a jacketed tubular reactor. J Process Control. 2012;22:1152–60.
19. Mesbah A, Streif S, Findeisen R, Braatz RD. Stochastic nonlinear model predictive control with probabilistic constraints. In: American Control Conference (ACC). American Automatic Control Council; 2014. p. 2413–9.
20. Galvanin F, Barolo M, Bezzo F. A backoff strategy for model-based experiment design under parametric uncertainty. AIChE J. 2010;56(8): 2088–102.
21. Asprey SP, Macchietto S. Designing robust optimal dynamic experiments. J Process Control. 2002;12(4):545–56.
22. Kadam JV, Schlegel M, Srinivasan B, Bonvin D, Marquardt W. Dynamic optimization in the presence of uncertainty: From off-line nominal solution to measurement-based implementation. J Process Control. 2007;17(5):389–98.
23. Podmajersky M, Fikar M, Chachuat B. Measurement-based optimization of batch and repetitive processes using an integrated two-layer architecture. J Process Control. 2013;23(7):943–55.
24. Diwekar UM, Kalagnanam JR. Efficient sampling technique for optimization under uncertainty. AIChE J. 1997;43:440–7.
25. Webster MD, Tatang MA, McRae GJ. Application of the Probabilistic Collocation Method for an Uncertainty Analysis of a Simple Ocean Model. Report 4 (MIT Joint Program on the Science and Policy of Global Change). 1996:32. http://globalchange.mit.edu/files/document/MITJPSPGC_Rpt4. pdf.
26. Haimes Y, Lasdon L, Wismer D. On a bicriterion of the problems of integrated system identification and system optimization. IEEE Trans Syst Man Cybernet. 1971;SMC-1:296–7.
27. Das I, Dennis JE. A closer look at drawbacks of minimizing weighted sums of objectives for Pareto set generation in multicriteria optimization problems. Structural optimization. 1997;14(1):63–9.
28. Messac A, Ismail-Yahaya A, Mattson CA. The normalized constraint method for generating the pareto frontier. Struct Multidiscip Optim. 2003;25(2):86–98.
29. Marler T, Arora J. Survey of multi-objective optimization methods for engineering. Struct Multidiscip Optim. 2010;41:853–62.
30. Logist F, Van Impe J. Novel insights for multi-objective optimisation in engineering using normal boundary intersection and (enhanced) normalised normal constraint. Struct Multidiscip Optim. 2012;45(3): 417–31.
31. Das I, Dennis J. Normal-boundary intersection: a new method for generating the pareto surface in nonlinear multicriteria optimization problems. Siam J Optim. 1998;8(3):631–57.
32. Bhaskar V, Gupta S, Ray A. Applications of multi-objective optimization in chemical engineering. Rev Chem Eng. 2000;16:1–54.
33. Reyes-Sierra M, Coello CAC. Multi-objective particle swarm optimizers: a survey of the state-of-the-art. Int J Comput Intell Res. 2006;2(3):287–308.
34. Suman B, Kumar P. A survey of simulated annealing as a tool for single and multiobjective optimization. J Oper Res Soc. 2006;57:1143–60.
35. Logist F, Telen D, Houska B, Diehl M, Van Impe J. Multi-objective optimal control of dynamic bioprocesses using acado toolkit. Bioprocess Biosyst Eng. 2013;36:151–64.
36. Andersson J. A General-Purpose Software Framework for Dynamic Optimization. PhD thesis. Belgium: Arenberg Doctoral School, KU Leuven, Department of Electrical Engineering (ESAT/SCD) and Optimization in Engineering Center, Kasteelpark Arenberg 10, 3001-Heverlee; 2013.
37. Wächter A, Biegler LT. On the implementation of an interior-point filter line-search algorithm for large-scale nonlinear programming. Math Programm. 2006;106:25–57.
38. Bartl M, Li P, Schuster S. Modelling the optimal timing in metabolic pathway activationÛuse of pontryagin's maximum principle and role of the golden section. Biosystems. 2010;101(1):67–77.
39. Bartl M, Li P, Schuster S. Just-in-time activation of a glycolysis inspired metabolic network - solution with a dynamic optimization approach. In: Proceedings 55nd International Scientific Colloquium. Ilmenau, Germany: Institute for Automation and Systems Engineering Ilmenau University of Technology; 2010.
40. Klipp E, Heinrich R, Holzhütter HG. Prediction of temporal gene expression. Eur J Biochem. 2002;269(22):5406–13.
41. Zaslaver A, Mayo AE, Rosenberg R, Bashkin P, Sberro H, Tsalyuk M, Surette MG, Alon U. Just-in-time transcription program in metabolic pathways. Nat Genet. 2004;36(5):486–91.
42. Wessely F, Bartl M, Guthke R, Li P, Schuster S, Kaleta C. Optimal regulatory strategies for metabolic pathways in escherichia coli depending on protein costs. Mol Syst Biol. 2011;7(1):1–13.
43. Vallerio M, Hufkens J, Van Impe J, Logist F. An interactive decision-support system for multi-objective optimization of nonlinear dynamic processes with uncertainty. Expert Syst Appl. 2015;42:7710–31.

12

Snoopy's hybrid simulator: a tool to construct and simulate hybrid biological models

Mostafa Herajy[1†], Fei Liu[2*†], Christian Rohr[3] and Monika Heiner[3]

Abstract

Background: Hybrid simulation of (computational) biochemical reaction networks, which combines stochastic and deterministic dynamics, is an important direction to tackle future challenges due to complex and multi-scale models. Inherently hybrid computational models of biochemical networks entail two time scales: fast and slow. Therefore, it is intricate to efficiently and accurately analyse them using only either deterministic or stochastic simulation. However, there are only a few software tools that support such an approach. These tools are often limited with respect to the number as well as the functionalities of the provided hybrid simulation algorithms.

Results: We present Snoopy's hybrid simulator, an efficient hybrid simulation software which builds on Snoopy, a tool to construct and simulate Petri nets. Snoopy's hybrid simulator provides a wide range of state-of-the-art hybrid simulation algorithms. Using this tool, a computational model of biochemical networks can be constructed using a (coloured) hybrid Petri net's graphical notations, or imported from other compatible formats (e.g. SBML), and afterwards executed via dynamic or static hybrid simulation.

Conclusion: Snoopy's hybrid simulator is a platform-independent tool providing an accurate and efficient simulation of hybrid (biological) models. It can be downloaded free of charge as part of Snoopy from http://www-dssz.informatik.tu-cottbus.de/DSSZ/Software/Snoopy.

Keywords: Hybrid simulation, Hybrid Petri nets, Hybrid biological models, Snoopy

Background

In order to study the dynamics of biological models, a simulation procedure is usually employed to emulate reaction firings. A vector representing the state of a system serves to track the species concentrations and/or the corresponding number of molecules as the simulation advances with respect to time. The chosen simulation procedure determines how the system state vector is updated as well as the progression of the simulation time. There are various approaches to capture reaction firings as well as their effects on the system state. However, all available algorithms can be grouped into four categories: stochastic, approximate stochastic, deterministic, and hybrid simulation approaches [1, 2].

Stochastic simulation methods [2–4] consider reaction firings as a random process and each reaction is executed individually. Therefore, stochastic simulation is very accurate compared to approximate approaches (e.g., approximate stochastic methods and deterministic ones). However stochastic simulation algorithms (SSA) are often referred to as computationally inefficient as they may consume much runtime to accomplish the discrete and individual firing of reactions. They can be used to simulate models with a moderate amount of reactions that do not fire too frequently, since, increasing the number of reactions could at the same time increase the number of stochastic events. As an improvement of the exact stochastic simulation, approximate stochastic simulation algorithms [5] group and fire multiple reactions at every step. Thus, they can save considerable

*Correspondence: feiliu@scut.edu.cn
†Equal contributors
2School of Software Engineering, South China University of Technology, 510006 Guangzhou, People's Republic of China
Full list of author information is available at the end of the article

runtime. Nevertheless, they will still require rather expensive computations.

On the contrary, deterministic simulation offers a completely different approach by considering reaction firing as a deterministic process which approximates reaction firings by constructing a system of ordinary differential equations (ODEs) or by using other approximation techniques (e.g., see [6]). Although deterministic simulation is computationally efficient, the results are not accurate for all kinds of computational models of biochemical reaction networks [2]. For instance, deterministic simulation is not applicable for many experiments, where molecular fluctuations of species with a few number of molecules drive the overall model behaviour (for examples see [7, 8]).

As a combined approach, hybrid simulation [9–15] merges exact stochastic and approximate algorithms. Thus, it takes advantage of computational efficiency, while avoiding result inaccuracy. Hybrid simulation works by first partitioning the set of reactions into stochastic and deterministic ones and correspondingly classifying the set of species into discrete and continuous ones. Afterwards, a system of ODEs is constructed for the deterministic regime using kinetic rate laws as specified (e.g., mass action). The system of ODEs is numerically integrated until a stochastic reaction is to occur and then the stochastic reaction takes place. The whole procedure is repeated until the end of the simulation time is reached.

However, the implementation of hybrid simulation is not a straightforward task compared with the comparably simple stochastic simulation methods, since it requires the interplay and integration of an ODE solver in addition to the SSA. Hence, it becomes intricate to write a dedicated and efficient simulation code for each model. Therefore efficient hybrid simulation software tools are required to accelerate the model development and execution. Unlike stochastic simulation, there are only a few software tools that currently support hybrid simulation (see e.g., [16, 17]). Furthermore, the original hybrid simulation algorithm introduced in [9] is not efficient to simulate all kinds of models. For example, a high frequency of reaction events leads to a performance drop. Therefore, recent hybrid approaches employ more sophisticated techniques in order to achieve a better performance (see e.g., [6, 12, 13]). Besides, hybrid simulation tools should continuously evolve and support the state of the art of hybrid simulation approaches such that they can cope with the continuously growing interest in systems biology.

In this paper, we present Snoopy's hybrid simulator, an efficient and generic (i.e., it does not assume a special kind of biochemical network models) hybrid simulator that supports state-of-the-art hybrid simulation approaches. Snoopy's hybrid simulator is deployed as a component of the Petri net tool Snoopy [18] and its steering server [19]. The latter tool permits different simulation scenarios

than the one discussed in this paper (please see [20] for more details). Snoopy's hybrid simulator has been recently restructured to support recent advances in hybrid simulation algorithms. Moreover, it admits a graphical representation of biochemical reactions by means of Petri nets (see below), while complex models that exhibit repeated components can be easily constructed as coloured Petri nets [15]. Snoopy's hybrid simulator is a free software tool that can run on many well-known platforms including MS Window, MacOSX and some Linux distributions. A comprehensive user manual is available at [21].

Modelling biochemical networks via Petri nets. Petri nets, as a discrete modelling approach, have been widely applied in many fields, including systems biology [22, 23]. In Petri nets, tokens on places represent discrete quantities of species such as the number of molecules or levels of species concentration. To accommodate different modelling scenarios, Petri nets have been extended in many ways [23]. For instance, stochastic Petri nets ($\mathcal{SPN}$) [22] were proposed by associating each transition with an exponentially distributed waiting time, and continuous Petri nets ($\mathcal{CPN}$) have been introduced to support continuous markings (cf., [22, 24]). The underlying semantics of a $\mathcal{CPN}$ model is a system of ODEs. However, there are different $\mathcal{CPN}$ interpretations. We adopt a special semantics of $\mathcal{CPN}$ called bio-semantics (cf., [25]). In the bio-semantics, we assume that transition rate equations are defined in terms of kinetic rate laws (e.g., mass action) that are commonly used to model biochemical networks. This assumption will considerably simplify the $\mathcal{CPN}$ simulation and its implementation for this particular application.

Furthermore, in order to allow discrete and continuous entities to coexist in one model, different types of hybrid Petri nets ($\mathcal{HPN}$) were proposed for different purposes [24]. We employ a special class of $\mathcal{HPN}$ called generalised hybrid Petri nets ($\mathcal{GHPN}$) [11] which is specifically tailored to the modelling of biochemical reaction networks. $\mathcal{GHPN}$ offer two types of places and five types of transitions, which permit together the convenient modelling of various kinds of (biological) processes. A detailed description of $\mathcal{GHPN}$ can be found in [11, 21].

Figure 1 presents an introductory example of using $\mathcal{GHPN}$ to model biochemical reaction networks. We follow a simplified scenario of the calcium dynamics detailed in [26]. Intracellular calcium dynamics is a complex process which requires hybrid modelling where channel opening and closing are stochastic processes while calcium diffusion is more efficiently modelled as a deterministic process [26]. In this example we assume the existence of only one channel which permits the flow of calcium to the cytoplasm when it is in the open state. We use two discrete places, *open* and *close*, to model the channel states, open and close, respectively. Likewise, the two stochastic

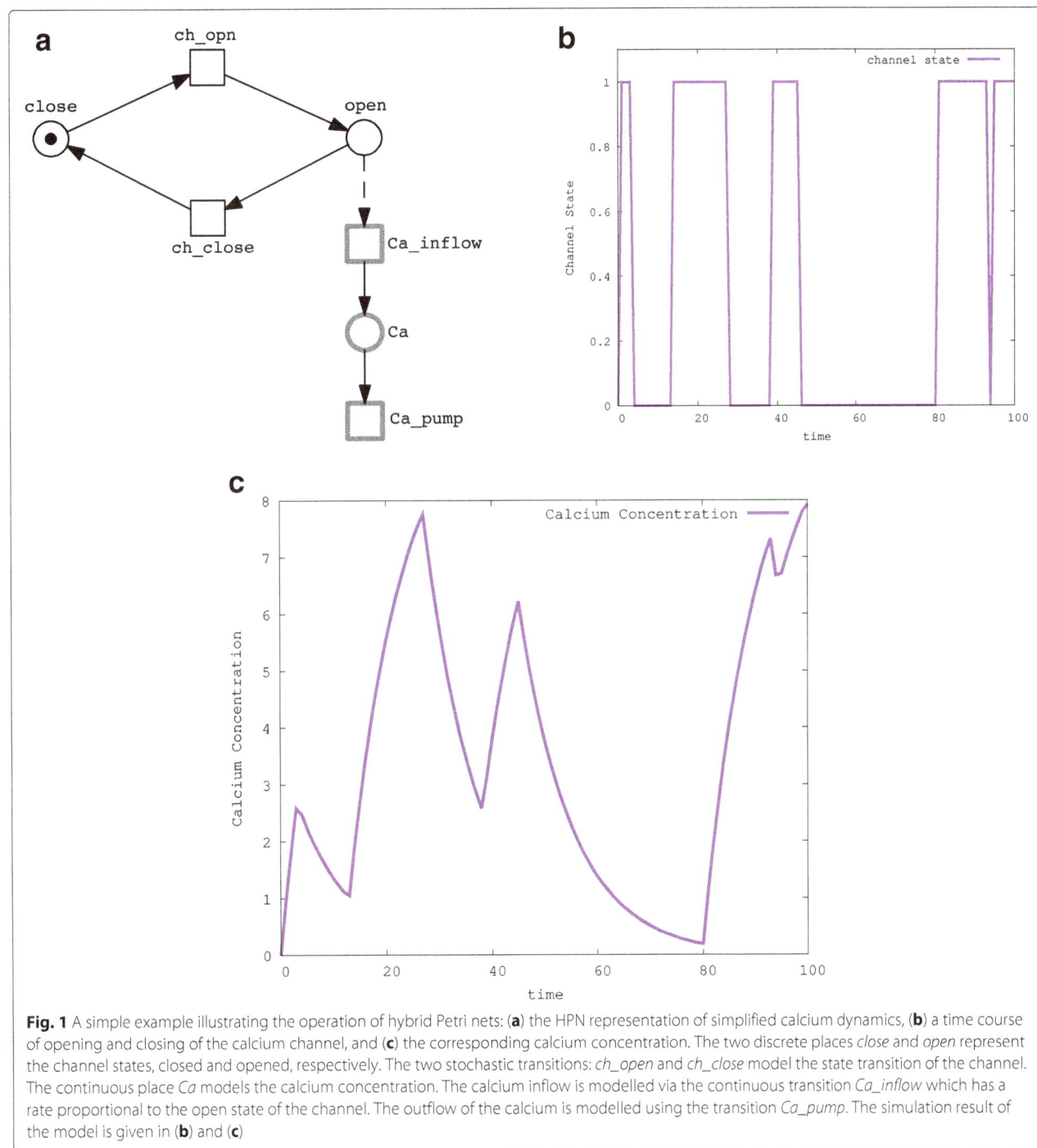

Fig. 1 A simple example illustrating the operation of hybrid Petri nets: (**a**) the HPN representation of simplified calcium dynamics, (**b**) a time course of opening and closing of the calcium channel, and (**c**) the corresponding calcium concentration. The two discrete places *close* and *open* represent the channel states, closed and opened, respectively. The two stochastic transitions: *ch_open* and *ch_close* model the state transition of the channel. The continuous place *Ca* models the calcium concentration. The calcium inflow is modelled via the continuous transition *Ca_inflow* which has a rate proportional to the open state of the channel. The outflow of the calcium is modelled using the transition *Ca_pump*. The simulation result of the model is given in (**b**) and (**c**)

transitions, *ch_open* and *ch_close*, model the processes of opening and closing the channel, respectively. When the channel is in the open state, the calcium can flow from the endoplasmic reticulum (not represented in this example) and enter the cytoplasm, which is represented by the continuous place *Ca*. The continuous transition *Ca_inflow* models this process. Finally, calcium can return back to the endoplasmic reticulum through a process called pump [26].

We model this process using the continuous transition *Ca_pump*. Figure 1b depicts the dynamics of channel opening and closing, while Fig. 1c provides the corresponding calcium concentrations. For the purpose of this example we have set the parameter values so that we can demonstrate the basic idea which has no immediate biological relevance. The corresponding Snoopy file is given in the Additional file 1: S1.

Beyond these extensions, Petri nets have also been extended in a parameterised way. Such an extension is called coloured Petri nets ($\mathcal{PN}^{\mathcal{C}}$) [27, 28]. In a $\mathcal{PN}^{\mathcal{C}}$, a group of similar components can be abstracted into one component (similar to a variable), each of which is defined as and thus distinguished by a colour (a specific value of the variable). In a $\mathcal{PN}^{\mathcal{C}}$, one or more colour sets have to be defined, and a colour set is assigned to each place. The tokens on a place are now distinguishable by colours. A guard, which is a Boolean expression, is assigned to each transition. For enabling a coloured transition, we not only check if the preplaces of the transition have sufficient and appropriate tokens, which is similar to what is done in standard Petri nets, but also have to evaluate the guard, which has to yield true. Each uncoloured Petri net

class can have a coloured counterpart. Thus by combining the parameterised modelling capability of $\mathcal{PN}^{\mathcal{C}}$ and the hybrid representation capability of $\mathcal{GHPN}$, we obtain coloured hybrid Petri nets ($\mathcal{GHPN}^{\mathcal{C}}$) [15], which can conveniently model a system having both multiple spatial and temporal scales. In what follows we refer to $\mathcal{GHPN}$ and $\mathcal{GHPN}^{\mathcal{C}}$ simply by $\mathcal{HPN}$ and $\mathcal{HPN}^{\mathcal{C}}$, respectively, unless explicitly stated otherwise.

To demonstrate the basic idea of $\mathcal{HPN}^{\mathcal{C}}$, we extend the example presented in Fig. 1 to include more than one channel arranged in one cluster and account for the spatial behaviour of calcium diffusion. This scenario will be much more realistic than the simple one presented in Fig. 1. Figure 2a shows a simple example of the calcium dynamics modelled as $\mathcal{HPN}^{\mathcal{C}}$. The corresponding

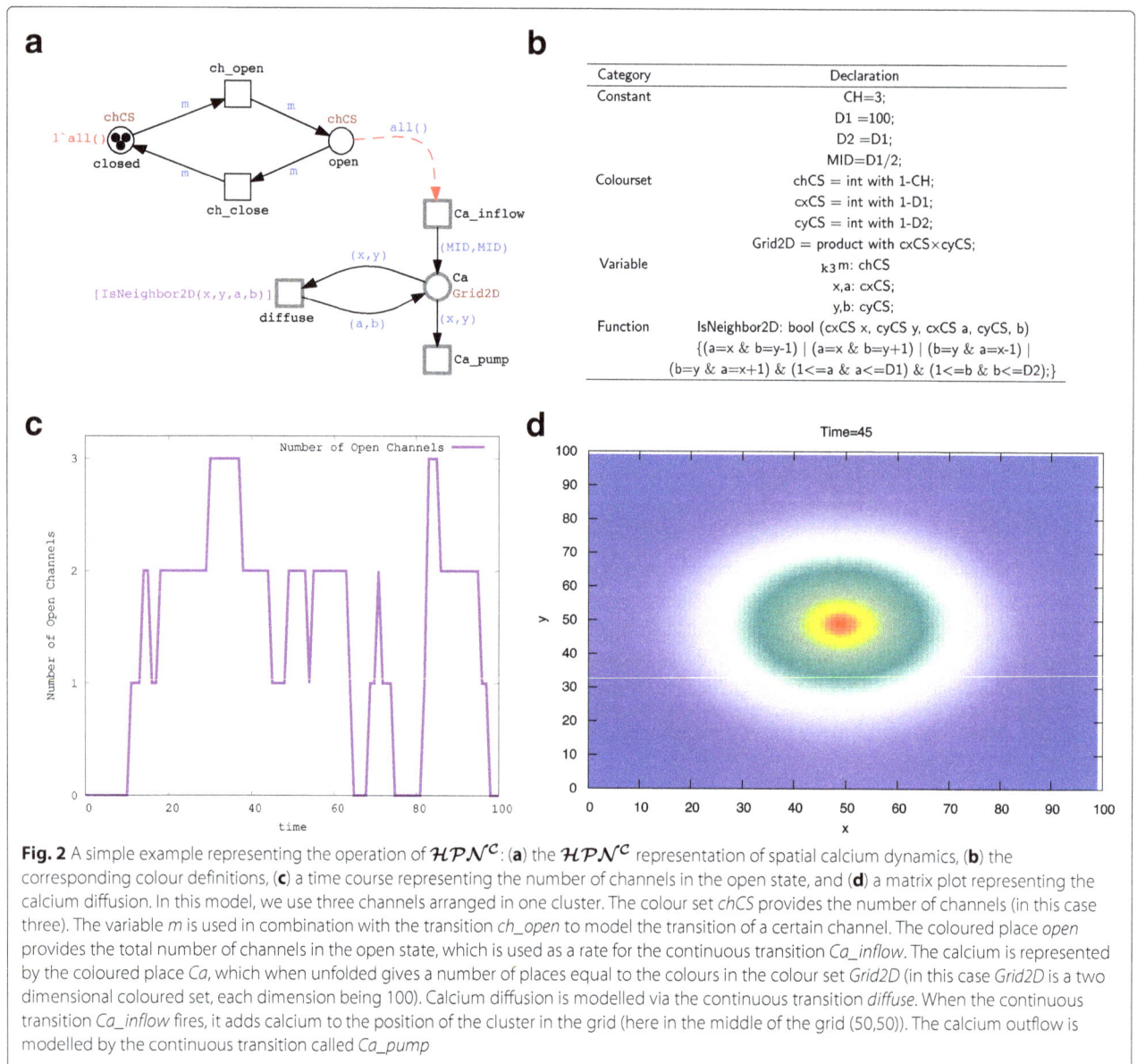

Fig. 2 A simple example representing the operation of $\mathcal{HPN}^{\mathcal{C}}$: (**a**) the $\mathcal{HPN}^{\mathcal{C}}$ representation of spatial calcium dynamics, (**b**) the corresponding colour definitions, (**c**) a time course representing the number of channels in the open state, and (**d**) a matrix plot representing the calcium diffusion. In this model, we use three channels arranged in one cluster. The colour set *chCS* provides the number of channels (in this case three). The variable *m* is used in combination with the transition *ch_open* to model the transition of a certain channel. The coloured place *open* provides the total number of channels in the open state, which is used as a rate for the continuous transition *Ca_inflow*. The calcium is represented by the coloured place *Ca*, which when unfolded gives a number of places equal to the colours in the colour set *Grid2D* (in this case *Grid2D* is a two dimensional coloured set, each dimension being 100). Calcium diffusion is modelled via the continuous transition *diffuse*. When the continuous transition *Ca_inflow* fires, it adds calcium to the position of the cluster in the grid (here in the middle of the grid (50,50)). The calcium outflow is modelled by the continuous transition called *Ca_pump*

colour declarations are given in Fig. 1b. Now the coloured discrete place *closed* is parameterised with the coloured set *chCS* which contains the colours from 1 to 3. Therefore it represents the state of three channels when they are closed, so does the coloured discrete place *opened*. In this coloured model version the two transitions *ch_open* and *ch_close* are bound with each colour in the colour set *chCS*. That is each transition has three different instances corresponding to the number of channels. Moreover, the calcium concentration is modelled by the continuous place *Ca*, which is associated with the colour set *Grid2D* to represent a two-dimensional grid of 100×100 cells (colours). Each of them represents a spatial calcium location. The calcium flow is modelled by the continuous transition *Ca_inflow* which adds calcium to the cluster location (here assumed to be in the middle of the grid: (50,50). The rate of the transition *Ca_inflow* is proportional to the total number of open channels in the cluster (see [26] for more details). The calcium diffusion is modelled by the continuous transition *diffuse* which diffuses the calcium to the four neighbouring cells of a calcium position. The calcium pump is done via the coloured continuous transition *Ca_pump* which positions a transition at each location in the grid. Figure 2c depicts the total number of channels in the open state, while Fig. 2d shows the calcium diffusion in the two dimension coordinates. In this example we can easily carry out different experiments by reconfiguring the model parameters. For instance, the number of channels in the cluster can be increased by just increasing the number of colours in the coloured set

chCS. Similarly, the grid coordinates can be adjusted by changing the colour set *Grid2D*. The Snoopy file for this introductory example is given in the Additional file 2: S2. A detailed discussion of simulating this coloured model is provided in the "Implementation" section.

Implementation

In this section we briefly describe the implementation of Snoopy's hybrid simulator by considering the architecture, available algorithms, export and import, and the deployed external libraries.

Architecture

Figure 3 presents the architecture of Snoopy's hybrid simulator. This architecture consists of three components: the user interface, which comprises the model editor and the simulation dialog; the simulator, which implements the simulation algorithms as well as storing the currently running models and the corresponding result views; and the Snoopy manager, which connects the user interface with the simulation module. Snoopy's hybrid simulator deploys a simple graphical user interface to permit a rapid configuration of the core simulation procedure. Figure 4 depicts the user interface; in the following, we discuss each of these components.

Model editor

The model editor permits the graphical construction of hybrid models using (coloured) hybrid Petri net notations defined in [11]. Reactions are represented by transitions,

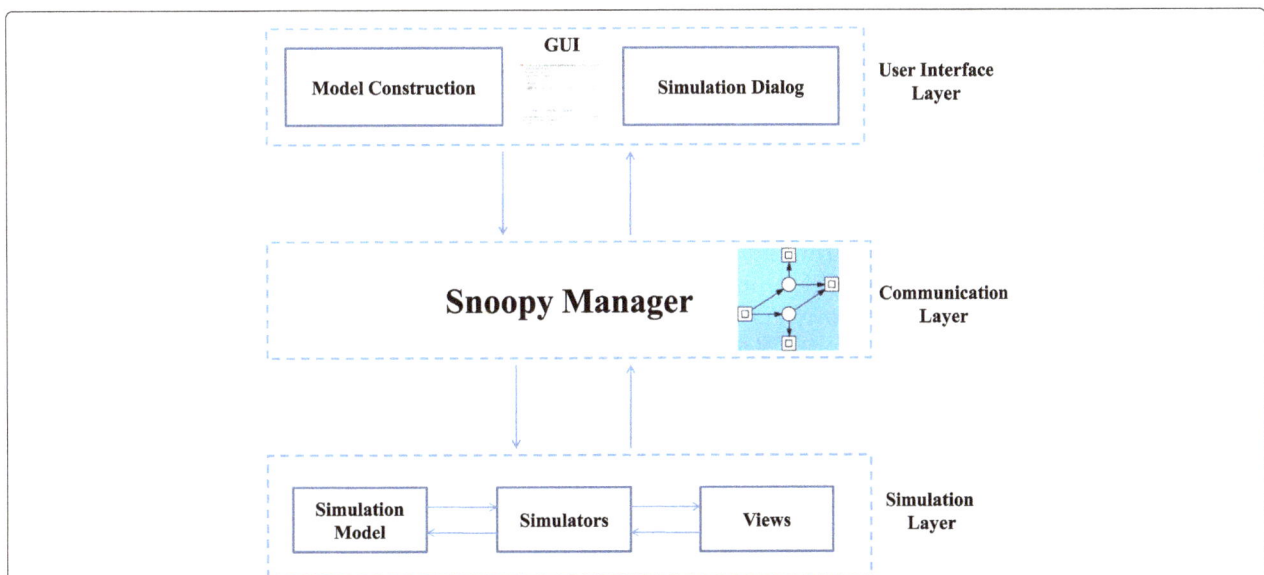

Fig. 3 Architecture of Snoopy's hybrid simulator. The different components of the architecture can be divided into three layers: user interface, communication, and simulation layer. The user interface is the user access point to construct and execute hybrid models. The simulation layer comprises the simulator as well as the simulation version of the constructed model definition. The communication layer connects the simulation component with the user interface

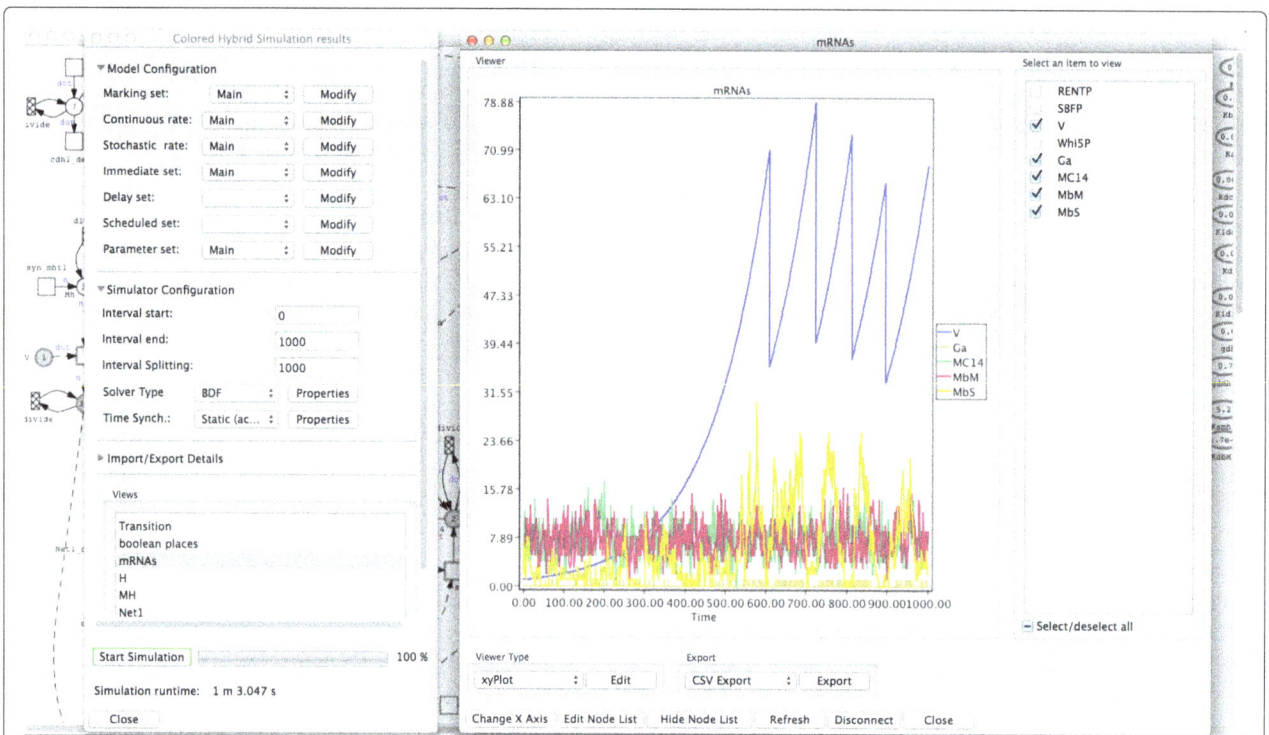

Fig. 4 Screenshot of Snoopy's graphical user interface. The simulation window is divided into two parts: configuration and viewer subwindows. The configuration window (*left*) permits to configure and control the simulation while the viewer window (*right*) is used to display the simulation results. Multiple viewer windows can be opened simultaneously to show the results from different perspectives. The viewer curves can be exported into CSV file for further analysis of the simulation results

while species are denoted by places. More information about hybrid Petri net notations can be found in [11, 19, 29] as well as in Snoopy's hybrid simulator user manual [21]. In addition to specifying all reactions, the model editor provides other features to configure the model parameters as well as the initial state. The model editor is applied as a pre-step before executing the simulation.

Simulation dialog

The simulation dialog is the user tool to run and manage the simulation. Through it the simulation experiment can be configured and then executed. Moreover, the simulation dialog provides access to the simulation algorithms that are implemented in Snoopy's hybrid simulator. Once a model has been constructed, a user can access the simulation dialog through Snoopy's menu bar. The simulation dialog consists of four parts: model configuration, simulator configuration, import/export, and simulation state (compare Fig. 4).

The model configuration section permits to adjust model settings including initial state, reaction rates and other similar parameters. The simulation configuration section deals with specifying the simulation options including the start and end time point of the simulation,

the type of the ODE solver, and the hybrid synchronisation method. The import/export section allows users to configure how Snoopy performs any export or import of simulation results. Finally, the simulation state section serves to start and stop the simulation as well as to monitor the simulation state.

The simulation results can be examined using views. Different result views can be defined to explore the simulation output from different perspectives. Each view has its own window to display the results using a dedicated result viewer. Result viewers permit to render the final data using different plotting techniques such as xy-plot. Finally, view curves can be exported into comma separated files (CSV) for further processing.

Simulation model

After a model is constructed, e.g. using the model editor, it can be sent to the simulator for execution. The simulation module takes a copy of the hybrid Petri net model, but ignoring the layout information. Usually, the model is a collection of species, reactions, stoichiometries as well as associated data such as kinetic rates and kinetic rate constants. Nevertheless, this information is mapped in terms of Petri net components. When the simulation model is partitioned into deterministic and stochastic parts, reactions and species are assigned to either of the two regimes. Unlike other implementations (e.g., in [10]), we consider only one version of the species: either discrete or continuous. Later, if a transformation is required (e.g., from number of molecules to concentration, or vice versa), this can be easily done at the position where such a conversion is required (e.g., in the rate equation). Similarly, when considering the simulation output, such a transformation can also be easily applied. However, if a species is manipulated by both a deterministic and a stochastic reaction, it needs to be represented as a continuous species.

Simulator

The simulation algorithms are implemented as a standalone, but built-in simulation library. The simulator module reads the model to be simulated and carries out the execution. A number of algorithms, which will be discussed in the next section, are available to execute a hybrid model. Please note, although the core simulator is implemented as a stand-alone library, it can currently not be used as a stand-alone application. However, we are working on this to achieve a stand-alone application (see the "Future improvements" section).

Views

Views are associated with models. Each view is defined over a set of places or transitions of which the dynamic behaviour shall be displayed when the simulation starts. These place/transition sets can be specified by a regular expression. Multiple views can be defined for the same model. A view is also associated with a viewer that displays the selected information. Views can be manipulated or removed after they were initially added to a model.

Snoopy manager

The communication between the user interface and the simulator is done via the Snoopy manager. The Snoopy manager acts as an intermediate agent that sends the GUI command to the simulator and gets the result back to visualise or export them to the chosen file format. As Snoopy is a stand-alone application, the communication between the user interface and the simulation module is done internally and not through a physical communication channel.

Available algorithms

Snoopy's hybrid simulator encompasses a set of simulation algorithms that together provide a convenient execution of hybrid biological models. The general idea of the hybrid simulation algorithms implemented in Snoopy is as follow. First, the synchronisation module (the hybrid algorithm) prepares the jump equation (see below). Afterwards, the ODE solver numerically integrates the system of ODEs due to the deterministic part until the

jump equation is fulfilled. At this point, the synchronisation module switches the control to the stochastic module to select and fire a stochastic reaction. The exact time point of the stochastic event is determined by the jump equation. In what follows, we outline each of these algorithms.

Haseltine and Rawlings algorithm

This is the realisation of the hybrid simulation idea proposed by Haseltine and Rawlings in [9]. According to this method, a system of ODEs is numerically solved until a stochastic event is to occur. The exact occurrence time of the stochastic event is captured through (1).

$$\int_t^{t+\tau} \sum_{j=0}^{N} a_j^s(\mathbf{x})dt = -log(r),\qquad(1)$$

where $\mathbf{x}$ is the state vector of the model at time t, τ is the firing time of the next slow reaction, r is a random number uniformly distributed in the interval $U(0,1)$, $a_j^s(\mathbf{x})$ is the propensity of the jth slow reaction, and N is the number of stochastic reactions.

In (1) we aim to determine the value of τ. This is achieved by first generating a random number from $U(0,1)$ and then integrating the propensity equations of all slow reactions together with the system of ODEs due to the deterministic part from the current simulation time t until (1) is satisfied. At this point we know that there is a stochastic event which needs to be fired.

Although this method is very accurate, it requires considerable time to switch from stochastic to deterministic simulation [13]. The performance of this method drops rapidly as soon as the number of stochastic events increases. Thus it is suitable only for simple models where the number of potential stochastic events is limited. Moreover, it can produce better results with ODE solvers that do not collect and use history information to advance the numerical integration time.

Accelerated Hybrid Simulation

To overcome the limitation of the Haseltine and Rawlings method, we follow an accelerated approach introduced in [13]. The accelerated algorithm takes advantage of the model structure to boost the overall simulation performance. According to this method, stochastic reactions are classified into two groups: dependent and independent. Dependent reactions affect the system state of the ODE solvers when they occur, while independent reactions have no effects. Therefore, the ODE solver is reinitialised only when a reaction in the dependent group is fired. Thus, the simulation performance becomes better than for the Haseltine and Rawlings method, particularly for bigger models. For instance, in [13] we compared the performance of the Haseltine & Rawlings method and the accelerated approach using three models. We found that there is a notable performance improvement for all three case studies and for certain models; the latter approach is ten times faster than the former one. This save in runtime is mainly due to the reduction of the number of times where the ODE solver is reinitialised. In order to achieve the better performance, the accelerated method approximates the exact capture of the stochastic event occurrence time given by (1) by another equation given in (2).

$$\sum_{j=0}^{N} a_j^s(\mathbf{x}) \cdot \Delta\tau = -log(r), \tag{2}$$

where $\Delta\tau$ is the time difference between the occurrence time of the previous event and the current event. Eqs. 1 or (2) has to be satisfied during the integration, when the Haseltine and Rawlings or accelerated method is used, respectively, until the ODE solver stops and returns the control back to the stochastic regime. Please note that although our approach mainly intends to reduce the reinitialisation of ODE solvers employing history information to advance the simulation time (e.g., multi-step ODE solver; see below), it can also be used with single step solvers (e.g., Runge-Kutta) to reduce the frequent recalculation of the step size after each firing of a discrete event.

Improved hybrid rejection-based stochastic simulation

The numerical integration of (1) as well as its approximation in (2) are computationally expensive to be satisfied. Therefore, in [12] a new hybrid simulation method was proposed based on the rejection-based stochastic simulation algorithm (RSSA) introduced in [30] which avoids the calculation of (1) and (2). The RSSA algorithm defines lower and upper bounds of the reaction propensities to minimise the propensity updates. The propensity lower and upper bounds are calculated based on a lower and upper bound of the system state values called fluctuation interval. Propensities are updated only when one or more of the system state entries move completely outside the defined fluctuation interval. The Hybrid Rejection-based Stochastic Simulation Algorithm (HRSSA) exploits this opportunity by switching from the deterministic to the stochastic regime only when the ODE solver reaches the time of a stochastic event or when any of the system state entries is outside the fluctuation interval. In the former case, the discrete regime does not affect the continuous one, while in the latter case the deterministic regime changes the state of the discrete species during the numerical integration. We apply an improved implementation of this method which combines the accelerated and hybrid rejection-based methods. Currently, the improved hybrid rejection stochastic simulation method is tested as the best hybrid algorithm implemented in our tool in terms of

performance (compare Table 2). In [12], the performance of the HRSSA algorithm has been compared with state of the art hybrid simulation algorithms using five benchmark models. It turned out that the HRSSA outperforms all competing algorithms.

Dynamic hybrid simulation

The previously discussed simulation approaches are based on static partitioning. Static partitioning adopts a predefined classification of the model reactions into stochastic and deterministic ones. The partitioning itself is usually performed by the user and exploited afterwards by the simulator during the whole simulation process. This approach is constructive for many applications with a clear cut between reactions which have to be simulated stochastically and those which should be simulated deterministically. For instance, in [31] and in many other similar publications that study cell fate, reactions related to the cell nucleus are considered as stochastic, while those happening inside the cytoplasm are considered as deterministic. However, such a clear cut is not always possible to be achieved for all models during the whole simulation period. Reactions can change their state from slow to fast and vice versa during the simulation. For example, in oscillating biological systems, reaction rates also oscillate with respect to time from fast to slow and the other way around. In this case, dynamic partitioning, where reactions are partitioned repeatedly during the simulation, can play a role in speeding up the whole simulation procedure. Our implementation of the dynamic hybrid simulation is based on the improved hybrid rejection method. Using this approach, reactions are repartitioned as soon as any of the state vector entries leaves the fluctuation interval. This will indeed eliminate the need for frequent checks of whether the set of reactions requires repartitioning.

Pure stochastic and pure deterministic simulation

To improve the comfort when simulating biological models with Snoopy's hybrid simulator, the user has the option to perform a pure stochastic or a pure deterministic simulation of a hybrid model. The direct method [3] is applied to implement the stochastic simulation, while the SUNDIAL CVODE [32] is used to carry out the deterministic simulation. This is a worthwhile feature during the experimentation phase to compare the hybrid results with the pure stochastic and pure deterministic ones. Using this feature, Snoopy's hybrid simulator ignores any reaction partitioning specified by the user and reads all model reactions as stochastic or deterministic ones, depending on the selected simulation algorithm.

Parallel multi-run simulation

Similar to stochastic simulation, hybrid simulation of biological models might require the execution of multiple runs to calculate average statistics. Snoopy's hybrid simulator permits the concurrent execution of different runs to take advantage of the existence of multiple cores in the user's machine.

Simulation of coloured models

A coloured model (such as $\mathcal{HPN}^{\mathcal{C}}$) with finite colour sets can be automatically unfolded to an uncoloured model (such as $\mathcal{HPN}$). See [33] for one of the unfolding algorithms deployed in Snoopy. Thus, the simulation of an $\mathcal{HPN}^{\mathcal{C}}$ model is done on the automatically unfolded $\mathcal{HPN}$ model. When the user starts the simulation of an $\mathcal{HPN}^{\mathcal{C}}$ model, an unfolding dialogue will be triggered, where the user can select an appropriate unfolding method to perform the unfolding. Afterwards, the simulation methods discussed in this section can be used to execute the unfolded model.

To better explain this idea, we consider the $\mathcal{HPN}^{\mathcal{C}}$ model presented in Fig. 2. First, the discrete subnet, consisting of the two places: *closed* and *open* as well as the two transitions: *ch_open* and *ch_close*, is unfolded into three identical subnets (because the colour set *chCS* consists of three colours). In other modelling scenarios where the unfolded subnets are not identical we can make use of transition guards to imply the required constraints. Similarly, the continuous subnet, consisting of the place *Ca* and the two transitions *Ca_pump* and *diffuse* is unfolded into 10,000 identical subsets (because the colour set *Grid2D* consists of 100×100 colours). However, because the transition *diffuse* has a guard expression, only transitions for the colours satisfying this Boolean expression are added. The transition *Ca_inflow* will have only one copy in the unfolded net because the input and output arcs contain a constant expression. Nevertheless, this procedure does not need to be implemented iteratively as we do in this small example. Instead, it can be viewed as a constraint satisfaction problem (CSP) which can be solved by a dedicated CSP solver (e.g. [34]).

Export and import

Snoopy supports the import/export of Petri net models from/to other tools and formats. First, Snoopy imports and exports the (C)ANDL format (cf. [21]) which is a human readable file format used by other software tools (e.g., Marcie [35] and Charlie [36]) which can be employed for a formal analysis of Petri net models (e.g., structure analysis, model checking, etc.). Moreover, Snoopy reads and writes SBML files [37] according to SBML level 2 version 4 by using libSBML [38]. However, we support only a subset of SBML elements that is compatible to our net classes; specifically we do not support any kind of rules or events. Snoopy passes all tests of the SBML Test Suite comprising supported elements. However, the partitioning of hybrid models is lost when exporting to SBML,

because SBML has no support for hybrid models yet. The user has to decide whether the exported model has to be treated stochastically or continuously. Furthermore, a coloured model is exported to SBML by first unfolding it into the low level representation and then performing the export.

Implementation language and external libraries

Snoopy's Hybrid Simulator has been implemented using standard C++. As a component of Snoopy, it adopts wxWidget [39] to implement the graphical user interface under different operating systems. Moreover, the stochastic and deterministic simulation components are implemented in a modular way such that different algorithms can be easily exchanged to execute the stochastic and deterministic regimes. Snoopy's hybrid simulator adopts internally an external library, SUNDIAL CVODE [32], to solve a system of ODEs. The ODE library provides two main algorithms: one for stiff and one for non-stiff ODEs. Additional ODE solver modules can easily be added in future releases. We also make use of the C++ library Boost [40] to carry out routine tasks such as input parsing and multithreading support.

Results

Snoopy's hybrid simulator provides a graphical and convenient way to construct hybrid models

Before using Snoopy's hybrid simulator, a model needs to be constructed by specifying reactions, species, stoichiometries, kinetic rates, etc. In Snoopy a model is usually constructed using Petri net notations. However, existing models can also be imported from other formats including the well known SBML. In what follows, we present two methods that permit the construction of hybrid models in Snoopy.

Simple models

For simple models which involve a limited number of reactions and species (e.g., both less than 100), we use $\mathcal{HPN}$ to construct them. Snoopy's hybrid simulator supports two types of places, five types of transitions and six arc types to facilitate the convenient modelling of hybrid biological systems. A complete description of these elements is provided in the user manual [21]. Unlike other hybrid Petri net tools and similar to the semantics discussed in [25], we apply the bio-semantics to execute the continuous part of the $\mathcal{HPN}$ (see also the "Background" section), which is more efficient than the adaptive semantics when simulating biological models.

Coloured models

For large-scale biological systems, the corresponding $\mathcal{HPN}$ models become difficult to manage. In this case, we may use $\mathcal{HPN}^{\mathcal{C}}$ (see the "Background" section) for model construction. These models may exhibit many repeated components as well as spatial behaviour (see Fig. 2 for an example). Colours have been successfully deployed to model many real biological applications (for examples see [28, 41, 42]).

Snoopy's hybrid simulator provides an efficient way to execute hybrid models

Once a model has been created, it can be simulated using one of the algorithms discussed in the previous section. The simulation dialog has been designed to be intuitive with many options to configure the simulation. Furthermore, the resulting time course data can either be viewed inside Snoopy or exported for further processing. In the following we summarise the required steps to execute a hybrid model. A more detailed discussion can be found in [21].

1. **Configure the constructed model:** Before running the simulation, you may need to adjust the model setting (see the model configuration section in Fig. 2), which includes choosing the initial state and/or the kinetic rate constants.

2. **Configure the chosen simulator:** To run a hybrid simulator you have to select an appropriate synchronisation algorithm, which is one of the discussed hybrid simulation methods, as well as the type of the ODE solver. Moreover, depending on the specific model, a user might need to adjust the options of the ODE solver. For many hybrid models, the default settings can be kept. However, Snoopy's hybrid simulator offers a wide range of other options for complex hybrid models that require special treatment (see the simulation configuration section in Fig. 2).

3. **Run the simulation and explore the results:** After the model and the simulator are configured, the simulation can be started. You may need to create a new result view to explore the simulation output.

4. **Export the simulation results:** As a final step, the resulting data can be exported to a CSV file for further post-processing.

Example of using Snoopy's hybrid simulator

To demonstrate how Snoopy's hybrid simulator can be used to deal with hybrid biological models, we include a sample application and show how this model can be constructed using $\mathcal{HPN}$ notations and then executed via a hybrid simulation algorithm.

When studying cell fate behaviour, where a cell decides either to undergo cell cycle arrest or commit apoptosis [43–45] in response to DNA damage, a model describing this phenomenon can be clearly divided into two parts: one with species exhibiting low number of molecules

and the other one involving species with high number of molecules. In such cases, it may not be feasible to apply stochastic simulation due to the huge number of stochastic events.

We deploy a recent model from [31] as a sample use case for illustrating Snoopy's hybrid simulator. This model permits to investigate the importance of various DDR (DNA Damage Response) elements after DNA damage induction during cell fate determination. More specifically, the model studies the ATM/p53/NF-κB pathway, consisting of four main modules: p53 (a tumor suppressor protein), ATM (ataxia telangiectasia mutated), NF-κB (a nuclear transcriptional factor) and Wip1 (a p53-induced protein phosphatase), and involves three different compartments: nucleus, cytoplasm, and extracellular matrix [31]. The main aim of the model is to explore the connection between these four key proteins and protein phosphates in order to understand cellular response to DNA damage which is important to understand cell fate determination. A key model component is Wip1, which is increased to a level that can block the corresponding cell apoptotic decision when DNA repair is successful [31]. However, the level of Wip1 should not stay high after DNA repair; otherwise the cell will not be sensitive to future damage. The model is divided into the following stochastic and deterministic parts. All the genes such as Wip and ATM are considered as discrete places, and all reactions related to genes (gene expression and degradation) are kept stochastic, while all the other species are considered as continuous places, and all other reactions, except those related to DSB (DNA double-strand breaks) creation and repair, are modelled as deterministic transitions. The output of the model consist of the levels of molecules with respect to time after irradiation and also the cell fate decision. Figure 5 depicts the Snoopy implementation of this model.

The accelerated hybrid simulation algorithm [13] is chosen to execute this model due to the weak coupling between the two reaction regimes. Figure 6 gives a screenshot of simulating the model using Snoopy's hybrid simulator which includes the time course behaviour of two versions of the gene Checkpoint kinase 2 (CHK2): inactive (CHK2_n,) and active (CHK2_pn,), in addition to the negative regulator protein of the p53 (denoted by MDM2), and the nuclear version of Wip1 (Wip1_n). The simulator output in Fig. 6 can also be exported to a CSV file for further processing. For example, in this model it may be required to count the number of cells that undergo apoptosis and those which exhibit cell cycle arrest. A threshold of the concentration of species p53, P21 (representing the p21 protein) and Bax (denoting the Bax protein) can be used to extract this information [31]. Such post-processing can be done by help of the exported simulation traces. Moreover, a performance comparison of three simulation algorithms when executing this model is

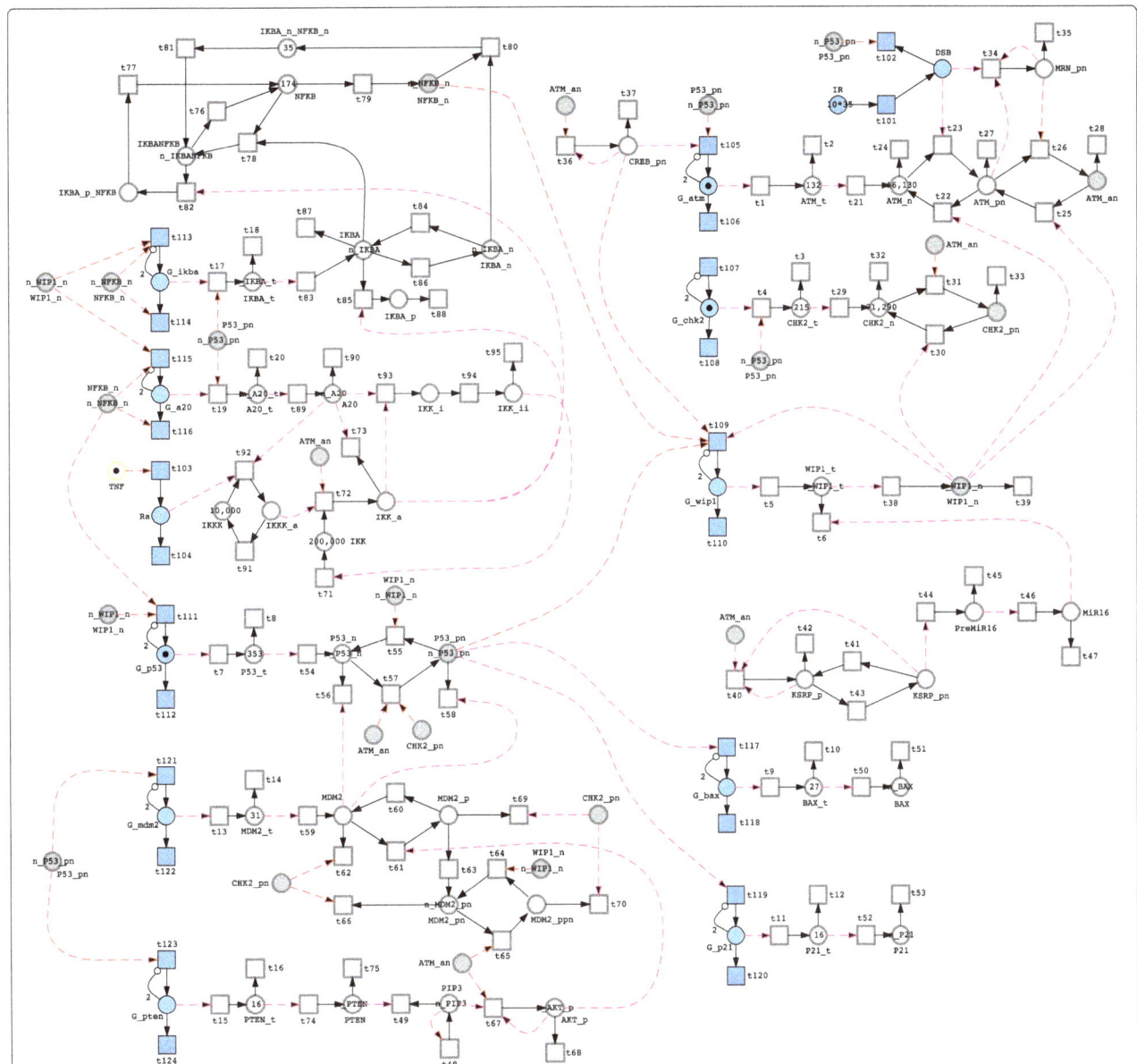

Fig. 5 Implementing the ATM/P53/NF-κB pathway model from [31] using Snoopy's hybrid simulator. *Circles* (places) represent species, *squares* (transitions) represent reactions, and arcs denote connections between the two node types. More information about these notations can be found in [11, 21]. *Coloured circles* represent discrete species, while the uncoloured ones represent continuous species. Similarly, *coloured squares* represent stochastic reactions, while uncoloured ones denote continuous reactions. *Solid black arcs* represent connections that consume molecules when the corresponding reaction fires, while *dashed coloured* ones just permit the use of substrates for defining reaction rates. *Grey nodes* are logical places that are repeated to simplify the network layout. Please note, inhibitor arcs (arc with *small circles*) enforce the number of genes to be at most two [31]. The complete Snoopy file is provided in the Additional file 2: S2.

provided in Table 2. The Snoopy file is included in the Additional file 3: S3, while a short description of how to execute the model is given in the Additional file 4: S4.

Discussion

Installation

Snoopy's hybrid simulation is installed as part of Snoopy. The Snoopy installation package can be run just by one click on a local computer with one of the three

well-known operating systems. No additional dependencies do exist. In other words, all dependent packages are installed with Snoopy's main package. A detailed procedure of how to install Snoopy on these platforms is given in its user manual.

Comparison with other tools

In this section we compare Snoopy's hybrid simulator with two of the well-known software tools that provide a hybrid

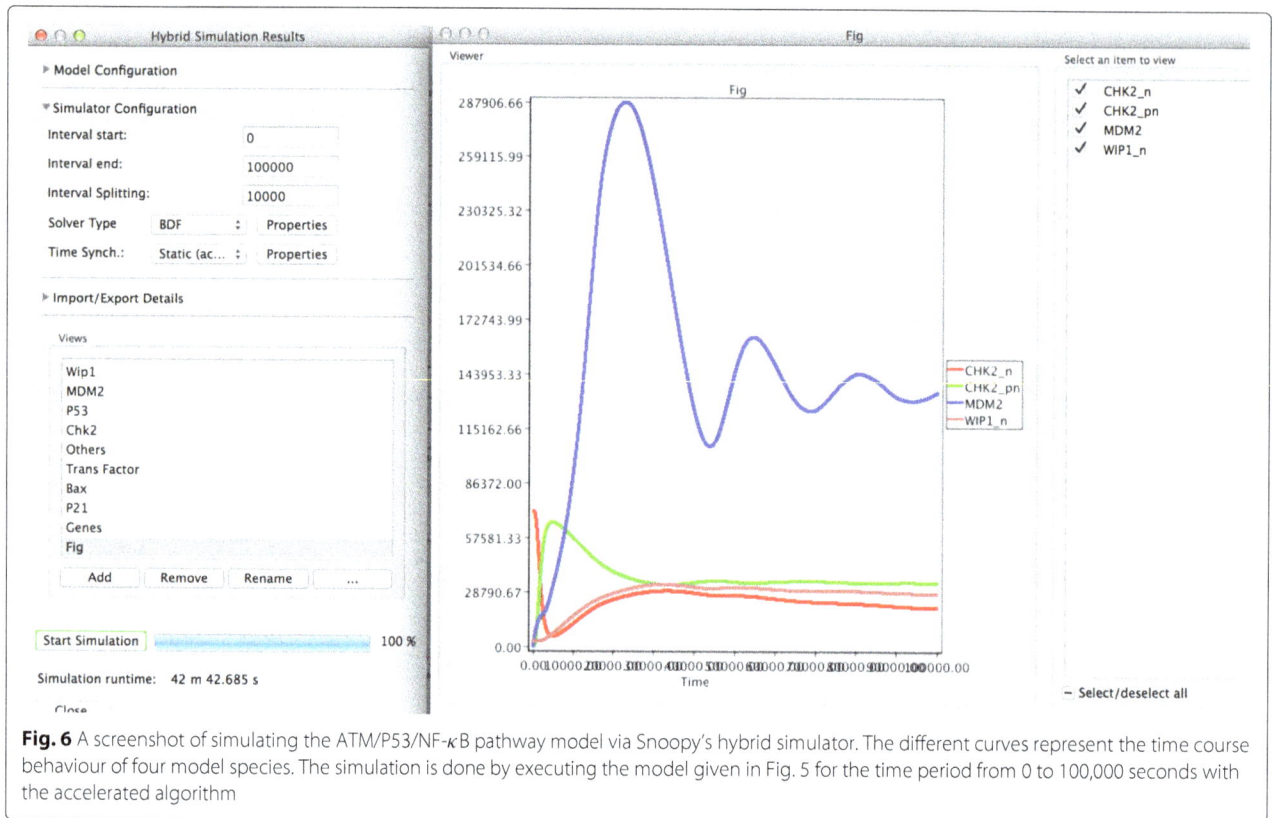

Fig. 6 A screenshot of simulating the ATM/P53/NF-κB pathway model via Snoopy's hybrid simulator. The different curves represent the time course behaviour of four model species. The simulation is done by executing the model given in Fig. 5 for the time period from 0 to 100,000 seconds with the accelerated algorithm

simulation module to systems biologists: COPASI [16] and Virtual Cell (VCell) [17]. Table 1 summarises the features of COPASI, Virtual Cell, and Snoopy's hybrid simulator with respect to the hybrid simulation procedures supported by the three tools.

COPASI [16] is a general-purpose software tool for constructing and executing computational biological models. It provides tables and widgets as user interface to specify compartments, reactions, species and other related parameters. It reads and writes models written in SBML.

For hybrid simulation, COPASI adopts a version similar to the Haseltine and Rawlings method, which has been independently developed at the same time [1, 16]. However, it deploys a tight coupling of the SSA and a specific ODE solver. To be precise, COPASI offers an hybrid Runge-Kutta/SSA algorithm, combining the classical Runge-Kutta ODE solver with the SSA algorithm, an LSODA/SSA, combining LSODA – a dynamic switching between stiff/nonstiff solvers – with the SSA algorithm, and recently it has been extended to support

Table 1 Comparison of Snoopy's hybrid simulator with two other similar tools

Features\Tools	Snoopy's hybrid simulator	COPASI [16]	VCell [17]
Use of graphical notations to specify model reactions	Yes	No	Yes
Use of a parameterised language to manage larger models	Yes	No	No
Support unstiff ODE solvers	Yes	Yes	Yes
Support stiff ODE solvers	Yes	Yes	Yes
Improving simulation performance by analysing the model structure	Yes	No	No
Interplay of stochastic and deterministic modules	Variable	Fixed	Fixed
Exact hybrid simulation	Yes	Yes	No
Platform-independent	Yes	Yes	Yes
Availability	Free	Free	Free

HybridRK-45. COPASI is platform-independent and is available free of charge.

The Virtual Cell [17] modelling and analysis tool also provides a module to execute hybrid models. Virtual Cell is deployed as a distributed application that can be downloaded free of charge. It uses the BioModel as well as VCell Markup Language to construct cell models. Three hybrid algorithms are supported: Hybrid (Gibson/Euler–Maruyama), Hybrid (Gibson/Milstein), and Hybrid (Adaptive Gibson/ Milstein).

Compared with COPASI and Virtual Cell, Snoopy's hybrid simulator offers a set of features that can improve the performance as well as the productivity of constructing and executing hybrid biological models. These include: analysing reaction networks to accelerate the simulation, implementing a modular design of the stochastic and deterministic procedures, implementing the state of the art of hybrid simulation algorithms, deploying accurate and efficient hybrid simulation algorithms, and utilising a parameterised language (coloured hybrid Petri nets) to construct large scale biological models.

On the one hand, Snoopy's hybrid simulator makes use of the structural information of the underlying reaction network to boost the overall simulation performance. For instance, the accelerated hybrid simulation algorithm, presented in [13], increases the performance of some hybrid models by ten times compared to the classical Haseltine and Rawlings method as it has been asserted in [13]. This improvement in the runtime is mainly due to detection of reaction dependencies between the deterministic and stochastic regime. In other words, Snoopy's hybrid simulator avoids unnecessary re-initialisations of the ODE solver when the system state of the ODE solver is not affected by the firing of the current discrete event.

On the other hand, Snoopy's hybrid simulator does not assume a fixed combination of the ODE solvers and the SSA algorithms as in COPASI and Virtual Cell. Instead, a user can select the appropriate type of the ODE solver, and the hybrid simulation algorithm acts as a time synchronisation module. Such modular design facilitates the support of new ODE solvers and SSA algorithms in the future with minimal efforts. Moreover, the user can take advantage of this modular design by selecting a different combination of the stochastic solver, ODE solver, and the hybrid time synchronisation procedure. This feature can be beneficial to address the issue that different models may have their own computational demands.

Furthermore, Snoopy's hybrid simulator implements the state of the art of hybrid simulation algorithms that have a better performance than the classical ones. For example, Snoopy's hybrid simulator implements the hybrid rejection-based stochastic simulation algorithm which has been recently introduced in [12]; it represents a promising direction to improve hybrid simulation when dynamic and static partitioning strategies are used.

To improve the simulation performance, the previous hybrid simulation algorithm implemented in COPASI does not include time-varying propensities in the slow subsystem [1] (i.e., there is no check for (1) or any similar exact methods, e.g., [12]). Although this approach can improve the simulation performance, it will affect the result accuracy. Recently, a new hybrid module (HybridRK-45,) has been added to COPASI to improve the simulation accuracy and overcome this limitation. On the contrary, Snoopy's hybrid simulator implements three exact versions of the algorithm in [9]. Moreover, recent advances of the theory of hybrid simulation (e.g,. in [12] and [13]) render it possible to overcome the computational overhead to check (1) or an alternative as (2).

Finally, unlike COPASI and Virtual Cell, Snoopy's hybrid simulator deploys a special parameterised language, coloured hybrid Petri nets, to deal with larger models which cannot be easily managed using traditional model construction approaches (see Fig. 2 for an example).

Compared with the different simulation approaches discussed in [1], Snoopy's hybrid simulator mainly supports three hybrid algorithms that consider time-varying propensities. That is the changes in the propensities of slow reactions, while the deterministically simulated reactions are evolving, are exactly captured using (1), (2), or using the approach introduced in [12]. The biochemical reaction networks can either be partitioned by the user (i.e., the net is drawn by the user as stochastic and deterministic subnets), or it can be partitioned online by Snoopy's hybrid simulator. In the latter case the reaction propensities as well as the number of molecules in the reaction substrates serve as criteria to carry out the partitioning.

Performance measures

To evaluate the performance of the three main algorithms implemented in Snoopy's hybrid simulator, we give the runtimes of four case studies as performance measures. The case studies range from simple to complex ones that involve many species and reactions. These include: a T7 phage model [8], a hybrid model of the eukaryotic cell cycle [29] based on the stochastic one in [7], the ATM/p53/NF-κB model [31] which has been discussed in the "Results" section, and the simple hybrid calcium model provided in Fig. 2. The simulation experiments have been conducted on a Mac Pro. with 3 GHz Core i7 processor and 8GB memory.

Table 2 summarises the number of species and reactions as well as the runtime of each example model when they are simulated using each of the three hybrid algorithms. For the T7 model we use the partitioning scheme discussed in [11], while for the eukaryotic cell cycle model

Table 2 Performance measures of the three implemented algorithms in Snoopy's hybrid simulator

Measures\Models	Model Information			Runtimes of Simulation Algorithms (sec.)		
	Species (Discrete)	Reactions (Stochastic)	Stochastic Events	Haseltine& Rawlings*	Accelerated Method	Improved HRSSA
T7 Phage (1,000 runs)	3 (2)	6 (4)	4,238,978	120.2	41.7 (281%)	26.6 (452%)
ATM/p53/NF-κB	62 (14)	119 (24)	3,726	6.3	1.9 (332%)	0.96 (656%)
Cell Cycle Model	26 (14)	51 (20)	762,612	870.4	625.7 (139%)	365.5 (238%)
Calcium Model (Fig. 2)	10,006 (6)	49,607 (6)	29	215.8	220.7 (-1.1%)	190 (1.13%)

*This refers to the algorithm version which exactly accounts for Eq. (1)

we apply the same partitioning as discussed in [29]. For the purpose of performance comparison we provide the number of stochastic events produced by each model. The percentage numbers given in parentheses represent the speed of the accelerated and improved algorithms compared to the Haseltine and Rawlings algorithm, and negative values mean the latter algorithm is faster.

In the Haseltine and Rawlings algorithm, the runtime required to simulate a model mainly depends on the number of corresponding stochastic events. This fact is illustrated by the four case studies. For instance, although the T7 phage model consists of only six reactions (four of them are simulated stochastically), it takes a considerable runtime compared to the calcium model where ten of thousands of reactions are involved. This is not the case in the accelerated and improved HRSSA algorithms, since not all of the stochastic reactions affect the deterministic solver. For instance, in the ATM/p53/NF-κB model, there is a substantial gain in terms of the runtimes because only very few stochastic events trigger a reinitialisation of the ODE solver (compare Fig. 5). Moreover, the runtime for the calcium model is comparable for all three algorithms because there are only a few stochastic events and the optimisation by the accelerated and improved HRSSA algorithms does not play a role.

Although the number of stochastic events in the cell cycle model is less than those in the T7 phage model, the latter model takes less runtime. The extra runtime is taken by the ODE solver, since the numerical integration of the cell cycle model exhibits more discontinuities, due to the volume division (cf. [7, 29]) than the T7 phage model where only two reactions are simulated deterministically.

The accuracy of the simulation results is the same for the three simulation algorithms since the core idea has not been changed. The accelerated and improved HRSSA approaches avoid the reinitialisation of the deterministic module for stochastic events which do not have an effect on the deterministic solver. This will not influence the simulation accuracy (please see [12, 13] for more details).

Future improvements
The development of Snoopy and its hybrid simulator is still active and new features and algorithms can be added

in the future to further enrich its simulation capabilities. We will continue to investigate how to further improve the performance of the hybrid simulation by exploiting the model structure. Moreover, we intend to support additional ODE solvers and other stochastic simulation algorithms to execute the semantics of different types of models. Currently, the simulation library depends on Snoopy's graphical user interface for reading a model. As a future extension of this scenario, we intend to create a command line application that reads SBML files or the Petri net file and simulates them directly using the simulation library. We will also continue to incorporate recent hybrid algorithms to Snoopy's hybrid simulator. The mailing list snoopy@informatik.tu-cottbus.de is dedicated to potential queries and bugs about Snoopy and its components.

Conclusions
In this paper we have presented Snoopy's hybrid simulator, a tool to execute hybrid biological models. Snoopy's hybrid simulator has been developed over the last five years, and reached recently a mature and reliable state. It employs a variety of hybrid simulation algorithms such that it can deal with various types of biological models that are usually encountered in systems biology. In addition to the simulation capabilities, the model can take advantage of the graphical representation via hybrid Petri nets notations when it is constructed and simulated via Snoopy's hybrid simulator.

Additional files

Additional file 1: An example $\mathcal{HPN}$ model. A Snoopy file implementing the calcium dynamics using $\mathcal{HPN}$ notations.

Additional file 2: An example $\mathcal{HPN}^{\mathcal{C}}$ model. A Snoopy file implementing the calcium spatial dynamics using $\mathcal{HPN}^{\mathcal{C}}$ notations.

Additional file 3: the ATM/p53/NF-κB HPN model. A Snoopy file implementing the ATM/p53/NF-κB.

Additional file 4: Description of the ATM/p53/NF-kB HPN model. A short description of how to open and simulate the Snoopy file of the ATM/p53/NF-κB HPN model.

Abbreviations

DSB: DNA double-strand breaks; $\mathcal{GHPN}$: Generalised hybrid Petri nets; $\mathcal{HPN}$: Hybrid Petri nets; $\mathcal{HPN}^{C}$: Coloured hybrid Petri nets; ODE: Ordinary differential equation; SBML: Systems biology markup language; SSA: Stochastic simulation algorithm

Acknowledgements

None.

Funding

This work was partially financed by National Natural Science Foundation of China (61273226).

Authors' contributions

MH proposed the idea. MoH (Mostafa Herajy), FL, and CR carried out the implementation. MH tested the implementation. MoH and FL wrote the draft. MH and CR read and improved the draft. All the authors read and approved the final version.

Competing interests

The authors declare that they have no competing interests.

Author details

[1]Department of Mathematics and Computer Science, Faculty of Science, Port Said University, 42521 Port Said, Egypt. [2]School of Software Engineering, South China University of Technology, 510006 Guangzhou, People's Republic of China. [3]Computer Science Institute, Brandenburg University of Technology, 10 13 44 Cottbus, Germany.

References

1. Pahle J. Biochemical simulations: stochastic, approximate stochastic and hybrid approaches. Brief Bioinform. 2009;10(1):53. doi:10.1093/bib/bbn050.
2. Gillespie D. Stochastic simulation of chemical kinetics,. Annu Rev Phys Chem. 2007;58(1):35–55. doi:10.1146/annurev.physchem.58.032806.104637.
3. Gillespie DT. A general method for numerically simulating the stochastic time evolution of coupled chemical reactions. J Comput Phys. 1976;22(4):403–34. doi:10.1016/0021-9991(76)90041-3.
4. Gillespie D. Exact stochastic simulation of coupled chemical reactions. J Phys Chem. 1977;81(25):2340–361. doi:10.1021/j100540a008.
5. Cao Y, Gillespie D, Petzold L. Adaptive explicit-implicit tau-leaping method with automatic tau selection. J Chem Phys. 2007;126(22):224101. doi:10.1063/1.2745299.
6. Duncan A, Erban R, Zygalakis K. Hybrid framework for the simulation of stochastic chemical kinetics. J Comput Phys. 2016;326:398–419. doi:10.1016/j.jcp.2016.08.034.
7. Kar S, Baumann WT, Paul MR, Tyson JJ. Exploring the roles of noise in the eukaryotic cell cycle. Proc Natl Acad Sci U S A. 2009;106(16):6471–476. doi:10.1073/pnas.0810034106.
8. Srivastava R, You L, Summers J, Yin J. Stochastic vs. deterministic modeling of intracellular viral kinetics. J theor Biol. 2002;218(3):309–21. doi:10.1006/jtbi.2002.3078.
9. Haseltine E, Rawlings J. Approximate simulation of coupled fast and slow reactions for stochastic chemical kinetics. J Chem Phys. 2002;117(15):6959–969. doi:10.1063/1.1505860.
10. Kiehl T, Mattheyses R, Simmons M. Hybrid Simul Cell Behav. Bioinformatics. 2004;20:316–22. doi:10.1093/bioinformatics/btg409.
11. Herajy M, Heiner M. Hybrid representation and simulation of stiff biochemical networks. J Nonlinear Anal Hybrid Syst. 2012;6(4):942–59. doi:10.1016/j.nahs.2012.05.004.
12. Marchetti L, Priami C, Thanh VH. HRSSA - efficient hybrid stochastic simulation for spatially homogeneous biochemical reaction networks. J Comput Phys. 2016;317:301–17. doi:10.1016/j.jcp.2016.04.056.
13. Herajy M, Heiner M. In: Cinquemani E, Donzé A, editors. Accelerated Simulation of Hybrid Biological Models with Quasi-Disjoint Deterministic and Stochastic Subnets. Cham: Springer; 2016, pp. 20–38. doi:10.1007/978-3-319-47151-8_2.
14. Herajy M, Heiner M. Modeling and simulation of multi-scale environmental systems with generalized hybrid Petri nets. Front Environ Sci. 2015;3:53. doi:10.3389/fenvs.2015.00053.
15. Herajy M, Liu F, Rohr C. Coloured hybrid Petri nets for systems biology. In: Proc. of the 5th International Workshop on Biological Processes & Petri Nets (BioPPN), Satellite Event of PETRI NETS 2014. CEUR Workshop Proceedings, vol. 1159. Tunisia: CEUR-WS.org; 2014. p. 60–76.
16. Hoops S, Sahle S, Gauges R, Lee C, Pahle J, Simus N, Singhal M, Xu L, Mendes P, Kummer U. Copasi—a complex pathway simulator. Bioinformatics. 2006;22(24):3067–074. doi:10.1093/bioinformatics/btl485.
17. Resasco DC, Gao F, Morgan F, Novak IL, Schaff JC, Slepchenko BM. Virtual cell: computational tools for modeling in cell biology. Wiley Interdiscip Rev Syst Biol Med. 2012;4(2):129–40. doi:10.1002/wsbm.165.
18. Heiner M, Herajy M, Liu F, Rohr C, Schwarick M. In: Haddad S, Pomello L, editors. Snoopy – A Unifying Petri Net Tool. Berlin: Springer; 2012, pp. 398–407. doi:10.1007/978-3-642-31131-4_22.
19. Herajy M, Heiner M. In: Ciardo G, Kindler E, editors. A Steering Server for Collaborative Simulation of Quantitative Petri Nets. Cham: Springer; 2014, pp. 374–84. doi:10.1007/978-3-319-07734-5_21.
20. Herajy M, Heiner M. Petri net-based collaborative simulation and steering of biochemical reaction networks. Fundamenta Informatica. 2014;129(1-2):49–67. doi:10.3233/FI-2014-960.
21. Herajy M, Liu F, Rohr C, Heiner M. (Coloured) Hybrid Petri Nets in Snoopy - User Manual. Technical Report 01-17. Brandenburg University of Technology Cottbus, Department of Computer Science; 2017. https://opus4.kobv.de/opus4-btu/files/4157/csr_01-17.pdf.
22. Heiner M, Gilbert D, Donaldson R. In: Bernardo M, Degano P, Zavattaro G, editors. Petri Nets for Systems and Synthetic Biology. Berlin, Heidelberg: Springer; 2008, pp. 215–64. doi:10.1007/978-3-540-68894-5_7.
23. Liu F, Heiner M. In: Chen M, Hofestädt R, editors. Petri Nets for Modeling and Analyzing Biochemical Reaction Networks. Berlin, Heidelberg: Springer; 2014, pp. 245–72. doi:10.1007/978-3-642-41281-3_9.
24. David R, Alla H. Discrete, Continuous, and Hybrid Petri Nets. Springer Berlin Heidelberg: Springer; 2010.
25. Gilbert D, Heiner M. From Petri Nets to Differential Equations – An Integrative Approach for Biochemical Network Analysis In: Donatelli S, Thiagarajan PS, editors. Petri Nets and Other Models of Concurrency - ICATPN 2006: 27th International Conference on Applications and Theory of Petri Nets and Other Models of Concurrency, Turku, Finland, June 26-30, 2006. Proceedings. Berlin, Heidelberg: Springer; 2006. p. 181–200. doi:10.1007/11767589_11.
26. Rüdiger S, Shuai JW, Huisinga W, Nagaiah C, Warnecke G, Parker I, Falcke M. Hybrid stochastic and deterministic simulations of calcium blips. Biophys J. 2007;93(6):1847–1857. doi:10.1529/biophysj.106.099879.
27. Jensen K. Coloured Petri nets and the invariant-method. Theor Comput Sci. 1981;14(3):317–36. doi:10.1016/0304-3975(81)90049-9.
28. Liu F. Colored Petri nets for systems biology. PhD thesis: Brandenburg University of Technology Cottbus; 2012.
29. Herajy M, Schwarick M, Heiner M. Transactions on Petri Nets and Other Models of Concurrency VIII. Hybrid Petri Nets for Modelling the Eukaryotic Cell Cycle. Berlin: Springer; 2013, pp. 123–41. doi:10.1007/978-3-642-40465-8_7.
30. Thanh VH, Zunino R, Priami C. On the rejection-based algorithm for simulation and analysis of large-scale reaction networks. J Chem Phys. 2015;142(24):. doi:10.1063/1.4922923.

31. Jonak K, Kurpas M, Szoltysek K, Janus P, Abramowicz A, Puszynski K. A novel mathematical model of ATM/p53/NF-κB pathways points to the importance of the DDR switch-off mechanisms. BMC Syst Biol. 2016;10(1): 75. doi:10.1186/s12918-016-0293-0.

32. Hindmarsh A, Brown P, Grant K, Lee S, Serban R, Shumaker D, Woodward C. Sundials: Suite of nonlinear and differential/algebraic equation solvers. ACM Trans Math Softw. 2005;31:363–96. doi:10.1145/1089014.1089020.

33. Liu F, Heiner M, Yang M. An efficient method for unfolding colored petri nets In: Laroque C, Himmelspach J, Pasupathy R, Rose O, Uhrmacher AM, editors. Proceedings of the 2012 Winter Simulation Conference (WSC 2012). 978-1-4673-4781-5/12. Berlin: IEEE; 2012. p. 3358–369. doi:10.1109/WSC.2012.6465203.

34. Tack G. Constraint propagation - models, techniques, implementation. phdthesis. Germany: Saarland University; 2009. http://www.gecode.org/paper.html?id=Tack:PhD:2009.

35. Heiner M, Rohr C, Schwarick M. MARCIE – Model Checking and Reachability Analysis Done Efficiently In: Colom J-M, Desel J, editors. Application and Theory of Petri Nets and Concurrency: 34th International Conference, PETRI NETS 2013, Milan, Italy, June 24-28, 2013. Proceedings. Berlin, Heidelberg: Springer; 2013. p. 389–99. doi:10.1007/978-3-642-38697-8_21.

36. Heiner M, Schwarick M, Wegener JT. Charlie – An Extensible Petri Net Analysis Tool In: Devillers R, Valmari A, editors. Application and Theory of Petri Nets and Concurrency: 36th International Conference, PETRI NETS 2015, Brussels, Belgium, June 21-26, 2015, Proceedings. Cham: Springer; 2015. p. 200–11. doi:10.1007/978-3-319-19488-2_10.

37. Keating SM, Le Novère N. In: Schneider MV, editor. Supporting SBML as a Model Exchange Format in Software Applications. Totowa: Humana Press; 2013, pp. 201–25. doi:10.1007/978-1-62703-450-0_11.

38. Bornstein BJ, Keating SM, Jouraku A, Hucka M. LibSBML: an API Library for SBML. Bioinformatics. 2008;24(6):880. doi:10.1093/bioinformatics/btn051.

39. wxWidgets website. http://www.wxwidgets.org/. Accessed: 8/3/2017.

40. Boost website. http://www.boost.org/. Accessed: 8/3/2017.

41. Liu F, Heiner M. Multiscale modelling of coupled Ca^{2+} channels using coloured stochastic Petri nets. IET Syst Biol. 2013;7(4):106–13. doi:10.1049/iet-syb.2012.0017.

42. Gao Q, Gilbert D, Heiner M, Liu F, Maccagnola D, Tree D. Multiscale modelling and analysis of planar cell polarity in the Drosophila wing. IEEE/ACM Trans Comput Biol Bioinforma. 2013;10(2):337–51. doi:10.1109/TCBB.2012.101.

43. Zhang XP, Liu F, Cheng Z, Wang W. Cell fate decision mediated by p53 pulses. Proc Natl Acad Sci. 2009;106(30):12245–12250. doi:10.1073/pnas.0813088106.

44. Elmore S. Apoptosis: A review of programmed cell death. Toxicol Pathol. 2007;35(4):495–516. doi:10.1080/01926230701320337.

45. Kracikova M, Akiri G, George A, Sachidanandam R, Aaronson SA. A threshold mechanism mediates p53 cell fate decision between growth arrest and apoptosis. Cell Death Differ. 2013;20(4):576–88. doi:10.1038/cdd.2012.155.

13

DMirNet: Inferring direct microRNA-mRNA association networks

Minsu Lee and HyungJune Lee[*]

Abstract

Background: MicroRNAs (miRNAs) play important regulatory roles in the wide range of biological processes by inducing target mRNA degradation or translational repression. Based on the correlation between expression profiles of a miRNA and its target mRNA, various computational methods have previously been proposed to identify miRNA-mRNA association networks by incorporating the matched miRNA and mRNA expression profiles. However, there remain three major issues to be resolved in the conventional computation approaches for inferring miRNA-mRNA association networks from expression profiles. 1) Inferred correlations from the observed expression profiles using conventional correlation-based methods include numerous erroneous links or over-estimated edge weight due to the transitive information flow among direct associations. 2) Due to the high-dimension-low-sample-size problem on the microarray dataset, it is difficult to obtain an accurate and reliable estimate of the empirical correlations between all pairs of expression profiles. 3) Because the previously proposed computational methods usually suffer from varying performance across different datasets, a more reliable model that guarantees optimal or suboptimal performance across different datasets is highly needed.

Results: In this paper, we present *DMirNet*, a new framework for identifying direct miRNA-mRNA association networks. To tackle the aforementioned issues, DMirNet incorporates 1) three direct correlation estimation methods (namely Corpcor, SPACE, Network deconvolution) to infer direct miRNA-mRNA association networks, 2) the bootstrapping method to fully utilize insufficient training expression profiles, and 3) a rank-based Ensemble aggregation to build a reliable and robust model across different datasets.
Our empirical experiments on three datasets demonstrate the combinatorial effects of necessary components in DMirNet. Additional performance comparison experiments show that DMirNet outperforms the state-of-the-art Ensemble-based model [1] which has shown the best performance across the same three datasets, with a factor of up to 1.29. Further, we identify 43 putative novel multi-cancer-related miRNA-mRNA association relationships from an inferred Top 1000 direct miRNA-mRNA association network.

Conclusions: We believe that DMirNet is a promising method to identify novel direct miRNA-mRNA relations and to elucidate the direct miRNA-mRNA association networks. Since DMirNet infers direct relationships from the observed data, DMirNet can contribute to reconstructing various direct regulatory pathways, including, but not limited to, the direct miRNA-mRNA association networks.

* Correspondence: hyungjune.lee@ewha.ac.kr
Department of Computer Science and Engineering, Ewha Womans
University, Seoul, South Korea

Background

MicroRNAs (miRNAs) are short endogenous non-coding RNAs that regulate their target mRNAs by promoting messenger RNA (mRNA) degradation or repressing translation [2]. It has been shown that miRNAs are involved in controlling a wide range of biological processes such as differentiation [3], cellular signalling [4], and several types of cancers [2]. Since miRNAs play crucial roles in regulating genes, the functional associations between miRNAs and mRNAs should be elucidated. However, experimental identification of miRNA-mRNA associations usually performs on a small-scale with a high cost. Therefore, various computational identification methods have been proposed [5].

MiRNAs regulate their target mRNAs post-transcriptionally by base paring to complementary sequences in the 3′-UTR of mRNAs [6]. Based on this property, several methods have been proposed to identify miRNA-target mRNA relationships using sequence data based on sequence complementarity or structural stability [7–9]. Even though the sequence-based computational methods work well with generating putative miRNA-target mRNA relationships, those methods suffer from high false positive rates and false negative rates [5].

To overcome the limitation of sequence-based computational methods, matched expression profiles have been incorporated to identify miRNA-mRNA association relationships. When a miRNA regulates a target mRNA, the expression level of its target mRNA should accordingly be changed. Therefore, there is a correlation between the expression levels of a miRNA and its target mRNA. Based on the premise, various computational methods have been proposed to identify miRNA-mRNA association relationships [10–12] or to build miRNA-mRNA regulatory networks [13–16] by incorporating the matched miRNA and mRNA expression profiles. The conventional approaches for identifying miRNA-mRNA associations using expression profiles are based on traditional correlation measures such as Pearson's linear correlation coefficient [17–19], Spearman's rank-based correlation coefficient [20] or mutual information [21]. These conventional correlation-based methods are valuable tools for generating putative miRNA-mRNA association relationships.

However, there remain some limitations to be resolved in inferring miRNA-mRNA associations from expression data. First, traditional correlation-based network analysis results in many spurious edges [22, 23]. Most of expression profile datasets come from high-throughput experiments, and the expression profiles include hundreds to thousands of variables. The inferred correlations from the observed expression profiles using conventional correlation-based methods contain indirect association relationships derived from transitive information flow

among direct associations [23]. In most cases, due to the limitations of information, it is hard to distinguish between direct associations and indirect associations among ten thousands of variables. Therefore, it is needed to suppress spurious associations from output results.

Second, the expression profiles from microarray experiments suffer from "High-dimension-low-sample-size (large p small n) problem" [24]. When we estimate the empirical correlation between all pairs of expression profiles or conditional dependencies among all variables to infer association relationships, a covariance matrix of size p × p has to be calculated. However, it is difficult to obtain an accurate and reliable estimate of the population covariance matrix from a dataset that has a large number of variables but includes few samples (n < <p) [24].

Third, it is impossible to know in advance which method will produce good results with user's datasets among various computational methods. It has been shown that there is no single computational method that performs well consistently across different datasets and different experimental environments [25]. Each method has been developed with a different premise and approach. Thus, different computational methods usually produce different outputs from the same input data, and one method usually shows different prediction performance across different datasets. As shown in the Result section, our empirical experiments on three datasets confirm the inconsistent performance of computational methods for identifying miRNA-mRNA association relationships. Therefore, a more reliable model that guarantees optimal or suboptimal performance across different datasets is highly needed.

In this study, we present a new framework for reconstructing direct miRNA-mRNA association networks from expression data. The main objectives of the proposed framework (called DMirNet) are as follows: 1) to identify direct associations between miRNA and mRNA, 2) to handle the large p small n problem in microarray expression data, and 3) to build a reliable and robust model across different datasets. To achieve the aforementioned objectives, we propose a direct miRNA-mRNA association network reconstruction method that adopts direct correlation identification methods, the bootstrapping, and an Ensemble approach. First, to suppress indirect associations from the observed expression profiles, we adopt three methods to identify direct relationships, namely partial correlation [24], sparse partial correlation [22], and network deconvolution [23] methods. Second, to overcome the high-dimension-low-sample-size problem, we reduce the dimension of a dataset by selecting the differentially expressed miRNA and mRNAs in an experiment. Also, we embed the bootstrapping approach to build a more accurate and reproducible network by fully utilizing the limited size of samples. Third, to improve the accuracy and reliability of the inferred association relationships, we select a non-parametric Ensemble approach. It has been shown that the ensemble methods that integrate different methods usually outperform individual methods [24, 25]. To aggregate bootstrapping results and different results from different methods, we choose a rank-product-based non-parametric Ensemble method.

We use experimentally confirmed miRNA-mRNA association datasets to evaluate the performance of DMirNet. The results of our empirical experiments on three matched miRNA and mRNA expression profiles show that DMirNet reconstructs a more accurate and reliable miRNA-mRNA association network by incorporating direct correlation methods, bootstrapping and Ensemble approach. We also compare the performance of DMirNet with the state-of-the-art Ensemble model [1] that combines Pearson's correlation, IDA [14], and Lasso [26] on the same datasets. The results of comparative experiment show that DMirNet performs better than the counterpart model with a factor of up to 1.29.

Methods
Framework for identifying direct miRNA-mRNA association relationships

In this section, we present an overview of the framework for identifying direct miRNA-mRNA association relationships as illustrated in Fig. 1. To infer direct miRNA-mRNA association relationships, a matched miRNA-mRNA expression data is needed. After pre-processing each sample, differentially expressed miRNAs and mRNAs are identified to reduce the dimension of data and to focus on the active miRNA-mRNA associations. Because miRNA and mRNA expression profiles are obtained from different platforms, their selected miRNA and mRNA expression profiles are integrated and then scaled.

To reconstruct base-direct microRNA-mRNA association networks, three bootstrapping-based direct correlation inference methods are applied to the integrated expression profiles. Notably, each direct correlation inference method produces a direct correlation model from the expression profiles as a form of a matrix that contains all combinations among miRNAs and mRNAs. Given the integrated expression profiles, the bootstrapping generates m new training data sets by resampling with replacement. For each direct correlation inference methods, m models are computed using the generated m bootstrap samples that are integrated by a rank-based aggregation method. Then, the bootstrapping outputs from the three methods are integrated using the rank-based aggregation method to produce a final direct correlation model. A direct miRNA-mRNA association network is reconstructed by thresholding the weights in the output correlation matrix.

Three direct association network inference methods

A conventional approach to reconstruct gene regulatory or association networks consists of computing the association weight among variables and inferring a link between the two variables by thresholding the association weight. However, the association weight also includes the confounding effect of other variables. By factoring out the dependency of other variables, a direct association network can be inferred. In this subsection, we introduce three methods that we have adopted for inferring direct association networks using expression profiles.

Partial correlation

A partial correlation measures the association weight between two random variables by suppressing the effect of a set of controlling random variables. The partial correlation-based methods can infer the conditional dependency by the non-zero entries in the concentration matrix which is the inverse of covariance matrix. When we apply the partial correlation-based method to identify a genetic network, the zero entries can be interpreted as two nodes that do not interact directly with each other.

Schafer and Strimmer [24] proposed a statistically efficient and computationally fast shrinkage estimator for the covariance and correlation matrix. We use the Corpcor package [24] to compute the partial correlations between selected miRNA and mRNA expression profiles. The resulting partial correlation coefficient between the two variables is regarded as an association weight between them.

Sparse partial correlation estimation (SPACE)

SPACE is another method to compute partial correlations under the large p and small n problem setting [22]. The main characteristics of SPACE are that it assumes that the partial correlation matrix is sparse, and most variable pairs are conditionally independent. Therefore, the output of space is a sparse matrix where many of the possible interactions are zeros. This method helps to select non-zero partial correlations. It estimates sparse partial correlation using sparse regression techniques and optimizes the results with a symmetric constraint and an L_1 penalization [22].

Network deconvolution

Network deconvolution is a direct dependency network inference method that eliminates an indirect weight from the inferred dependency network from the observed data [23]. The network deconvolution method assumes that the measured association weights from the observed data represent the sum of direct and indirect weights. Moreover, the method assumes that the indirect information flow can be approximated as the product of

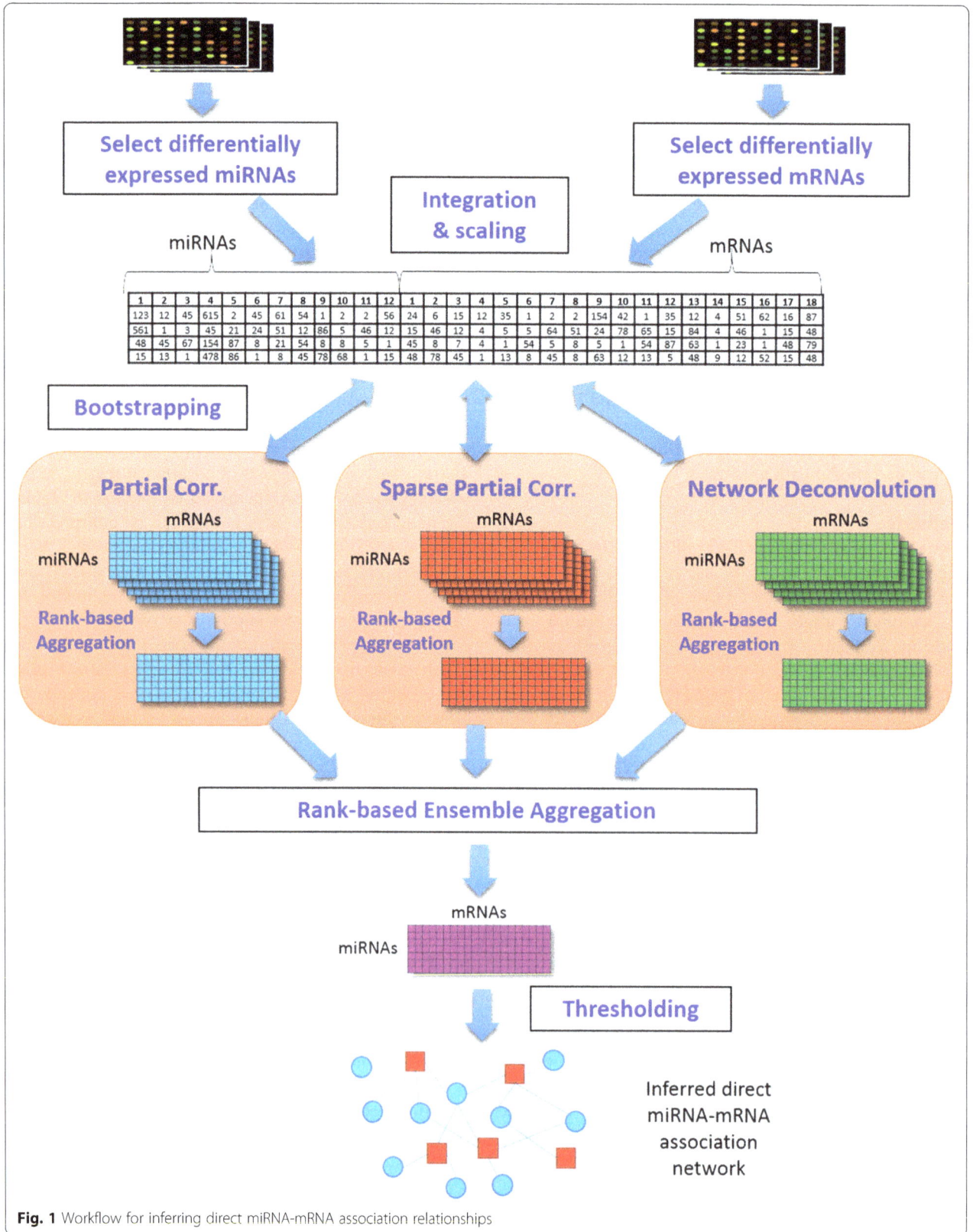

Fig. 1 Workflow for inferring direct miRNA-mRNA association relationships

direct association weights. Let G_{obs} be an observed dependency network, G_{tru} a true direct dependency network, and G_{ind} an indirect dependency network. Then, the indirect network can be expressed in terms of all indirect effects along paths of increasing length, and we can express the observed network (G_{obs}) in terms of the true network (G_{tru}) and the indirect network (G_{ind}) as follows:

$$G_{obs} = G_{tru} + G_{ind}$$
$$= G_{tru} + \left(G_{tru}^2 + G_{tru}^3 + G_{tru}^4 + ...\right)$$
$$= G_{tru}(I - G_{tru})^{-1} \qquad (1)$$

Therefore, the network deconvolution method [23] infers true direct dependency network by reversing the effect of transitive information flow across all possible indirect paths. That is, the true direct network can be calculated using the observed network as follows:

$$G_{tru} = G_{obs}(I + G_{obs})^{-1} \qquad (2)$$

The network deconvolution method can be applied with various correlation measures. In this study, we compute the pair-wise observed correlations between miRNA and mRNA expression profiles using mutual information, and then apply the network deconvolution method to suppress indirect correlation relationships from the observed correlations.

Bootstrapping

Bootstrapping is a method for generating multiple versions of a model, and using these to generate an aggregated model. It is designed to improve accuracy and stability [27]. Given a training set D, bootstrapping generates m new training data sets D_i by sampling from D uniformly and with replacement. The m models are computed using the generated m bootstrap samples and combined by aggregating the outputs.

Because the bootstrap aggregation usually reduces variance and helps to avoid overfitting, the bootstrap procedure works well when the sample size is insufficient for straightforward model inference. Therefore, we adopt the bootstrapping procedure to reconstruct multiple networks from a single original dataset using a single direct association network inference method, which can then be aggregated into a more accurate and reproducible association network.

Rank-based Ensemble aggregation

Because computational methods often show varying performances across different datasets [25], it is necessary to improve the reliability and accuracy of the inferred networks using computational methods. In this case, the Ensemble methods that integrate different methods can be used because they have shown better performances

than individual methods [1, 25]. Also, the Ensemble methods may be useful to capture nonlinear relationships as well as linear relationships among variables by integrating results from linear or nonlinear correlation inference methods.

When several results from computational methods are integrated, the distribution of the weights between two elements usually varies considerably among computational methods. It is difficult to directly integrate real-valued weights between two variables from individual methods. Thus, it is challenging to aggregate real-valued weights of inferred association networks from different methods or datasets.

To aggregate different output networks from various methods, we adopt a non-parametric approach based on ranking. Because a rank-based Ensemble aggregation method only considers the rank of the weight and does not assume specific distribution of the source data, the rank-based method does not depend on the actual distribution of weights derived from different methods [28]. The characteristic of rank-based aggregation is the ability to combine lists from different sources and platforms. Hence, we employ a rank-based Ensemble approach to aggregate the outputs from bootstrapping iterations and different methods. The conventional rank-based aggregation methods include the rank-sum-based approach, average-rank-based approach, and Borda count election [1]. In this study, we use an inverse-rank-product method [29] to combine networks reconstructed from the same set of genes, after empirically comparing the performances of the Borda count election method and the normalized-weight-sum method with the inverse-rank-product method. The rank of a particular weight between a miRNA and an mRNA in the aggregated network is calculated by taking the product of the ranks of the same edge across all networks. Then, to assign a lower rank to a higher weight, the inverse of rank-product is used as a representative association weight between the miRNA and the mRNA. Let G be a set of association networks to be integrated, and let r_{ij} be a rank of association weight between node i and j. Then, the association weight of an integrated graph using the inverse-rank-product strategy (r'_{ij}) can be calculated as follows:

$$r'_{ij} = \frac{1}{\log\left(\Pi_{m \in G}\left(r_{ij}^m + 1\right)\right)} \qquad (3)$$

We apply the inverse-rank-product method to aggregate bootstrapping outputs from the single direct association identification method and to integrate the outputs from different methods.

Experiments for performance evaluation

To evaluate our proposed DMirNet, we performed empirical experiments with three matched miRNA and mRNA expression profiles. First, we analysed the effect of bootstrapping and Ensemble to identify miRNA-mRNA association relationships. Second, we compared the performance of DMirNet with a best-performed Ensemble model [1] for inferring miRNA-mRNA regulatory relationships from expression data.

Experimental datasets

To avoid the biased or intentional selection of experimental data, we used the same three matched miRNA and mRNA expression profiles used in a recently published comparative study [1, 30]. The three processed datasets were obtained from [30].

Epithelial to Mesenchymal Transition (EMT) data includes the matched miRNA-mRNA expression profiles of epithelia class (11 samples) and mesenchymal class (36 samples). Multi-Class Cancer (MCC) data includes 60 samples from normal and cancerous tissues from eight organs. Breast Cancer (BR) data has 50 samples from basal and luminal groups. After applying the differentially expressed gene (DEG) analysis with *limma* package of Bioconductor and a false discovery correction process at a significant level (adjusted p-value <0.05), 35 miRNAs and 1154 mRNAs were identified as DEGs of the EMT data; additionally, 108 miRNAs and 1860 mRNAs were identified as DEGs of the MCC data. Regarding the BR data, 92 miRNAs (adjusted p-value <0.2) and 1500 mRNAs (adjusted p-value <0.0001) were identified as DEGs. The selected and integrated miRNA and mRNA expression profiles were standardized across samples before applying our DMirNet.

Implementation of DMirNet

To identify a direct miRNA-mRNA association network, its base association networks were reconstructed using the three direct association relationships inference method with bootstrapping. For each method, the base miRNA-mRNA association networks were iteratively built using randomly resampled data with replacement. To get the bootstrapping results, we randomly selected 95% of the dataset with replacement and iteratively rebuilt association networks 100 times for each dataset.

To utilize three direct association network identification methods, we use *corpcor* and *space* R packages [31, 32] from Bioconductor and an existing network deconvolution algorithm [33]. Aggregations of the results from bootstrapping of a single method and Ensembles of different methods were performed using equation (3).

Performance evaluation method

Currently, 1,881 miRNA precursors and 2,588 mature sequences in the Human genome are listed in miRBase (GRCh38), and the number of human genes is estimated at 20,000-25,000 [34]. Several manually curated miRNA-target mRNA databases show that one miRNA may regulate many genes as its targets, while one gene may be targeted by many miRNAs. This indicates that the relationships between miRNAs and their target mRNAs may not be one-to-one. However, the number of experimentally validated miRNA-mRNA interactions for evaluating a computational model has been very limited until now. Since there is no complete ground-truth for evaluating performances, the union of public miRNA-target mRNA databases, which include both experimentally verified relationships and some predicted relationships, has been used to evaluate performance and to compare different computational methods [1, 30, 35, 36]. The union of Tarbase v.6.0 [37], miRecords v2013 [38], miR-Walk v2.0 [39] and miRTarBase v.4.5 [40] includes 62,858 unique miRNA-target mRNA interactions among 693 miRNAs and 16,091 genes. We use the union of these four databases [30] as a ground-truth dataset.

Based on the ground-truth data, the performance of each method was evaluated by checking the number of overlaps between top k high-ranked mRNAs of each miRNA on an inferred network and the ground-truth miRNA-mRNA pairs. Even though the number of ground-truth is very limited, the fraction of inferred correlations that are experimentally validated pairs may be regarded as a measure of the precision of the computational method. Since the total number of selected miRNA-mRNA correlations is same across all the methods in the comparative study, a higher number of overlaps can be regarded as higher precision on inferring direct miRNA-mRNA association network.

Results

Performance evaluation of DMirNet

To investigate the performance of DMirNet and to examine the effects of all components of the framework, we performed comparative empirical experiments using EMT, MCC, and BR datasets and three direct correlation inference methods: Corpcor, SPACE, and mutual information-based network deconvolution (MIND). For bootstrapping execution, the number of bootstrapping iterations was set to 100, and the sampling rate was set to 95%. Additionally, an inverse-rank-product method was applied for aggregating bootstrapping results and integrating results from different methods. For each method, the number of experimentally confirmed miRNA-mRNA associations was evaluated as a measure of precision by computing the overlaps between ground-truth pairs and inferred top 100 mRNAs per a miRNA. Table 1

Table 1 Number of experimentally confirmed miRNA-mRNA associations by the ground-truth data

		Single Method			Ensemble Method			
		Corpcor	Space	MIND	C&S	C&M	S&M	C&S&M
EMT	Whole	35	45	24	45	34	35	41
	Bootstrap	32	38	25	40	24	37	40
MCC	Whole	200	183	210	204	206	201	209
	Bootstrap	211	204	207	201	217	220	216
BR	Whole	98	83	95	90	94	97	102
	Bootstrap	107	95	99	99	102	100	105

The Top 100 correlations for each miRNA were selected from each experiment for performance comparison. To evaluate the effect of three direct correlation inference methods, bootstrapping and Ensemble approach, we performed a comparative study using EMT, MCC and BR datasets. Corpcor (denoted as C) is the partial correlation estimation method, SPACE (denoted as S) is the sparse partial correlation estimation method, and MIND (denoted as M) is the mutual information-based network deconvolution method. 'Whole' means that the whole expression profiles were used to infer a direct correlation matrix, and 'Bootstrap' means that 100 direct correlation matrices were computed using 100 bootstrapped samples and then aggregated based on an inverse-rank-product method

summarizes the precisions of all combinations of DMirNet component.

First, we investigated each single direct correlation estimation method across three datasets. The results of empirical experiments confirm that there is no single inference method that performs optimally across all datasets. Corpcor (C) shows the best precision with the BR dataset, but it ranks the medium with the EMT and the MCC datasets. SPACE (S) performs best with the EMT dataset, but has the worst performance with BR and MCC datasets. On the other hand, even though MIND (M) performs worst with the EMT dataset, it shows good performance with both MCC and BR dataset. The results indicate that each method has its own limitation on inferring direct correlations; thus, it is difficult to identify the whole direct miRNA-mRNA correlations using any single method. In such cases, the Ensemble aggregation of different methods can improve the accuracy and stability of an inferred correlation network.

We also determined the effects of bootstrapping in DMirNet framework. By applying a bootstrapping strategy, the precision of three methods was strictly increased within MCC and BR datasets. However, regarding the EMT dataset, bootstrapping does not lead to any performance improvement. The results imply that the bootstrapping procedure does not guarantee an increase in the fraction of experimentally validated pairs among inferred pairs.

Although an Ensemble method that combines three inference methods (C&S&M) shows good performance, on occasion, single methods (SPACE with EMT whole and Corpcor with BR bootstrap) or Ensembles of two inference methods (S&M with MCC bootstrap) outperforms C&S&M. This phenomenon was derived by combining the worst-performed model to the Ensemble. For example,

MIND shows the worst performance with the EMT dataset but the Ensemble method excluding MIND (i.e. C&S) with the EMT dataset performs best. It should be noted that although C&M, S&M, and C&S&M perform relatively worse because they are integrated with MIND, the combined ensemble models turn out to outperform MIND itself. Additionally, when the number of aggregated methods increases from two to three, the precision of Ensemble methods also increases. The experimental results show that the Ensemble aggregation approach helps to relieve the effect of the worst model and achieves a relatively higher performance.

We also investigated the combinatorial effect of bootstrapping and Ensemble aggregation on DMirNet framework. Regarding the EMT dataset, there was no improvement in the precision using bootstrapping. However, the Ensemble aggregation of different methods reduced the effect of the worst-performed MIND. In the MCC and BR dataset, the results show performance improvements by bootstrapping across almost all experiments, as well as a relief of the effect of the worst model (SPACE) and improved precision by Ensemble aggregation. Regarding the BR dataset, each method with the combination of bootstrapping and Ensemble aggregation turns out to be effective.

The effect of bootstrapping and Ensemble approaches can be quantified using a paired t-test. Figure 2 demonstrates the average number of confirmed miRNA-mRNA correlations using each method. Additionally, in order to assess the statistical significance of difference on the precision between two methods, the p-values using the paired t-test were calculated.

We summarize the performance evaluation on precisions for all combinations of DMirNet component using the limited number of ground-truth pairs as follows: 1) The performance of each direct correlation estimation method slightly varies across the three datasets. 2) Applying the bootstrapping procedure generally improves the precision of the model. 3) If an Ensemble model aggregates a poorly performed model, the Ensemble approach guarantees at least the average performance of aggregated methods. 4) The balanced combination of three direct correlation inference methods, bootstrapping and Ensemble approach, strictly reduces the effect of the worst-performed model and achieve the best or the second best precision. Therefore, we demonstrate that the use of both bootstrapping and Ensemble approaches helps to build a more reliable and robust model across different expression datasets, while tackling the large p small n problem.

Performance comparison between DMirNet and the state-of-the-art Ensemble-based model

DMirNet framework adopts the three direct correlation network inference methods to identify direct miRNA-

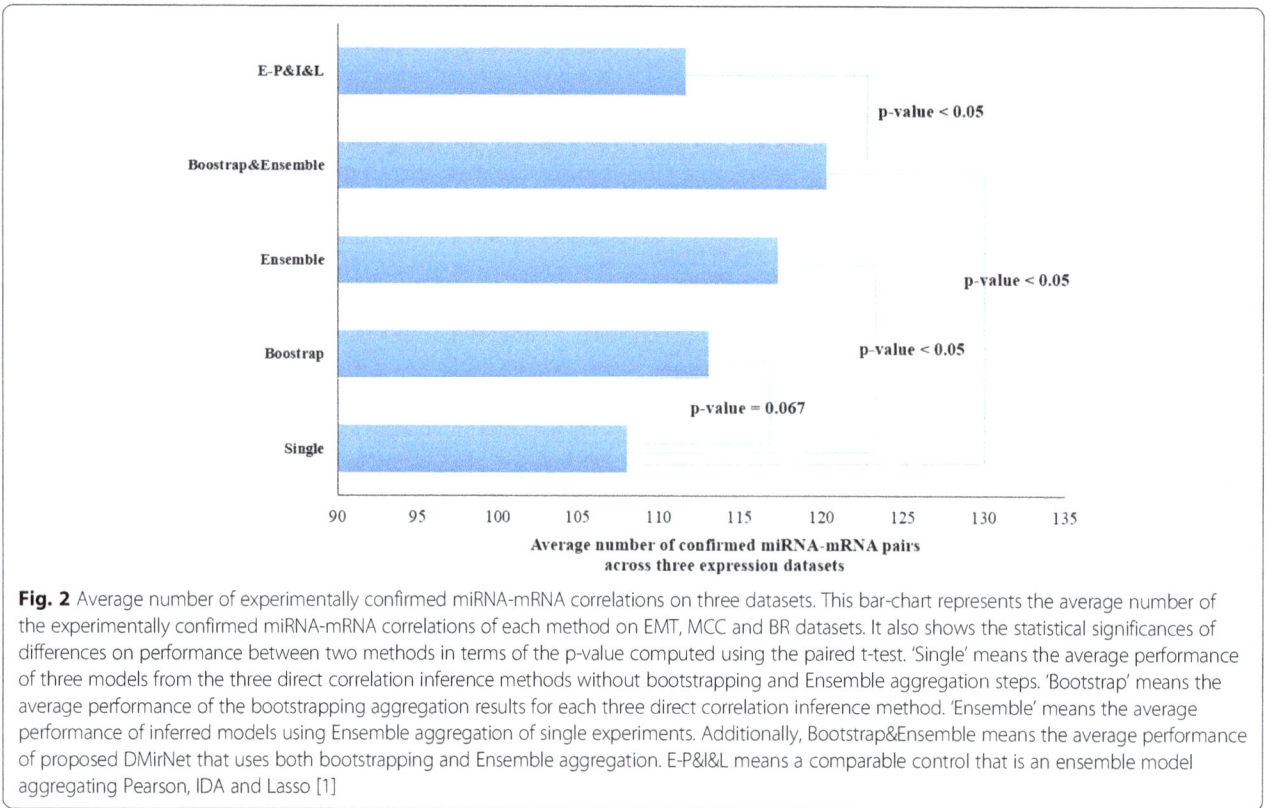

Fig. 2 Average number of experimentally confirmed miRNA-mRNA correlations on three datasets. This bar-chart represents the average number of the experimentally confirmed miRNA-mRNA correlations of each method on EMT, MCC and BR datasets. It also shows the statistical significances of differences on performance between two methods in terms of the p-value computed using the paired t-test. 'Single' means the average performance of three models from the three direct correlation inference methods without bootstrapping and Ensemble aggregation steps. 'Bootstrap' means the average performance of the bootstrapping aggregation results for each three direct correlation inference method. 'Ensemble' means the average performance of inferred models using Ensemble aggregation of single experiments. Additionally, Bootstrap&Ensemble means the average performance of proposed DMirNet that uses both bootstrapping and Ensemble aggregation. E-P&I&L means a comparable control that is an ensemble model aggregating Pearson, IDA and Lasso [1]

mRNA association network. It embeds the bootstrap aggregation for fully utilizing the limited training expression profiles and the Ensemble approaches for improving reliability and performance. To show the effectiveness of DMirNet on identifying direct miRNA-mRNA interactions, we compare the performance of it with the state-of-the-art Ensemble-based model [1]. The Ensemble-based model integrates Pearson's correlation (denoted as P), IDA (denoted as I) [14], and Lasso (denoted as L) [26] using the Borda count election aggregation method. Through a rigorous comparative study using EMT, MCC, and BR dataset and eight correlation inference methods, the ensemble of P&I&L was selected as a best-performed model across the three datasets [1]. Table 2 shows the number of experimentally confirmed miRNA-mRNA correlations inferred from

combinations of components in DMirNet framework and the P&I&L Ensemble model. Table 2 shows interesting results of the comparative study. The solo use of Corpcor, Space, and MIND methods usually does not outperform Pearson, IDA, and Lasso methods. Moreover, Regarding the BR dataset, Pearson, IDA and Lasso rather considerably outperform Corpcor, Space, and MIND with the current ground-truth data. However, when three direct correlation estimation methods are bootstrapped and aggregated, the integrated model considerably performs better. The p-value of the difference on performance between DMirNet (Bootstrap&Ensemble) and P&I&L is less than 0.05 (p-value = 0.040) as shown in Fig. 2. This implies that the difference of the above two methods is statistically significant, and thus, DMirNet is a better choice than P&I&L in a statistical sense.

Table 2 Performance comparison of DMirNet with the state-of-the-art Ensemble model

Dataset	Direct correlation inference methods					the state-of-the-art method			
	Corpcor	Space	MIND	E-C&S&M	B&E-C&S&M	Pearson	IDA	Lasso	E-P&I&L
EMT	35	45	24	41	40	30	29	29	31
MCC	200	183	210	209	216	205	198	187	203
BR	98	83	95	102	105	114	124	120	101

To compare the performance of our method with a related work, we investigate the number of experimentally confirmed miRNA-mRNA associations of the state-of-the-art Ensemble model. It combines Pearson's correlation (denoted as P), IDA (denoted as I), and Lasso (denoted as L) using the Borda count election and was reported as the best-performed Ensemble model on the three datasets [1]. 'E' denotes the Ensemble approach, and 'B&E' denotes the DMirNet with both bootstrapping and Ensemble aggregation

Network analysis of inferred direct miRNA-mRNA association networks

Based on the proposed DMirNet framework, we reconstructed direct miRNA-mRNA association networks for each dataset. Through the procedures described in Method section with 100 bootstrapping iterations, the output miRNA-mRNA correlation matrix was generated. We selected top 1000 miRNA-mRNA association relationships to reconstruct association networks for each dataset. The top 1000 miRNA-mRNA pairs for each dataset are listed in the Additional file 1.

We visualized the reconstructed networks from the top 500 pairs using the Cytoscape [41] environment, and analysed their network structure using the ModuLand plug-in [42]. The ModuLand can determine overlapping network modules and community centrality. Since the outputs of the ModuLand represent representative community centralities and connections among network modules, the results of the ModuLand effectively show an abstraction of the whole network structure. For each dataset, the reconstructed network and its key network structure are shown in the Additional file 2. Among them, Fig. 3 shows the core network structure of the

inferred Top 500 miRNA-mRNA associations of the MCC data. Additionally, the network of modules represents the indirect associations among miRNAs mediated by the mRNAs.

To interpret the related biological pathway of inferred miRNA-mRNA association network, we analyse the functions of mRNAs listed in the Top 500 and Top 1000 pairs based on KEGG pathway [43]. We used the ClueGO [44] Cytoscape plug-in to extract the biological pathways for associated mRNAs, and to visualize the selected KEGG pathway terms in a functionally grouped network. The overall results of identifying significant KEGG pathway across three dataset are summarized in the Additional file 3. Figure 4 demonstrates the KEGG biological pathways related to Top 1000 pairs of the MCC dataset. The size of the nodes reflects the statistical significance of the terms. The degree of connectivity between terms (edges) is calculated using kappa statistics. The calculated kappa score is also used to define functional groups. A node having more than two colours is a term that can be included in several groups. It should be noted that the MCC dataset is the expression profiles of normal and cancerous tissues from eight

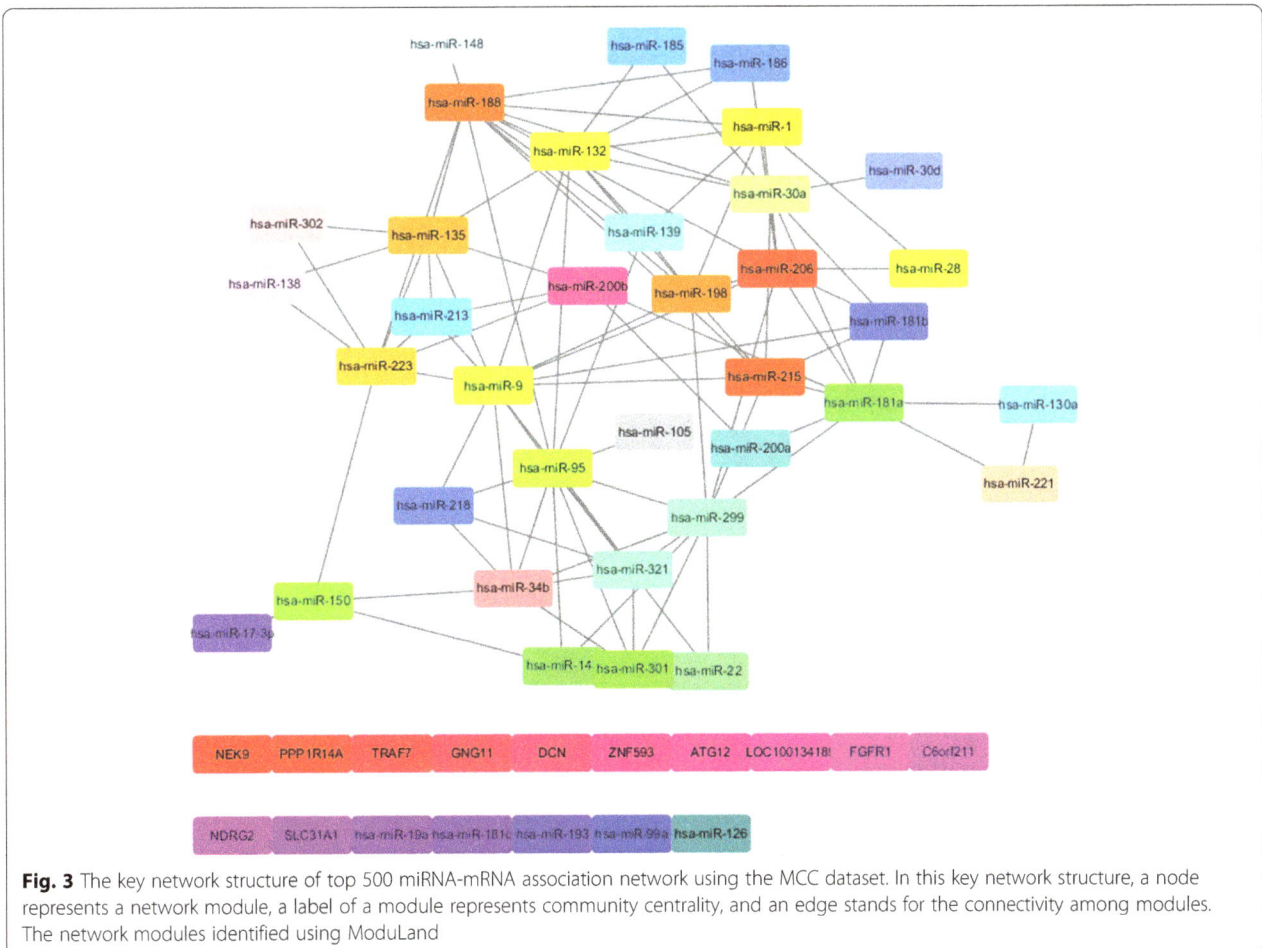

Fig. 3 The key network structure of top 500 miRNA-mRNA association network using the MCC dataset. In this key network structure, a node represents a network module, a label of a module represents community centrality, and an edge stands for the connectivity among modules. The network modules identified using ModuLand

organs. The top 1000 pairs of the MCC dataset consists of 103 miRNAs and 572 mRNAs. The biological pathway analysis was performed on the 572 mRNAs. Among various biological pathways, there are three cancer-related categories; namely 'Transcriptional misregulation in cancer,' 'MicroRNAs in cancer,' and 'Choline metabolism in cancer.' The three cancer-related categories are associated with 44 miRNA-mRNA pairs among 33 miRNAs and 27 mRNAs. The list of the 44 miRNA-mRNA pairs is shown in Additional file 4, and Fig. 5 shows the list in a network form.

To investigate putative novel multi-cancer-related miRNA-mRNA pairs, we checked the overlaps between the 44 multi-cancer-related miRNA-mRNA pairs and ground-truth data which is a union of the four manually curated database. Our DMirNet found out a strong miRNA-mRNA association between hsa-miR-181a and BMPR2 as top 809 out of 200,880 pairs (upper 0.4% percentile). This miRNA-mRNA relationship has already been confirmed in [45] such that the hsa-miR-181a plays a direct role in down-regulating the BMPR2. This means that our DMirNet inference provide a consistent result with pre-known miRNA-mRNA relationships.

Regarding hsa-miR-299::CDKN2C (top 479) and hsa-miR-301::BCL6 (top 593) in the 44 multi-cancer-related pairs, they are not listed in the ground-truth data.

However, the ground-truth data includes closely related pairs (namely, has-miR-299-5p::CDKN1A and hsa-miR301a::BCL2L11) of which mRNA is from the same gene family. In many cases, genes in a family have a similar structure of function, or proteins produced from these genes work together as a unit or participate in the same process. Therefore, the existence of similar miRNA-mRNA pair may support the plausibility of the inferred pairs by DMirNet.

After excluding the known miRNA-mRNA pair (hsa-miR-181a::BMPR2), 43 among 44 miRNA-mRNA pairs can accordingly be regarded as the putative novel multi-cancer-related miRNA-mRNA pairs.

Discussion

By investigating the combinatorial effect of the bootstrapping and the Ensemble aggregation on DMirNet framework, the performance enhancement factors of DMirNet are demonstrated. The bootstrapping procedure helps to build a more accurate and reproducible network by fully utilizing the limited size of samples. Additionally, the Ensemble model helps to avoid the worst performance by guaranteeing at least the average performance of aggregated methods. The balanced combination of three direct correlation inference methods, bootstrapping and Ensemble approach, strictly reduces

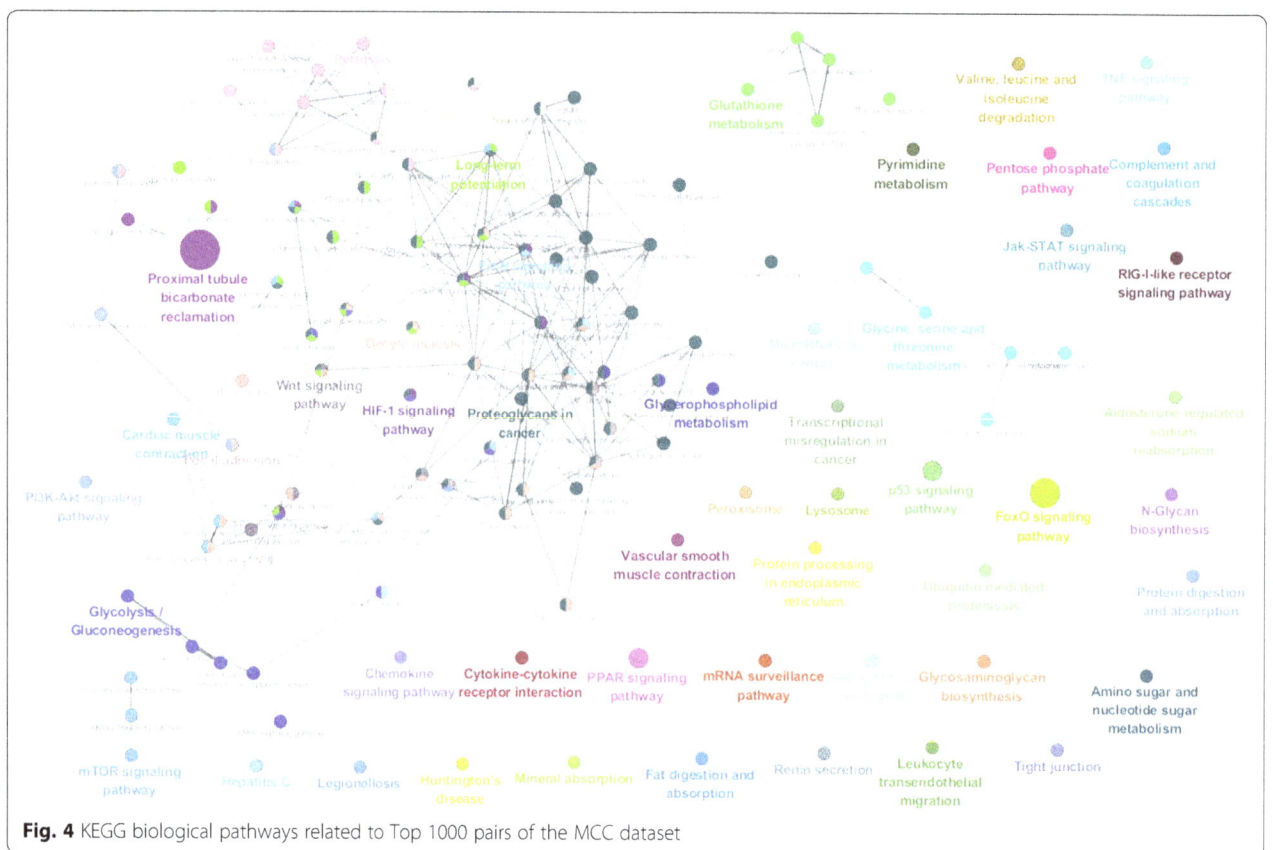

Fig. 4 KEGG biological pathways related to Top 1000 pairs of the MCC dataset

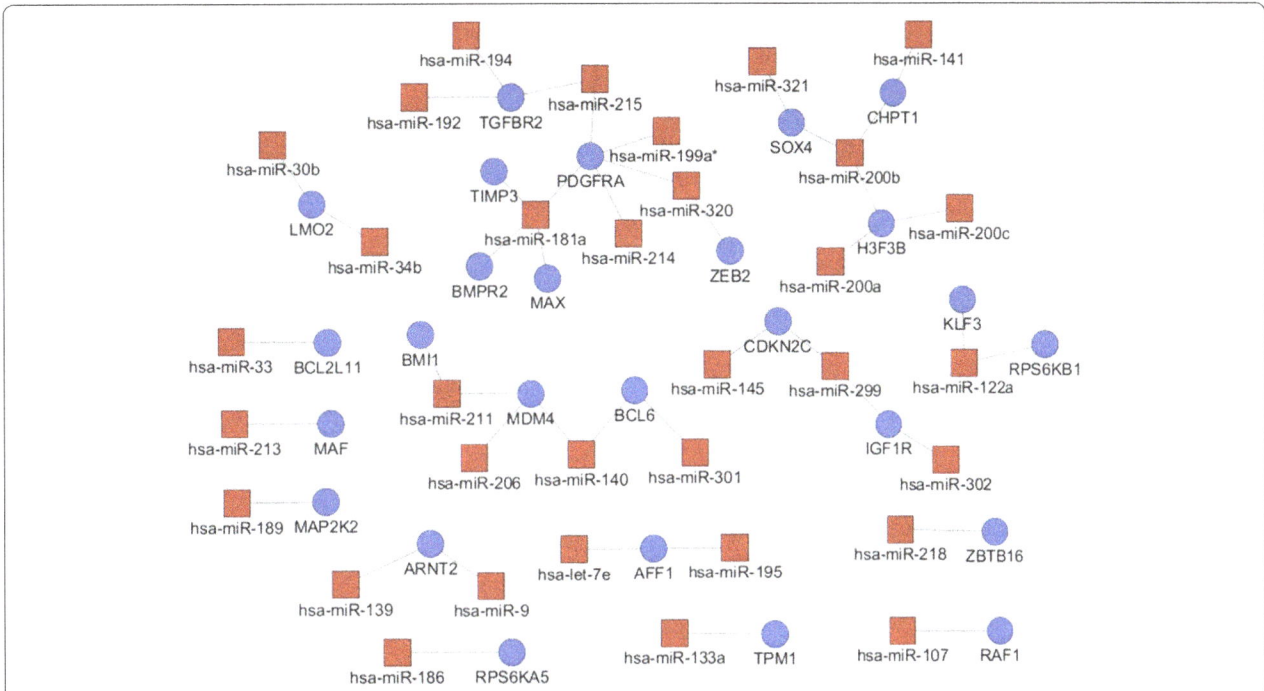

Fig. 5 Cancer-related miRNA-mRNA association networks among Top 1000 pairs of the MCC dataset. The red rectangle nodes are mRNAs and the blue circle nodes are mRNAs

the effect of the worst-performed model and achieves a better precision.

Additionally, when we compare the performance of DMirNet with P&I&L, three single direct correlation inference methods do not show good performance compared to Pearson, IDA, and Lasso. This result indicates that even though each direct correlation estimation method suppresses its indirect information from an observed data in some degree, they are still incomplete. However, by incorporating the bootstrapping and Ensemble aggregation, DMirNet outperforms the best-performed P&I&L across three datasets. These results demonstrate the effectiveness of DMirNet procedure in terms of accuracy and robustness. Although the three direct correlation inference methods cannot perfectly suppress the whole indirect relationships from the observed data, we can effectively focus on the direct associations through incorporating the bootstrapping and the Ensemble approach. We expect that if we can integrate more direct correlation inference methods to DMirNet, the performance of DMirNet would be more improved. Also, if Pearson, IDA, and Lasso methods can be

integrated with additional information such as sequence-based miRNA-mRNA target prediction result, the indirect associations might be filtered, and it may further improve the performance of the Ensemble model.

We would like to discuss the limitation of the ground-truth dataset which was used at the experiments. The number of pairs in ground-truth data is significantly smaller than the expected number of miRNA-mRNA correlation pairs in a genome. Moreover, the miRNA-mRNA relationships are dynamically changed according to the experimental method, sample, and experimental condition. For example, 'hsa-miR-19a-3p' sometimes directly down-regulates the mRNA of RAB14 on the Kidney tissue with the PAR-CLIP experiment [46], whereas 'hsa-miR-19a-3p' sometimes does not regulate RAB14 on the lung tissue with the Luciferase reporter assay method results [47] as shown in Table 3. Therefore, it is difficult to fully estimate the performance of computational inference methods based only on the overlap with the limited size of the ground-truth data. We expect that the number of experimentally confirmed pairs will increase as miRNA mediated gene regulation research

Table 3 Conflict between experimental results on hsa-miR-19a-3p and RAB14

Publication	Method	Tissue	Cell line	Tested cell line	Result	Regulation
Hafner M. et al. 2010 [46]	PAR-CLIP	Kidney	HEK293	N/A	Positive	Down
Kanzaki H et al. 2011 [47]	Luciferase Reporter Assay	Lung	SBC3	HEK293	Negative	?

A manually curated miRNA-target database includes conflict experimental results for some miRNA-mRNA pairs. As an example, this table shows a conflict experimental result on hsa-miR-19a-3p and RAB14(hsa) from TarBase 6.0 [37]

in this field becomes more mature and flourished. More extensive ground-truth findings may confirm our false negative inference cases as true positive ones.

Regarding the MCC datasets, we identify putative novel multi-cancer-related miRNA-mRNA pairs by utilizing KEGG pathway analysis and ground-truth data. After excluding previously known one pair and similar two pairs with the ground-truth data, 43 out of 44 miRNA-mRNA association pairs are reported.

Although our DMirNet improves the performance by incorporating the bootstrapping and Ensemble approach, the bootstrapping procedure may come with computational overhead. The bootstrapping procedure generates m training datasets using sampling with replacement, computes m direct correlation matrices, and aggregates the m models. If the bootstrapping procedures are combined with Ensemble approach that aggregates n different methods, we have to run the bootstrapping procedure n times. However, in many bioinformatics applications, there is a trade-off between performance improvement and computation complexity. Also, we can accelerate the bootstrapping and ensemble procedure by utilizing the MPI.

Conclusions

We have presented the DMirNet framework that identifies direct miRNA-mRNA association networks from expression profiles. DMirNet takes full advantage of three direct association estimation methods, the bootstrapping and the Ensemble approach based on an inverse-rank-product method. The performance evaluation has shown a substantial effectiveness of DMirNet in terms of the number of the matched miRNA-mRNA cases with a ground-truth data. Our proposed DMirNet framework outperforms the state-of-the-art Ensemble model with a factor of up to 1.29 with the EMT data in terms of precision. These empirical experimental results show the effectiveness of the combinatorial effects of the direct association estimation, the bootstrapping, and the Ensemble approaches in DMirNet. This paper demonstrates that our DMirNet can be a promising alternative to other existing methods to identify direct and novel miRNA-mRNA relationships more extensively. We expect that DMirNet can contribute to reconstructing various direct regulatory pathways, including, but not limited to, the direct miRNA-mRNA association networks.

Additional files

Additional file 1: Top1000 miRNA-mRNA association relationships for each dataset.

Additional file 2: Reconstructed miRNA-mRNA association networks and their key structures with top 500 miRNA-mRNA association relationships. The

reconstructed network was visualized with Cytoscape and the key modular structure of the network was analysed using ModuLand.

Additional file 3: Functional analysis of reconstructed miRNA-mRNA association networks based on KEGG pathway. To interpret the functions of inferred miRNA-mRNA association network, related KEGG pathway in Top 500 and Top 1000 pairs were analysed using ClueGO

Additional file 4: The list of multi-cancer-related 44 miRNA-mRNA pairs. The putative novel multi-cancer-related pairs are coloured with yellow.

Declaration

This article has been published as part of BMC Systems Biology Volume 10 Supplement 5, 2016. 15th International Conference On Bioinformatics (INCOB 2016): systems biology. The full contents of the supplement are available online http://bmcsystbiol.biomedcentral.com/articles/supplements/volume-10-supplement-5

Funding

The publication charge was funded by the "Convergence Female Talent Education Project for Next Generation Industries" through the MSIP and NRF(2015H1C3A1064579) to HL. This work was supported by National Research Foundation of Korea grants funded by the Korean government (MSIP) (KW-2014PPD0053 and NRF-2015R1C1A1A01054305) to ML, and also funded by the Ministry of Education (NRF-2015R1D1A1A01057902) and MSIP (NRF-2015H1C3A1064579) to HL.

Authors' contributions

ML conceived the study, designed and implemented the proposed framework, performed empirical experiments, and wrote the manuscript. HL participated in the design and coordination of the study, and wrote the manuscript. Both authors read and approved the final manuscript.

Competing interests

The authors declare that they have no competing interests.

References

1. Le TD, Zhang J, Liu L, Li J. Ensemble methods for miRNA target prediction from expression data. PLoS One. 2015;10(6):e0131627.
2. Bartel DP. MicroRNAs: target recognition and regulatory functions. Cell. 2009;136:215–33.
3. Esquela-Kerscher A, Slack FJ. Oncomirs-microRNA with a role in cancer. Nat Rev Cancer. 2006;6:259–60.
4. Cui Q, Yu Z, Purisima EO, Wang E. Principles of microRNA regulation of human cellular signalling network. Mol Syst Biol. 2006;2:1–7.
5. Rajewsky N. microRNA target prediction in animals. Nat Genet. 2006;38: S8–S13.
6. Bartel DP. MicroRNAs: genomics, biogenesis, mechanism, and function. Cell. 2004;116:281–97.
7. Lewis BP, Burge CB, Bartel DP. Conserved seed paring, often flanked by adenosines, indicates that thousands of human genes are microRNA targets. Cell. 2005;120:15–20.
8. Enright AJ, John B, Gaul U, Tuschl T, Sander C, et al. MicroRNA targets in Drosophila. Genome Biol. 2004;5:R1.
9. Kim SK, Nam JW, Rhee JK, Lee JW, Zhang BT. miTarget: microRNA target gene prediction using a support vector machine. BMC Bioinformatics. 2006;7(1):411.

10. Van der Auwera I, Limame R, van Dam P, Vermeulen PB, Dirix LY, Van Laere SJ. Integrated miRNA and mRNA expression profiling of the inflammatory breast cancer subtype. Br J Cancer. 2010;103:532–41.

11. Diaz G, Zamboni F, Tice A, Farci P. Integrated ordination of miRNA and mRNA expression profiles. BMC Genomics. 2015;15:767.

12. Joung JG, Hwang KB, Nam JW, Kim SJ, Zhang BT. Discovery of microRNA-mRNA modules via population-based probabilistic learning. Bioinformatics. 2007;23:1141–7.

13. Liu B, Li J, Tsykin A, Liu L, Gaur AB, Goodall GJ. Exploring complex miRNA-mRNA regulatory networks by splitting-average strategy. BMC Bioinformatics. 2009;10:408.

14. Le TD, Liu L, Tsykin A, Goodall GJ, Liu B, Sun BY, Li J. Inferring microRNA-mRNA causal regulatory relationships from expression data. Bioinformatics. 2013;29(6):765–71.

15. Zhang Y, Liu W, Xu Y, Li C, Wang Y, Yang H, Zhang C, Su F, Li X, Li X. Identification of subtype specific miRNA-mRNA functional regulatory modules in matched miRNA-mRNA expression data: Multiple myeloma as a case. Biomed Res Int. 2015;501262.

16. Kim SK, Ha JW, Zhang BT. Constructing higher-order miRNA-mRNA interaction networks in prostate cancer via hypergraph-based learning. BMC Syst Biol. 2013;7:47.

17. Fu J, Tang W, Du P, Wang G, Chen W, Li J, Zhu Y, Gao J, Cui L. Identifying microRNA-mRNA regulatory network in colorectal cancer by a combination of expression profile and bioinformatics analysis. BMC Syst Biol. 2012;6:68.

18. Zhuang X, Li Z, Lin H, Gu L, Lin Q, Lu Z, Tzeng CM. Integrated miRNA and mRNA expression profiling to identify mRNA targets of dysregulated miRNAs in non-obstructive azoospermia. Nature Science Reports. 2015;5:7922.

19. Li Y, Xu J, Chen H, Bai J, Li S, Zhao Z, Shao T, Jiang T, Ren H, Kang C, Li X. Comprehensive analysis of functional microRNA-mRNA regulatory network identifies miRNA signatures associated with glioma malignant progression. Nucleic Acids Res. 2013;41(22):e203.

20. Jacobsen A, Silber J, Harinath G, Huse JT, Schultz N, Sander C. Analysis of microRNA-target interactions across cancer types. Nat Struct Mol Biol. 2013;20:1325–32.

21. Jung D, Kim B, Freishtat RJ, Giri M, Hoffman E, Seo J. miRTarVis: an interactive visual analysis tool for microRNA-mRNA expression profile data. BMC proceedings. 2015;9 suppl 6:S2.

22. Peng J, Wang P, Zhou N, Zhu J. Partial correlation estimation by joint sparse regression models. J Am Stat Assoc – Theory and Methods. 2009;104(486):735–46.

23. Feizi S, Marbach D, Médard M, Kellis M. Network deconvolution as a general method to distinguish direct dependencies in networks. Nat Biotechnol. 2013;31:726–33.

24. Schäfer J, Strimmer K. A shrinkage approach to large-scale covariance matrix estimation and implications for functional genomics. Statist Appl Genet Mol Biol. 2005;4:32.

25. Marbach D, Costello JC, Kuffner R, Vega NM, Prill RJ, et al. Wisdom of crowds for robust gene network inference. Nat Methods. 2012;9:796–804.

26. Tibshirani R. Regression shrinkage and selection via the lasso. J R Stat Soc B Methodol. 1996;267–288.

27. Breiman L. Bagging predictors. Machine Learning. 1996;24(2):123–40.

28. Pihur V, Datta S, Datta S. RankAggreg, an R package for weighted rank aggregation. BMC Bioinformatics. 2009;10:62.

29. Zhong R, Allen JD, Xiao G, Xie Y. Ensemble-based network aggregation improves the accuracy of gene network reconstruction. PLoS One. 2014;9(11):e016319.

30. Le TD, Zhang J, Liu L, Liu H, Li J. miRLAB: an R based dry lab for exploring miRNA-mRNA regulatory relationships. PLoS One. 2015;10(12):e0145386.

31. Corpcor R package: https://cran.r-project.org/web/packages/corpcor/index.html. Accessed 14 Nov 2016.

32. Space R package: https://cran.r-project.org/web/packages/space/index.html

33. Network deconvolution matlab code: http://compbio.mit.edu/nd/code.html. Accessed 14 Nov 2016.

34. Pennisi E. ENCODE project writes eulogy for junk DNA. Science. 2012; 337(6099):1159–61.

35. Zhang J, Le TD, Liu L, Liu B, He J, Goodall GJ, Li J. Identifying direct miRNA-mRNA causal regulatory relationships in heterogeneous data. J Biomed Inform. 2014;52:438–47.

36. Karim SMM, Liu L, Le TD, Li J. Identification of miRNA-mRNA regulatory modules by exploring collective group relationships. BMC Genomics. 2015; 17 suppl 1:7.

37. Vergoulis T, Vlachos IS, Alexiou P, Georgakilas G, Maragkakis M, Reczko M, et al. TarBase 6.0: capturing the exponential growth of miRNA targets with experimental support. Nucleic Acids Res. 2012;40(D1):D222–9.

38. Xiao F, Zuo Z, Cai G, Kang S, Gao X, Li T. miRecords: an integrated resource for microRNA–target interactions. Nucleic Acids Res. 2009;37 suppl 1:D105–10.

39. Dweep H, Sticht C, Pandey P, Gretz N. miRWalk–database: prediction of possible miRNA binding sites by walking the genes of three genomes. J Biomed Inform. 2011;44(5):839–47.

40. Hsu SD, Tseng YT, Shrestha S, Lin YL, Khaleel A, Chou CH, et al. miRTarBase update 2014: an information resource for experimentally validated miRNA-target interactions. Nucleic Acids Res. 2014;42(D1):D78–85.

41. Shannon P, Markiel A, Ozier O, Baliga NS, Wang JT, Ramage D, Amin N, Schwikowski B, Ideker T. Cytoscape: a software environment for integrated models of biomolecular interaction networks. Genome Res. 2003;13:2498.

42. Szalay-Beko M, Palotai R, Szappanos B, Kovacs IA, Papp B, Csermely P. ModuLand plug-in for Cytoscape: determination of hierarchical layers of overlapping network modules and community centrality. Bioinformatics. 2012;28:2202–4.

43. Kanehisa M, Goto S, Kawashima S, Nakaya A. The KEGG databases at GenomeNet. Nucleic Acids Res. 2002;30(1):42–6.

44. Bindea G, Mlecnik B, Hackl H, Charoentong P, Tosolini M, Kirilovsky A, Fridman WH, Pagès F, Trajanoski Z, Galon J. ClueGO: a Cytoscape plug-in to decipher functionally grouped gene ontology and pathway annotation networks. Bioinformatics. 2009;25(8):1091–3.

45. Karginov FV, Hannon GJ. Remodeling of Ago2-mRNA interactions upon cellular stress reflects miRNA complementarity and correlates with altered translation rates. Genes Dev. 2013;27(14):1624–32.

46. Hafner M, Landthaler M, Burger L, Khorshid M, Hausser J, Berninger P, Rothballer A, Ascano Jr M, Jungkamp AC, Munschauer M, Ulrich A, Wardle GS, Dewell S, Zavolan M, Tuschl T. Transcriptome-wide identification of RNA-binding protein and microRNA target sites by PAR-CLIP. Cell. 2010; 141(1):129–41.

47. Kanzaki H, Ito S, Hanafusa H, Jitsumori Y, Tamaru S, Shimizu K, Ouchida M. Identification of direct targets for the miR-17-92 cluster by proteomic analysis. Proteomics. 2011;11(17):3531–9.

14

Quantifying differences in cell line population dynamics using CellPD

Edwin F. Juarez[1,2*], Roy Lau[1], Samuel H. Friedman[1], Ahmadreza Ghaffarizadeh[1], Edmond Jonckheere[2], David B. Agus[1], Shannon M. Mumenthaler[1] and Paul Macklin[1*]

Abstract

Background: The increased availability of high-throughput datasets has revealed a need for reproducible and accessible analyses which can quantitatively relate molecular changes to phenotypic behavior. Existing tools for quantitative analysis generally require expert knowledge.

Results: CellPD (cell phenotype digitizer) facilitates quantitative phenotype analysis, allowing users to fit mathematical models of cell population dynamics without specialized training. CellPD requires one input (a spreadsheet) and generates multiple outputs including parameter estimation reports, high-quality plots, and minable XML files. We validated CellPD's estimates by comparing it with a previously published tool (cellGrowth) and with Microsoft Excel's built-in functions. CellPD correctly estimates the net growth rate of cell cultures and is more robust to data sparsity than cellGrowth. When we tested CellPD's usability, biologists (without training in computational modeling) ran CellPD correctly on sample data within 30 min. To demonstrate CellPD's ability to aid in the analysis of high throughput data, we created a synthetic high content screening (HCS) data set, where a simulated cell line is exposed to two hypothetical drug compounds at several doses. CellPD correctly estimates the drug-dependent birth, death, and net growth rates. Furthermore, CellPD's estimates quantify and distinguish between the cytostatic and cytotoxic effects of both drugs—analyses that cannot readily be performed with spreadsheet software such as Microsoft Excel or without specialized computational expertise and programming environments.

Conclusions: CellPD is an open source tool that can be used by scientists (with or without a background in computational or mathematical modeling) to quantify key aspects of cell phenotypes (such as cell cycle and death parameters). Early applications of CellPD may include drug effect quantification, functional analysis of gene knockout experiments, data quality control, minable big data generation, and integration of biological data with computational models.

Keywords: Phenotype digitizer, Growth rate, Net birth rate, Phenotype comparison, Cell population dynamics, Parameter estimation, Computational modeling, Mathematical models, Open source, User friendly, MultiCellDS

Background

The growing adoption of systems biology and high-throughput experimental techniques increasingly demonstrates the need for quantitative and dynamic measurements to better characterize the complexity of biological systems [1, 2]. Measurements from a single experimental snapshot in time (e.g., an endpoint analysis) can often be misleading. For example, cell growth dynamics are influenced by cell density and nutrient availability, which are continually in flux. It is therefore better to use data across multiple time points when measuring cell phenotypes. On a broader scale, dynamical measurements can help to compare data across labs and identify protocol errors/discrepancies that may go unnoticed if data are only collected at a single time point.

With the increasing availability of high-throughput microscopy and high content screening (HCS), one can measure cell counts in many different environmental contexts with high precision over several time points [3]. These platforms can be used to link studies on molecular biology to observable, quantitative changes in cell behavior [4, 5]. However, as these experimental platforms have advanced,

* Correspondence: juarezro@usc.edu; Paul.Macklin@MathCancer.org
[1]Lawrence J. Ellison Institute for Transformative Medicine, University of Southern California, Los Angeles, California, USA
Full list of author information is available at the end of the article

they have allowed the generation of vast amounts of data which, in turn, require sophisticated bioinformatics tools for analysis [6]. In order to leverage these bioinformatics tools, a scientist needs to learn how to use complex computer programs [7] or work in bioinformatics-oriented programming environments (e.g., R, MATLAB, or Python), often without the benefit of graphical interfaces to assist them [8]. The need for specialized knowledge is compounded when the user, for example, may want to test several mathematical models of cell population dynamics (e.g., birth, death, and clearance rates) and choose one among them. Furthermore, the many ways in which a software package, the operating system in which it runs, and its required dependencies can be configured lead to challenges in data reproducibility [9–11]. While built-in functions in popular spreadsheet software such as Microsoft Excel can perform basic analyses on small, simple datasets (e.g., total cell counts for a few replicates), the functions cannot easily be extended to fit more sophisticated mathematical models that are better suited to analyzing more complex datasets. Furthermore, in the process of implementing more sophisticated mathematical models, a scientist can inadvertently introduce elusive bugs in their calculations [12, 13]. Therefore, tools which facilitate the replication and implementation of new analyses should be used in scientific computing.

To expedite the quantitative analysis of cellular phenotypes from experimental data while promoting data reproducibility, we introduce CellPD: a user-friendly cell line phenotype digitizer which obtains best-fit parameters and uncertainty estimates for cell birth, death, and population carrying capacity, based upon well-established "canonical" mathematical forms (e.g., exponential and logistic growth, with either net birth rates or separate birth, death, and dead cell clearance rates). CellPD has been designed to comply with the MultiCellDS data standard [14], therefore it can be expanded to record additional phenotype parameters, such as pharmacodynamics and cell motility. CellPD uses Microsoft Excel-compatible spreadsheets containing cell counts and experimental metadata as its sole input. The spreadsheets are also compatible with open source office suites such as LibreOffice. It is packaged as both a Python script (for those with existing Python 3 or Python 2.7 installations) and standalone executables for Windows and OSX, eliminating the need for installing and learning any additional software. Finally, CellPD generates locally-stored webpage outputs to clearly and intuitively present parameter estimation results with publication-quality tables and graphics as well as machine-readable XML outputs. These webpage reports also rank the quality of each fitted model to help the user choose the appropriate results without specialized mathematical knowledge. (See Additional file 1 for two examples of CellPD outputs.) CellPD is a beneficial tool for experimentalists, especially for those without a computational background or an existing partnership with a trained biostatistician or mathematician, as it provides a uniform and precise method for analyzing cellular dynamics data. Furthermore, CellPD not only computes growth rates from time series data, but also fits mathematical models in order to gain further insights from time series data, such as discerning between cytostatic and cytotoxic drug effects (as shown in the Results and Discussion section). While all the tasks that CellPD performs (automatic analysis by multiple models, uncertainty quantification, automatic ranking of fitted models by quality, user-friendly interfaces, publication-quality graphics, open data standards-compliant outputs for future data sharing, and utility in high-content screening experiments) are in principle possible today with significant custom scripts (e.g., in R, Python, or Matlab), no tools available today have already been tailored to these tasks and shared with the scientific community in a user-friendly format. In this article we describe some applications of CellPD and link to its source code (which will be updated as new features are added) so the scientific community can use it and build upon it.

Implementation

CellPD was designed with the following goals in mind:

- *Utility for experimental biologists:* The main goal of CellPD is to facilitate a quantitative description and analysis of cell population dynamics, using mathematical models that are powerful enough to make full use of increasingly detailed datasets.
- *Ease of learning and ease of use:* A scientist with no training in mathematical/computational modeling can learn how to use CellPD in an hour or less.
- *Robustness to sparsity in data:* CellPD can fit mathematical models to irregularly and sparsely sampled data requiring a minimum of two data points to fit the most basic mathematical model.
- *Accessibility and Shareability:* CellPD is open source and free to use with an unrestrictive license.
- *Extensibility:* We have planned extensions to CellPD's capabilities. In addition, its source code may be modified by any member of the scientific community, provided they follow the guidelines of the (permissive) MIT License.
- *Portability:* CellPD's Python code is packaged with a Python interpreter and all the required libraries; therefore, a computer running Windows, OSX, or Linux can run it without installing any software.

Previous work

There have been numerous efforts to compare and standardize cell line data across labs to ensure reproducibility and accuracy [15–17]. For example, an early effort by Osborne et al. characterized MCF7 breast cancer cells grown in four different laboratories [18]. Their investigation exposed substantial differences in the four labs' cell population doubling times. However, it can be difficult to discern any irregularities between cell cultures from different labs using doubling times for comparison, especially if those doubling times have not been computed to account for confluency effects.

Many tools have been developed specifically to estimate cell line growth parameters. Several were written in R such as cellGrowth [4, 19], grofit [20], and minpack.lm [21, 22]; MATLAB packages include PHANTAST [23] and SBaddon [24]; Ruby packages include BGFit [25]; and Python packages include ABC-SysBio [26] and GATHODE [27]. Although these packages are very useful, they are difficult to use for those without formal programming or bioinformatics expertise; moreover, the MATLAB-based packages require additional, costly software licenses. Some of these packages require data to be formatted in an inflexible format, for example requiring the data to be the output of a specific high content screening microscope. None of these tools and software packages are designed for regular use by scientists without extensive training with computational tools (i.e., they do not incorporate user-friendly inputs and outputs). They are also primarily designed for single-lab use. For example, they create outputs with lab-specific formats, rather than a standardized, well-annotated format suitable for curation and meta-analysis. These output formats make it challenging to compare different datasets from multiple laboratories. Thus, they do not answer the call for (big) data sharing [28]. While spreadsheet software such as Microsoft Excel can be used for some basic calculations (such as computing the net growth rate of an exponentially growing cell culture), fitting more sophisticated models is much more difficult and potentially error-prone. Hand-coded spreadsheet formulas and macros can hide subtle but critical errors (e.g., incorrect row/cell numbers), sometimes invalidating results or requiring paper retractions [12, 13].

CellPD aims to fill these workflow gaps by providing a user-friendly tool to estimate some key cell phenotype parameters from data acquired using common experimental platforms. In this paper, we lay the groundwork for a suite of tools that can be shared among different labs, that will help to facilitate data sharing and cross-lab meta-analyses. While the first release of CellPD is focused on quantifying cell cycle and death parameters, it has been built from the ground up to allow future extensions to quantify other phenotypic parameters, such as motility and pharmacodynamics, and to leverage anticipated advances in single-cell tracking to test hypotheses-driven phenotype parameters (e.g., an S-phase duration that depends upon glucose availability and cell size). Some early applications of CellPD may include quantification of drug effects on cancer cells (using data from assays containing varying drug doses), functional analysis of gene and other knockout experiments (such as the ones used in Gagneur et al. [4]), quantifying the effect of the microenvironment on cell phenotype (such as described in Garvey et al. [5]), cell culture quality control (by comparing estimated growth rates with a curated database), data mining (by extracting data from databases and analyzing it with CellPD) and generation of minable big data (by creating digital cell lines for each experiment that CellPD analyzes), and integration of biological data into computational models (by using CellPD's estimates as parameters for computational models).

In order to simplify the user interface, the primary input for CellPD is a Microsoft Excel spreadsheet that contains the experimental data (e.g., total cell counts at different time points), metadata related to the experimental setup (e.g., the name of the cell line and user notes), and the user information (e.g., the name and contact information of the CellPD user/data creator). Every CellPD download includes template spreadsheets to guide users through the spreadsheet layout and data formatting. In order to bring mathematical modelling expertise to biologists, the user-supplied data are then parsed and used to calibrate several mathematical models (for example exponential and logistic growth); the models were designed for extracting biologically-meaningful cell parameters from typical experimental data, such as the cell population growth rate and cell cycle information (such as population doubling times) with adjustment for confluence effects. See the Methods section for details on the mathematical models, their underlying assumptions, and a layperson's description of the mathematical models.

CellPD automatically selects one or more mathematical models for fitting based upon the type and quantity of data supplied by the user. For example, if the user provides cell counts at different time points but no cell viability, then CellPD (without extra input from the user) will calibrate models which describe cell counts but will omit models which describe cell death. Finally, a series of locally-stored webpage outputs are created to report and rank the fitted models (by quality of fit) and their parameters. Each fitted model includes a layperson's description of the underlying model assumptions and the biological meaning of each parameter. The results are reported in publication-quality figure (PNG, JPG, SVG, EPS, and PDF files) and table formats (XLSX and CSV files). For a list of all the outputs, see the Additional file 1.

CellPD has been designed to run on both finely-sampled data (measured at many time points, for example every 15 min, such as in yeast and bacteria experiments where dynamics have shorter timescales) and more sparsely (measured once per day, as is common in cancer cell culture experiments). We intend

CellPD to be used in a wide range of applications so it is robust to sparse data, but its accuracy improves when given more data points (as shown in Fig. 1). For a model with n parameters, CellPD requires at least n data points to estimate those parameters and at least $n + 1$ data points to estimate standard error of the mean for each parameter. The simplest model that CellPD can fit is the exponential model (a two-parameter model; see the methods section for more details); CellPD can analyze data so as long as there are at least 2 data points, in which case CellPD will assume exponential growth in between those two points.

Downloading CellPD

CellPD is hosted at SourceForge.net. Its source code, Windows and OSX executables can be downloaded at http://CellPD.sf.net [29]. Alternatively, tutorials and the most recent version of CellPD can also be downloaded at http://MultiCellDS.org/CellPD/ [30].

Results and discussion

We first validate CellPD's parameter estimates by comparing its results against another previously published cell growth parameter estimator, cellGrowth, and Excel 2016's built in functions (see Additional file 2 for a list of other tools that we examined). After validating the code, we demonstrate some key applications of CellPD by (1) evaluating its use in cell culture quality control by comparing two cultures of the same cell line (HCT 116, a colon cancer epithelial-like cell line) grown in two different media, (2) demonstrating its utility in HCS experiments

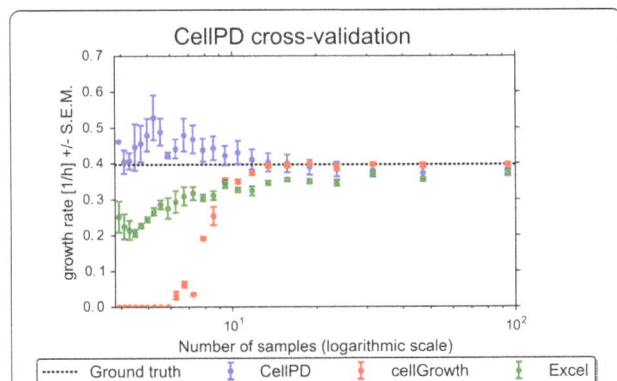

Fig. 1 Cross-validation of growth rates. Growth rates ± Standard error of the mean (SEM) of Yeast strain seg_07A grown in YPD media computed by cellGrowth (*red*), Excel (*green*) and CellPD (*blue*) using different number of sampling time points (i.e., at different sampling rates). All three tools correctly estimated the maximum growth rate for high sampling rates. For low sampling numbers (approximately less than 10 samples), the three tools become less accurate; cellGrowth lacks the necessary number of data points to perform data smoothing, Excel becomes inaccurate, but CellPD continues to estimate reasonable growth rates. Even at the limit case of only 3 sampled data points, CellPD provides a reasonable estimate (although it can no longer estimate SEM of the parameter)

by calculating dose-dependent cell birth and death parameters in a simulated drug screening experiment, and (3) using CellPD to determine whether these drugs are cytostatic or cytotoxic.

To test and quantify user-friendliness, we subjected CellPD, cellGrowth, and Excel to a series of timed use cases: installing all necessary software (first time setup for a new user), formatting data (6 replicates of yeast growth data from [4, 19] sampled every 15 min) running the software, and analyzing output. The lead author performed this test and recorded them (see Table 1 for links to the videos), and a group of 12 biologists volunteered to perform this test using CellPD either as individual testers (two participants) or in either pairs or groups of three. We also tested robustness by repeating the use cases for sparser data samples.

CellPD comparison and cross-validation
Advantages of CellPD over Excel's built-in fitting functions
Some of the computations that CellPD performs automatically can be replicated, albeit not automatically, using a spreadsheet. Spreadsheet software can use built-in regression functions to fit an exponential curve to experimental data. While other, more complex, dynamics can be modeled using a spreadsheet (such as logistic curves), these approaches push the limits of spreadsheet software by requiring hand-coded formulas, macros, or VisualBasic coding. Such calculations tend to be less reusable and more error prone, occasionally invalidating study results or even contributing to retractions [12, 13]. Hence, spreadsheets should only be used for minor calculations; more complex applications are best left to well-designed, purpose-built open source scientific software. CellPD is designed to estimate parameters accounting for expected behavior in cell population growth (such as logistic limitations). Additionally, CellPD is modular and extensible, thus allowing the user to fit multiple mathematical models at once, modify its current models, or even code new custom models. CellPD also creates high resolution figures which can be used in scientific publications, as well as minable XML reports and intuitive webpage reports.

Cross-validating CellPD
In order to cross-validate CellPD's parameter estimates, we utilized a publicly-available dataset from [4]. From this dataset, we selected the strain *seg_07A* grown in "YPD" medium (high in glucose) and computed the population growth rates using CellPD, Excel's linear regression function (linest), and cellGrowth. Not all of the replicates from that dataset have the same number of measurements, so we truncated the longer-time replicates so that they all have the same number of time points. We first computed the maximum growth rate of the data using all three tools, and defined cellGrowth's

Table 1 Comparison between CellPD, cellGrowth, and Excel

	CellPD	cellGrowth	Spreadsheet (Excel)
0.25 h sampling rate (95 samples) max_growth_rate ($\pm$SEM) h^{-1}	0.375 ($\pm$0.00963)	0.3971 ($\pm$0.0045)	0.3751 ($\pm$0.0045)
3 h sampling rate (7 samples) max_growth_rate ($\pm$SEM) h^{-1}	0.438 ($\pm$0.0326)	0.1916 ($\pm$0.0045)	0.3044 ($\pm$0.0096)
6 h sampling rate (3 samples) max_growth_rate ($\pm$SEM) h^{-1}	0.462 (N/A)	Breaks down	0.2519 ($\pm$0.0435)
Usability benchmark: T_{total}, lead author Usability benchmark: T_{total}, lead author	6 m 55 s	6 m 35 s	2 m 27 s
Usability benchmark: $T_{analysis}$, lead author Usability benchmark: $T_{analysis}$, lead author	5 m 43 s	3 m 17 s	2 m 27 s
Usability benchmark: T_{total}, 12 scientists unfamiliar with CellPD	Approximate range, in minutes [20, 30]	N/A	N/A
Usability benchmark: $T_{analysis}$, 12 scientists unfamiliar with CellPD	Approximate range, in minutes [14, 26]	N/A	N/A
Run time	~30 s	<1 s	<1 s
Video of timed use case	https://youtu.be/3xR9x_2pBKs	https://youtu.be/DO-LkVVgIIg	https://youtu.be/YCyCfzl7yFY
Comments on ease of use	Tutorial available, drag and drop option	Good tutorial to use custom data	Present on (viritually) every computer, many tutorials available online
Comments on input user interface	Executable file, command line option	Command-line in R	Manual input of formulas within the GUI
Comments on output user interface	Easy to read webpages with downloadable plots	Option to display and save an informative plot	Easy to create simple graphs
Typical UI	https://youtu.be/3xR9x_2pBKs?t=276	https://youtu.be/DO-LkVVgIIg?t=272	https://youtu.be/YCyCfzl7yFY?t=12
Typical output	https://youtu.be/3xR9x_2pBKs?t=406	https://youtu.be/DO-LkVVgIIg?t=391	https://youtu.be/YCyCfzl7yFY?t=158
Feature comparison matrix:			
Uncertainty quantification	Yes	If user computes it	If user computes it
Parametric growth models	Yes	Yes	If user creates them
Nonparametric growth models	No	Yes	If user creates them
Publication quality graphs	Yes	No	No
Fully annotated results in a standardized markup language	Yes	No	No
Open Source	Yes	Yes	No
Language written	Python	R	C/C++, C++/Java/Python
Required software to run	Spreadsheet editor (Excel, LibreOffice), Web browser (internet Explorer will suffice)	R	Excel, LibreOffice
Required computational expertise	No specialized experience	Working knowledge of R	Familiarity with spreadsheets

All three tools correctly estimate the growth rate when provided with a large number of samples. cellGrowth is more precise than CellPD for higher number of samples (i.e., shorter sampling intervals). However, even with fewer samples (i.e., larger sampling intervals), CellPD correctly estimates the growth rate (within the 95 % confidence interval). For fewer samples (i.e., larger sampling intervals), both cellGrowth and Excel become unreliable. CellPD is slower than cellGrowth or Excel for an experienced user, but CellPD does not require prior programming knowledge (unlike cellGrowth) and it also creates multiple useful outputs (Excel does not generate publication-quality graphs and cellGrowth has the option of creating a single graph which the user can export). CellPD is quicker to set up than cellGrowth, but it takes longer to run in order to create the multiple outputs. Excel usually requires no set up (beyond installing Microsoft Office), and it is often already installed in a research computer. The lead author computed the usability benchmark running a fixed, "clean" Windows 7 configuration on a Virtual Machine (VM). This VM included an installation of LibreOffice 5.1.4 and was run in a Lenovo ThinkPad Yoga with an Intel Core i7-4600U CPU with 8GB of Ram running Windows 10 (64-bit)

estimate (0.397 h^{-1} which corresponds to a population doubling time of 1.75 h) as the ground truth (the true value). We then used the three tools to estimate the growth rate using only a subset of the data (to simulate an experiment where samples are taken less frequently). We first fitted to the original data, which corresponds to 95 samples (and a sampling time interval of 0.25 h), then we reduced the number of samples roughly by half (the equivalent of sampling every 0.5 h), then we used roughly 1/3rd of the number of samples (the equivalent of sampling every 0.75 h), and so on, until we fitted to only 3 data points (i.e., sampling every 6 h). Figure 1 shows the estimated growth rates as the number of samples is varied (for the same experiment) using the three tools (see Additional file 3 for a figure where the x-axis represents sampling interval). CellPD, Excel, and cellGrowth correctly estimate the maximum growth rate for largest number of samples. CellPD generates reasonable growth parameter estimates over a broad range of data sampling rates, while both Excel and cellGrowth rapidly lose accuracy as the number of samples decreases. In particular, cellGrowth fails altogether when there are fewer than 6 samples. cellGrowth relies on smoothing methods arising from signal processing methods in order to provide accurate growth rates. Thus, it requires a substantial number of data samples. Neither CellPD nor Excel perform data smoothing, so they can estimate parameters with fewer data samples. While Excel can still compute the net growth rate with very few sampling points, its approach is prone to user bias (because users must choose which points to include in the linear regression and which to omit, e.g., due to confluence effects) and replication errors due to the manual nature of this computation. Hence there is a need for tools like CellPD which perform these analyses systematically and automatically. However, even tools which do not perform data smoothing are susceptible to problems caused by low number of data samples. Figure 1 shows that the uncertainty of the parameter estimates reduced for all three tools as the number of samples is increased, as described in Harris et al. [31], two data points are not enough to accurately model the dynamics of cell population growth.

Usability comparison testing

In order to quantitatively assess the usability of software package we used the following usability benchmark:

Usability benchmark

Measures how easily a new user can set up the package, starting from a "clean slate" using data formatted as outputted by a generic high content screening microscope (i.e., raw cell counts or optical densities at different times, each replicate recorded in its own file).

Use case:

Step 1: Install and setup software (included required dependencies)
Step 2: Reformat data for the software
Step 3: Run software
Step 4: Compute means and standard deviations of maximum growth rate

Total time (T$_{total}$) = Step 1 to Step 4.
T$_{analysis}$ = Step 3 and Step 4.

We recorded times measured by the usability benchmark as rows of a feature comparison matrix described in Table 1. We also recorded videos while the lead author performed the usability benchmark. Links to the videos are also listed in Table 1. To minimize user experience differences between the methods, the lead author spent time learning R and cellGrowth before performing the benchmark tests. Thus, the times reported are the minimum times to perform an analysis. For a novice user with no computational expertise, the differences would be larger. In particular, such users would require at least 1–10 days to learn introductory R before completing the benchmark use case, whereas users can complete the benchmark use case with CellPD without any additional training. The usability benchmark was repeated for CellPD by multiple volunteers with a wide range of computational experience, these times are also recorded in Table 1. The volunteers did not repeat the benchmarks on cellGrowth because using it requires familiarity with R.

Using CellPD to compare two cultures of the same cell line

A significant issue in biological experimentation is inter- and intra-laboratory variability [10, 15–17, 32]. For example, cells are typically grown in various media irrespective of the initial culturing methods. Moreover, even when culturing conditions are standardized, the use of biological reagents that are inherently variable in composition (such as fetal bovine serum (FBS)) can dramatically impact cell growth [33]. As result, it is important to devise tools such as CellPD to assess cell growth and perform quantitative quality control. We used CellPD and Excel (there are not enough sample points for cellGrowth to process these data) to compute the growth rate of two HCT116 cultures grown in two different media. In this paper, cells grown in McCoy's media are labeled "USC" and those grown in DMEM are labeled "WFU". Figure 2 shows the growth rates and the 95 % confidence interval (CI) as computed by CellPD and Excel. With either tool, the 95 % CI of the same HCT116 cells grown using these two different media do not overlap. This experiment was designed to observe different growth

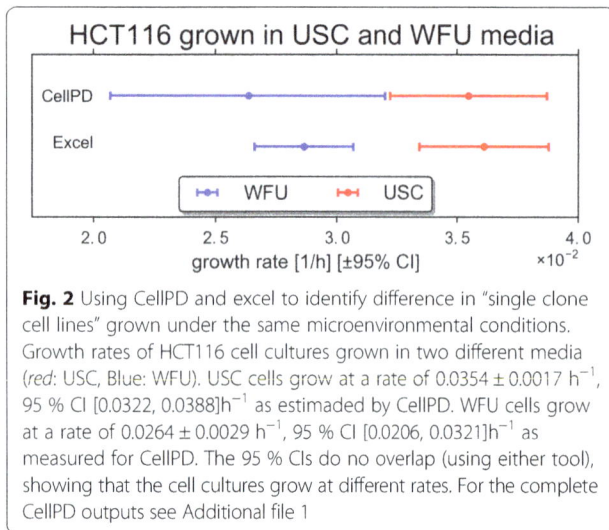

Fig. 2 Using CellPD and excel to identify difference in "single clone cell lines" grown under the same microenvironmental conditions. Growth rates of HCT116 cell cultures grown in two different media (*red*: USC, Blue: WFU). USC cells grow at a rate of 0.0354 ± 0.0017 h^{-1}, 95 % CI [0.0322, 0.0388]h^{-1} as estimaded by CellPD. WFU cells grow at a rate of 0.0264 ± 0.0029 h^{-1}, 95 % CI [0.0206, 0.0321]h^{-1} as measured for CellPD. The 95 % CIs do no overlap (using either tool), showing that the cell cultures grow at different rates. For the complete CellPD outputs see Additional file 1

dynamics when culturing the same cell line (HCT116) in two different media due to differences in glucose concentrations (McCoy's – 16 mM; DMEM – 25 mM) and other nutrients and growth factors [33]. CellPD allows for detection and quantification of such differences. These quantifications can be used to identify potential deviations from the protocol, such as in this case (where we intentionally used the wrong medium). This result highlights the importance of standardizing experimental conditions within and across laboratories. Such a large discrepancy in growth rates as a result of culture media could significantly alter the interpretation of standard tumor cell growth and their response to stimuli or inhibitors, such as chemotherapies.

Using CellPD to analyze (synthetic) high-content drug screening data

CellPD can be used to quantitate cell phenotype under multiple drug conditions using data generated in high content screening platforms. To demonstrate this feature, we first generated a synthetic drug screening dataset typical of high-content screening platforms and tested CellPD against these synthetic data. Specifically, we generated synthetic live and dead cell counts for two drugs at 5 doses, with 5 biological replicates for each experimental condition (a specific drug at a specific dose). Each simulated experimental measurement included normally-distributed noise to approximate both instrument and biological variability. See Additional files 4 and 5 for full details on the synthetic dataset, generating code, and the synthetic data themselves. CellPD was able to quantitate the net birth rate for each experimental condition, along with uncertainty estimates, and plot the responses. See Fig. 3a. Note that these analyses would be difficult to automate with cellGrowth, and would require substantial manual effort when using Excel.

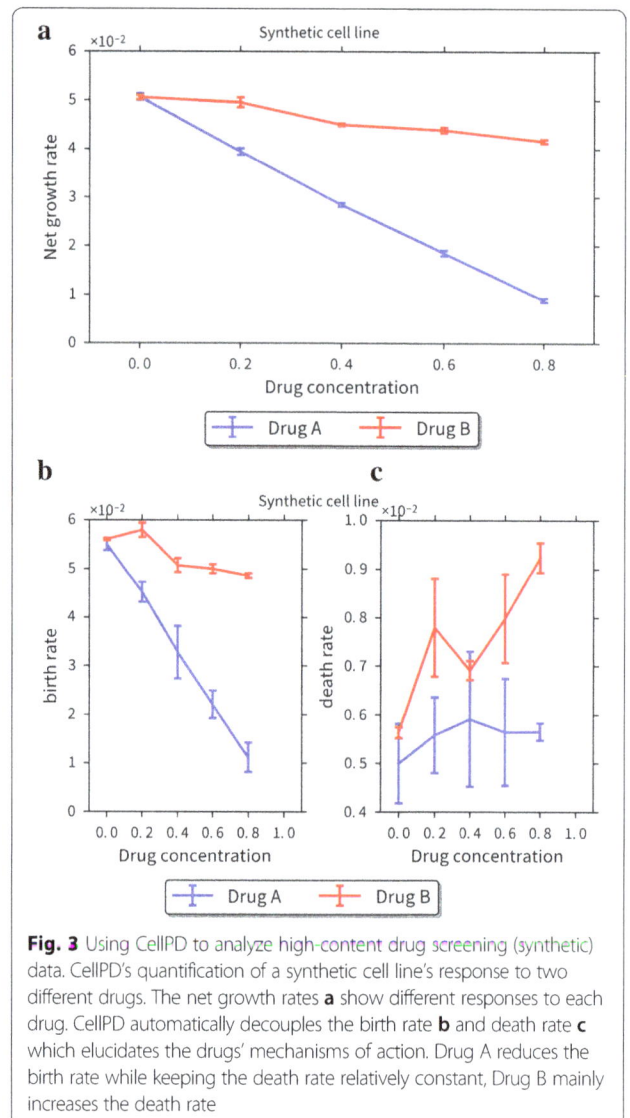

Fig. 3 Using CellPD to analyze high-content drug screening (synthetic) data. CellPD's quantification of a synthetic cell line's response to two different drugs. The net growth rates **a** show different responses to each drug. CellPD automatically decouples the birth rate **b** and death rate **c** which elucidates the drugs' mechanisms of action. Drug A reduces the birth rate while keeping the death rate relatively constant, Drug B mainly increases the death rate

We note that because we used a synthetic dataset with known "true" values, we can assess CellPD's accuracy and to test its robustness to measurement noise. We found that even with 10 % noise, CellPD was able to recover the correct parameter values for both simulated drugs at all doses, with a mean error of 2.6 % for the birth rate, 6.7 % for the death rate, and 3.0 % for the net growth rate. See Additional files 4 and 5 for more details.

Using CellPD to differentiate between cytotoxic and cytostatic drug effects

Continuing with the prior analysis, we used CellPD to separately quantitate the cell birth and death rate parameters for each experimental condition. See Fig. 3b-c. This additional analysis correctly reproduced the known birth and death rate parameters. Moreover, CellPD found that drug A was primarily cytostatic (it mainly influenced the cell birth rate) and drug B was cytotoxic (it chiefly affected the cell death rate). Because cellGrowth and Excel only fit the total population curve (i.e., they determine the net birth rate = birth − death), they cannot easily repeat this analysis nor help distinguish between cytostatic and cytotoxic drug effects here. In fact, by only examining the dose-dependent net birth rate, it can be (wrongly) inferred that drug A is cytotoxic because increasing the dose of drug A rapidly decreases its net birth rate to negative values, whereas drug B has a very shallow dose-net birth rate curve, which could be (wrongly) interpreted as arresting cell birth to maintain a zero or small net birth rate. Here, a more detailed analysis (facilitated by CellPD) was necessary to discern that counter to intuition from the net birth rate graphs, drug A is cytostatic, and drug B is cytotoxic. More sophisticated mathematical analyses—made possible by the broader class of models encoded in CellPD with straightforward usability—are necessary to discern the mechanism of action of each simulated drug compound.

Limitations

CellPD can currently only estimate parameters for predefined models (see Methods section for a list of the models); it does not currently support user-defined mathematical models except by directly modifying the source code. CellPD can directly analyze a dataset in which multiple environmental conditions are changing, e.g. if a dataset includes experiments where both the oxygenation and the media are changed independently. However, this functionality is under development. Currently publication-quality plots can only be generated for a single environmental condition with multiple values (e.g., different levels of oxygenation). When more than one environmental condition is changing at the same time, CellPD will perform the quantitative analysis and it will create plots, but those plots may not be as intuitive to read as the rest of the output files that CellPD generates. CellPD is designed to analyze population-level phenotypic data and is currently not equipped to provide single-cell dynamic information, although this is a feature that could be added in the future.

Future versions of CellPD

CellPD is open source: anyone may modify the code under the terms of the MIT License (MIT). We plan to update CellPD and release future versions which will include:

- Implement other common mathematical models of cell growth (e.g., Gompertzian).
- Implement additional cell cycle models (e.g., cells transitioning between G_0/G_1, S, G_2, and M phases) suitable to flow cytometry data.

- Automatically handle multiple-condition datasets for a single experimental factor (e.g., varying a drug level or the oxygenation). This function is in active development and testing. A beta version is now included with the main code of CellPD, with a fully-supported version anticipated soon.
- Automatically handle multiple-condition datasets for multiple experimental factors (e.g., varying a drug level and the oxygenation).
- Implement pharmacodynamics (drug response) models.
- Interactive web version.
- Interface with ORCID's API [34] to pull in user details automatically.
- Black-box optimization, allowing the user to define a custom mathematical model and estimate its parameters.
- Alternative minimization techniques such as cross-validation, bootstrap, genetic algorithms, and different heuristics for global optimization.

Conclusions

CellPD is a useful tool for biologists to analyze, quantify, and share phenotypic data. It can be used for data quality control and to identify unexpected changes in cell population dynamics. It can help automate analysis of high-content screening data, while distinguishing between cytostatic and cytotoxic drug effects. In all of its analyses, it makes use of biological and technical replicates to help assess uncertainty. CellPD facilitates integration of experimental data into computational models, rapidly quantifying critical phenotypic characteristics such as a cell line's net birth rate and producing consistent publication quality data.

Methods
Cell culture and reagents
HCT116 cells were acquired from ATCC and maintained in McCoy's media (USC HCT116). HCT116 cells cultured in DMEM media were also gifted to us by the Soker laboratory at Wake Forest University (WFU HCT116). All culture media was supplemented with 10 % fetal bovine serum (Gemini) and 1 % penicillin/streptomycin solution and cells were kept under standard tissue culture conditions (5 % CO_2, 37 °C).

Live/dead cell counts
Cells were seeded at 1,000 cells per well in 96 well plates (Corning #3904). Live and dead cell counts were determined at 0, 48, and 72 h using the Operetta high content screening (HCS) platform by PerkinElmer. Briefly, cells were stained with 5 μg/mL Hoechst 33342 (Invitrogen #H21492) and 7.5 nM Sytox Red (Life Technologies S34859) prior to imaging to identify cells as live or dead, respectively. Using the Harmony 3.5.2 (PerkinElmer)

image analysis software, individual cells were quantified as live or dead using nuclear segmentation algorithms and intensity thresholds. Each assay condition was performed in triplicate. All data points used in the analysis were taken before any confluence effects were apparent. Raw data can be found in Additional file 6.

Software implementation

CellPD is written in Python 3 (but a version compatible with Python 2.7 was created, in part, using 3to2), and it can be downloaded as source code to run in scripted python, or as a downloadable, self-standing executable (Windows and Mac). As shown in the graphical abstract, CellPD takes in a Microsoft Excel file as an input (or a compatible XLSX/XLS spreadsheet created or edited with open source software such as LibreOffice [35]). It creates a collection of webpage (HTML) reports as primary outputs, including a summary and ranking of fitted models and publication-quality graphics. CellPD also generates multiple supplementary outputs to facilitate data extraction. Supplementary outputs include: log-linear plots, black and white plots, figure captions, model descriptions, model equations in latex, table of estimated parameters in multiple formats (TEX, XLSX, and CSV), and a digital cell line XML file (a standardized, hierarchical reporting of cell phenotype and metadata; see the MultiCellDS project website for more details [14]). The HTML-based report is saved locally for later access.

In order to run properly, CellPD requires the following common Python libraries (note that they are included in the executable files):

- LMFIT: Non-Linear Least-Square Minimization and Curve-Fitting for Python. CellPD requires it to perform the minimization required in parameter estimation and to store the parameter values in LMFIT's Parameter's structure [36]. MIT license [37].
- NumPy: A basic numerical library for Python. CellPD requires it through to create numerical arrays and to use various mathematical functions [38]. BSD license [39].
- SciPy: A scientific computing library for Python. CellPD requires it to complement NumPy's mathematical functions and for its numeric integration algorithms [40]. Custom BSD compatible license [41].
- matplotlib: A common and versatile plotting library for Python. CellPD requires it to generate publication quality plots [42]. Custom BSD compatible license: [43].
- tzlocal: A library with time and locale tools for Python. CellPD requires it for generating time-zone sensitive time stamp [44]. License CC0 1.0 Universal [44].

- tabulate: A library for handling tables in Python. CellPD requires it for generating table of parameters for the reports [45]. MIT License [45].
- OpenPyXL: A library to read/write Excel files in Python. CellPD requires it to read the input files and to create the excel files that are supplemental outputs [46]. MIT/Expat license: [46].
- PyInstaller: A software to create stand-alone executable files for Windows and Mac. CellPD does not invoke PyInstaller, rather, we use PyInstaller to package CellPD with Python interpreters into a single executable file [47]. Modified GPL license to "have no restrictions in using PyInstaller as-is" [47].
- 3to2: A python script that converts most Python 3 code into Python 2. CellPD does not invoke 3to2, rather, we use 3to2 to translate most of CellPD's Python 3 code into Python 2.7 code, the rest of the code that is not translated by 3to2 is translated manually [48]. Apache Software License [49].

Further algorithmic detail

- Levenberg–Marquardt algorithm (LMA): CellPD uses LMFIT [50], which uses the LMA to minimizes an error metric to obtain an optimal, best fit between a supplied mathematical and data. LMA is a generalization of the steepest gradient descent method designed to solve smooth nonlinear problems (by using a "damping" on the gradient). A good explanation can be found in the book Numerical Recipes, The Art of Scientific Computing [51]. In our application of LMA, we used the following error metric:

$$\text{Error}_i = \text{Data}_i - \text{Simulation}_i$$

$$\text{WSSE} = \sum_i^N \frac{1}{\sigma_i} \frac{[\text{Error}_i]^2}{\text{Data}_i}$$

where sigma is the standard deviation of the ith data point. Thus, data with the largest uncertainty carries the least weight in the optimization (i.e., LMA prioritizes data with higher certainty). WSEE is the Weighted Sum of Squared Errors.

- MAPE and Reduced χ_v^2: In order to compare different models, the Mean Absolute Percentage Error (MAPE) and the reduced chi squared are computing using the formulas:

$$\text{MAPE} = \frac{1}{N} \sum_i^N 100 \frac{|\text{Error}_i|}{\text{Data}_i} \quad \chi_v^2 = \frac{\text{WSSE}}{N_{\text{data}} - N_{\text{parameters}}}$$

where N_{data} is the number of data samples and $N_{\text{parameters}}$ is the number of parameters of the model being evaluated. MAPE gives an intuitive sense of how well the model fit the individual data time points, on average, expressed as a percentage of the fitted data. χ_ν^2 adjusts this metric to account for the complexity of the fitted model (the difference between the number of measurements and the number of model parameters). It is meant to find a balance between fitting the data and simplifying the model, to avoid overfitting. (E.g., with enough parameters, a model could be made to fit every data point, even if it were a very poor model of the underlying biological system.) Both MAPE and χ_ν^2 are appropriate scores for ranking models applied to a given dataset so we provide both to allow the user decide which metric they prefer.

- Estimates of standard error of the mean (SEM): To estimate the SEM of the estimated parameter i, LMFIT uses the formula:

$$SEM_i = \sqrt{\chi_\nu^2 Cov(i, i)}$$

where $Cov(i, i)$ is the element of the covariance matrix in the ith row and the ith column. A good explanation of these numerical methods can be found in LMFIT's website [36, 50] and can be supplemented by [52].

Mathematical models implemented

- live: This is an exponential model that describes the growth of the live cells:

$$\frac{d[\text{Live}]}{dt} = \text{growth\_rate}[\text{Live}]$$

  Here, [Live] is the total number of live cells, and growth_rate is the net rate of live cell population growth (cell birth minus death).
- live_logistic: This modifies the exponential growth model to account for logistic growth effects (e.g., depletion of a growth substrate, or approaching cell confluence):

$$\frac{d[\text{Live}]}{dt} = \text{growth\_rate}\left(1 - \frac{[\text{Live}]}{L_{\text{cap}}}\right)[\text{Live}]$$

  [Live] is the total number of live cells, and growth_rate is the net rate of live cell population growth (cell birth minus death). L_{cap} is the total cell population carrying capacity (the maximum number of live cells).
- live_dead: This extends the exponential model to describes the changes of the live and dead cell populations:

$$\begin{aligned} \frac{d[\text{Live}]}{dt} &= \text{birth\_rate}[\text{Live}] - \text{death\_rate}[\text{Live}] \\ \frac{d[\text{Dead}]}{dt} &= \text{death\_rate}[\text{Live}] - \text{clearance\_rate}[\text{Dead}] \end{aligned}$$

[Live] is the total number of live cells, [Dead] is the total number of dead cells, birth_rate is the cell birth rate, death_rate is the cell death rate, and clearance_rate is the rate at which dead cells are cleared from the system (or the rate at which they become undetectable/unrecognizable to cell segmentation, cell counter, or other measurement techniques). 1/birth_rate is the mean time between cell divisions, and 1/clearance_rate is the mean time required for dead cells to degrade and/or cease to be recognized as cells by cell detection software.

- live_dead_logistic: This model modifies the live_dead model to account for logistic population effects:

$$\frac{d[\text{Live}]}{dt} = \text{birth\_rate}\left(1 - \frac{[\text{Live}]}{L_{\text{cap}}}\right)[\text{Live}] - \text{death\_rate}[\text{Live}]$$

$$\frac{d[\text{Dead}]}{dt} = \text{death\_rate}[\text{Live}] - \text{clearance\_rate}[\text{Dead}]$$

  [Live] is the total number of live cells, [Dead] is the total number of dead cells, birth_rate is the cell birth rate, death_rate is the cell death rate, and clearance_rate is the rate at which dead cells are cleared from the system (or the rate at which they become undetectable/unrecognizable to cell segmentation, cell counter, or other measurement techniques). L_{cap} is the total cell population carrying capacity (the maximum number of live cells). 1/birth_rate is the mean time between cell divisions, and 1/clearance_rate is the mean time required for dead cells to degrade and/or cease to be recognized as cells by cell detection software.
- total: This is an exponential model that describes the growth of the live and dead cells combined:

$$\frac{d[\text{Total}]}{dt} = \text{growth\_rate}[\text{Total}]$$

  [Total] = [Live] + [Dead] is the total number of cells, and growth_rate is the net rate of cell population growth (cell birth minus death).
- total_logistic: This modifies the exponential growth model to account for logistic growth effects (e.g., depletion of a growth substrate, or approaching cell confluence) in the total cell population:

$$\frac{d[\text{Total}]}{dt} = \text{growth\_rate}\left(1 - \frac{[\text{Total}]}{T_{\text{cap}}}\right)[\text{Total}]$$

  [Total] = [Live] + [Dead] is the total number of cells, and growth_rate is the net rate of cell population growth (cell birth minus death). T_{cap} is the total cell population carrying capacity (the maximum number of total cells).

Additional files

Additional file 1: Example of CellPD's outputs. This folder contains two examples of the outputs generated by CellPD (using the data from Fig. 2 and Additional file 6).

Additional file 2: Other tools that we examined. This file contains a list of five other tools we attempted to use and the reasons why we did not use them.

Additional file 3: Growth rate estimates versus sampling frequency. This file contains an alternative representation of the same data from Fig. 1. In this representation, the x-axis is the sampling interval (instead of the log 10 of the number of samples).

Additional file 4: Creation of synthetic data. This file contains a description of how the synthetic data for Fig. 3 was created, as well as the calculation of the estimation error.

Additional file 5: Synthetic data. This folder contains the synthetic data used for Fig. 3 and Additional file 4. Additionally, it contains the python scripts that were used to make the synthetic data, Fig. 3, and Additional file 4: Figure S9-1.

Additional file 6: Raw data. This spreadsheet contains the raw data used for Fig. 2, as well as the experimental setup.

Additional file 7: CellPD Tutorial. This file contains a step by step tutorial detailing how to add experimental data and metadata to CellPD's input file, as well as how to run CellPD in Windows, OSX, or Linux.

Additional file 8: Cuantificación de Dinámicas Poblacionales de Cultivos de Líneas Celulares Utilizando CellPD. This file is a Spanish summary of the manuscript.

Additional file 9: Source code. This folder contains the Python code with the version of CellPD used throughout this manuscript.

Additional file 10: Windows and OSX distributions of CellPD. This file contains simple instructions and links to download CellPD and use it on a Windows or an OSX computer.

Abbreviations

CellPD: Cell phenotype digitizer; CI: Confidence interval; FBS: Fetal bovine serum; HCS: High content screening; LMA: Levenberg-Marquardt algorithm; MAPE: Mean absolute percentage error; SEM: Standard error of the mean; VM: Virtual machine; WSSE: Weighted sum of squared errors;

Acknowledgments

Thanks to Kian Kani, Dan Ruderman, and Jonathan Katz for their helpful discussions and recommendations during development of CellPD and for their edits of this manuscript.
Thanks to the Soker group at Wake Forest University for supplying HCT116 cells and the media in which they were grown.
Thanks to Ruth Alvarez, Joanna Chen, Patrick Chiang, Chi-li Chiu, Carolina Garri, Collen Garvey, Ahyoung Joo, Sonya Liu, Katherin Patsch, Christine Solinsky, Anjana Soundararajan, Kiran Sriram, Kat Tiemann for their help in testing and timing the usage of CellPD.
Thanks to Jaime Juarez and David Juarez for their revisions of the Spanish translation of the abstract.

Funding

We thank the USC Center for Applied Molecular Medicine for generous resources, the National Institutes of Health (Physical Sciences Oncology Center grant 5U54CA143907 for Multi-scale Complex Systems Transdisciplinary Analysis of Response to Therapy (MCSTART), and 1R01CA180149), the Breast Cancer Research Foundation, the USC James H. Zumberge Research and Innovation Fund, and USC Provost's PhD fellowship for their generous financial support.

Authors' contributions

Conceptualization, EFJ, PM, SMM, and DBA; Methodology, EFJ, PM, SHF, AG, and EJ; Software, EFJ, SHF, AG, and PM; Investigation, EFJ and RL; Writing – Original Draft, EFJ, PM, and SMM, Writing – Review & Editing, EFJ, PM, SMM, SHF, AG, EJ, and DBA; Supervision, PM, SMM, EJ, and DBA. All authors read and approved the final manuscript.

Competing interests

The authors declares that they have no competing interests.

Author details

[1]Lawrence J. Ellison Institute for Transformative Medicine, University of Southern California, Los Angeles, California, USA. [2]Department of Electrical Engineering, Viterbi School of Engineering, University of Southern California, Los Angeles, California, USA.

References

1. Barbolosi D, Ciccolini J, Lacarelle B, Barlési F, André N. Computational oncology - mathematical modelling of drug regimens for precision medicine. Nature Reviews Clinical Oncology. 2016;13(4):242-54.
2. Karr JR, Williams AH, Zucker JD, Raue A, Steiert B, Timmer J, et al. Summary of the DREAM8 parameter estimation challenge: toward parameter identification for whole-cell models. PLoS Comput Biol. 2015;11:e1004096.
3. Zanella F, Lorens JB, Link W. High content screening: seeing is believing. Trends Biotechnol. 2010;28:237–45.
4. Gagneur J, Stegle O, Zhu C, Jakob P, Tekkedil MM, Aiyar RS, et al. Genotype-environment interactions reveal causal pathways that mediate genetic effects on phenotype. PLoS Genet. 2013;9:e1003803.
5. Garvey CM, Spiller E, Lindsay D, Chiang C-T, Choi NC, Agus DB, et al. A high-content image-based method for quantitatively studying context-dependent cell population dynamics. Sci Rep. 2016;6:29752.
6. Kitano H. Computational systems biology. Nature. 2002;420:206–10.
7. Gilbert D. Bioinformatics software resources. Brief Bioinform. 2004;5:300–4.
8. Hall BG, Acar H, Nandipati A, Barlow M. Growth rates made easy. Mol Biol Evol. 2014;31:232–8.
9. D. James, N. Wilkins-Diehr, V. Stodden, D. Colbry, C. Rosales, M. Fahey, et al. Standing together for reproducibility in large-scale computing: Report on reproducibility@ XSEDE. arXiv preprint arXiv:1412.5557. 2014.
10. Sandve GK, Nekrutenko A, Taylor J, Hovig E. Ten simple rules for reproducible computational research. PLoS Comput Biol. 2013;9:e1003285.
11. Soergel DA. Rampant software errors may undermine scientific results. F1000Research. 2014;3:303.
12. Baggerly KA, Coombes KR. Deriving chemosensitivity from cell lines: Forensic bioinformatics and reproducible research in high-throughput biology. The Annals of Applied Statistics. 2009;3(4):1309–34.
13. Herndon T, Ash M, Pollin R. Does high public debt consistently stifle economic growth? A critique of Reinhart and Rogoff. Camb J Econ. 2014;38:257–79.
14. Macklin P, Friedman SH. MultiCellDS MultiCellular Data Standard Project. Available: http://MultiCellDS.org. (Accessed 15 Sept 2015)
15. Facilitating reproducibility. Nat Chem Biol. 2013; 9: 345.
16. Begley CG, Ellis LM. Drug development: Raise standards for preclinical cancer research. Nature. 2012;483:531–3.
17. Mobley A, Linder SK, Braeuer R, Ellis LM, Zwelling L. A survey on data reproducibility in cancer research provides insights into our limited ability to translate findings from the laboratory to the clinic. PLoS One. 2013;8:e63221.

18. Osborne CK, Hobbs K, Trent JM. Biological differences among MCF-7 human breast cancer cell lines from different laboratories. Breast Cancer Res Treat. 1987;9:111–21.

19. Gagneur J, Neudecker A. cellGrowth: Fitting cell population growth models. R package version. 2012. Available Online from:https://www.bioconductor.org/packages/release/bioc/manuals/cellGrowth/man/cellGrowth.pdf. (Accessed 27 Sept 2015).

20. Kahm M, Hasenbrink G, Lichtenberg F. grofit: fitting biological growth curves with R. J Stat Softw. 2010;33:1–21.

21. Elzhov TV, Mullen KM, Bolker B. minpack. lm: R Interface to the Levenberg-Marquardt Nonlinear Least-Squares Algorithm Found in MINPACK. R package version. 2009

22. Elzhov TV, Mullen KM, Bolker B. minpack. lm: R Interface to the Levenberg-Marquardt Nonlinear Least-Squares Algorithm Found in MINPACK. R package version. 2009. Available Online from: https://cran.rproject.org/web/packages/minpack.lm/minpack.lm.pdf. (Accessed 30 Dec 2015).

23. Jaccard N, Griffin LD, Keser A, Macown RJ, Super A, Veraitch FS, et al. Automated method for the rapid and precise estimation of adherent cell culture characteristics from phase contrast microscopy images. Biotechnol Bioeng. 2014;111:504–17.

24. Schmidt H, Jirstrand M. SBaddon: high performance simulation for the Systems Biology Toolbox for MATLAB. Bioinformatics. 2007;23:646–7.

25. Veríssimo A, Paixão L, Neves AR, Vinga S. BGFit: management and automated fitting of biological growth curves. BMC bioinformatics. 2013;14:1.

26. Liepe J, Kirk P, Filippi S, Toni T, Barnes CP, Stumpf MPH. A framework for parameter estimation and model selection from experimental data in systems biology using approximate Bayesian computation. Nat Protoc. 2014;9:439–56.

27. Jung PP, Christian N, Kay DP, Skupin A, Linster CL. Protocols and programs for high-throughput growth and aging phenotyping in yeast. PLoS One. 2015;10:e0119807.

28. Sagiroglu S, Sinanc D. Big data: a review. 2013. p. 42–7.

29. Macklin P, Juarez EF. CellPD: Cell Phenotype Digitizer. Available: http://CellPD.sf.net. (Accessed 8 Feb 2016)

30. Macklin P, Juarez EF. MultiCellDS/CellPD: Cell Phenotype Digitizer. Available: http://MultiCellDS.org/CellPD/. (Accessed 8 Feb 2016).

31. Harris LA, Frick PL, Garbett SP, Hardeman KN, Paudel BB, Lopez CF, et al. An unbiased metric of antiproliferative drug effect in vitro. Nat Methods. 2016; 13(6):497–500.

32. Bell AW, Deutsch EW, Au CE, Kearney RE, Beavis R, Sechi S, et al. A HUPO test sample study reveals common problems in mass spectrometry-based proteomics. Nat Methods. 2009;6:423–30.

33. Masters JR, Stacey GN. Changing medium and passaging cell lines. Nat Protoc. 2007;2:2276–84.

34. ORCID. ORCID Connecting Research and Researchers. Available: http://orcid.org/. (Accessed 1 Oct 2015).

35. LibreOffice.org. LibreOffice The Document Foundation. Available: https://www.libreoffice.org/. (Accessed 2 Nov 2015)

36. Newville M. LMFIT Non-Linear Least-Square Minimization and Curve-Fitting for Python. Available: http://cars9.uchicago.edu/software/python/lmfit/. (Accessed 15 Sept 2015)

37. Newville M. LMFIT License. Available: http://cars9.uchicago.edu/software/python/lmfit/installation.html#license. (Accessed 15 Sept 2015).

38. N. developers. NumPy. Available: http://www.numpy.org/. (Accessed 15 Sept 2015)

39. N. developers. Numpy license. Available: http://www.numpy.org/license.html. (Accessed 15 Sept 2015)

40. S. developers. SciPy library. Available: http://www.scipy.org/scipylib/index.html. (Accessed 15 Sept 2015)

41. S. developers. SciPy license. Available: http://www.scipy.org/scipylib/license.html. (Accessed 15 Sept 2015)

42. Hunter J, Dale D, Firing E, Droettboom M, Matplotlib-development-team. matplotlib. Available: http://matplotlib.org/. (Accessed 15 Sept 2015).

43. Hunter J, Dale D, Firing E, Droettboom M and Matplotlib-development-team. matplotlib license. Available: http://matplotlib.org/users/license.html. (Accessed 15 Sept 2015)

44. Regebro L. tzlocal 1.2.2. Available: https://pypi.python.org/pypi/tzlocal. (Accessed 2 Dec 2015)

45. Astanin S. tabulate 0.7.5. Available: https://pypi.python.org/pypi/tabulate. (Accessed 26 Nov 2015)

46. Gazoni E, Clark C. openpyxl - A Python library to read/write Excel 2010 xlsx/xlsm files. Available: https://openpyxl.readthedocs.org/en/2.3.3/. (Accessed 29 Sept 2015)

47. Zibricky M, Goebel H, Cortesi D, Vierra D. PyInstaller. Available: http://www.pyinstaller.org/. (Accessed 30 Sept 2015)

48. Amenta J. Joe Amenta's Blog. Available: http://www.startcodon.com/wordpress/category/3to2/. (Accessed 16 Jan 2016)

49. Amenta J. 3to2 1.1.1. Available: https://pypi.python.org/pypi/3to2/1.1.1. (Accessed 16 Jan 2016)

50. Newville M, Stensitzki T, Allen DB, Ingargiola A. LMFIT: Non-linear least-square minimization and curve-fitting for python. 2014.

51. Press WH. Numerical recipes 3rd edition: The art of scientific computing. Cambridge: University press; 2007.

52. Gavin H. The Levenberg-Marquardt method for nonlinear least squares curve-fitting problems. 2011.Availble Online from: http://people.duke.edu/~hpgavin/ce281/lm.pdf.. (Accessed 15 Sept 2015).

15

Kinetic stability analysis of protein assembly on the center manifold around the critical point

Tatsuaki Tsuruyama

Abstract

Background: Non-linear kinetic analysis is a useful method for illustration of the dynamic behavior of cellular biological systems. To date, center manifold theory (CMT) has not been sufficiently applied for stability analysis of biological systems. The aim of this study is to demonstrate the application of CMT to kinetic analysis of protein assembly and disassembly, and to propose a novel framework for nonlinear multi-parametric analysis. We propose a protein assembly model with nonlinear kinetics provided by the fluctuation in monomer concentrations during their diffusion.

Results: When the diffusion process of a monomer is self-limited to give kinetics non-linearity, numerical simulations suggest the probability that the assembly and disassembly oscillate near the critical point. We applied CMT to kinetic analysis of the center manifold around the critical point in detail, and successfully demonstrated bifurcation around the critical point, which explained the observed oscillation.

Conclusions: The stability kinetics of the present model based on CMT illustrates a unique feature of protein assembly, namely non-linear behavior. Our findings are expected to provide methodology for analysis of biological systems.

Keywords: Protein assembly, Nonlinear kinetics, Fluctuations

Background

Numerical simulation based upon multi-parametric kinetic equations is the principal methodology for the analysis of the behavior of biological systems. Researchers often encounter a number of parameters in the governing equations of the system. Here, we introduce the center manifold theory (CMT) for simplification of the study of dynamic biological systems. CMT provides mathematical prescription for carrying out reduction of the number of parameters near the steady state, as well as information regarding the stability of the steady state. As a result, simulation is oriented to illustrate behavior around the critical point, at which system behavior drastically changes in the qualitative structure. The observable change is termed bifurcation, and the threshold values of the parameters are referred to as critical values or bifurcation values. The aim of this study was to provide a simple algorithm for the application of CMT to multi-parametric kinetic

equations, in order to clearly illustrate the behavior of the biological system. The CMT has been applied to the Lotka-Volterra model of predator–prey system to provide important simulation results [1, 2]. In addition, several pioneering studies have applied CMT to neural network analysis [3]. Time-delay and diffusive effects play important roles in bifurcation phenomena [1, 4]. However, to date, there are few applications of the CMT to biochemical reaction models. We previously reported a model of cell signaling systems using non-linear kinetics and demonstrated the phase transition phenomenon via a numerical simulation [5].

Pivotal protein-protein interactions during cytoskeleton formation were selected as the application model for the present CMT method. Among the interactions between protein monomers, tubulin and actin polymerization are well-known events that have been analyzed using the numerical method [6–10]. The physical robustness of the cytoskeleton is based on the biophysical properties of actin and tubulin. In particular, various mathematical models have been proposed to explain the kinetic behavior of tubulin assembly [6–11]. A

Correspondence: tsuruyam@kuhp.kyoto-u.ac.jp
Department of Pathology, Kyoto University, Graduate School of Medicine, Yoshida-Konoe-Cho 1, Kyoto, Kyoto Prefecture 606-8501, Japan

theory of polymerization of macromolecules has been established on the basis of the kinetic model of aggregation [12, 13]. Oosawa and Asakura previously reported that polymerization is similar to micelle formation or crystallization, and that there is a critical monomer concentration above which monomers effectively polymerize. The authors additionally suggested that the nucleation step represents the rate-limiting step for polymerization. Nucleation and growth occur in parallel during the progression of polymerization. There is a gap in free energy change between initial nucleation and progression of linear polymerization [13]. The stable nucleus for polymerization consists of trimers or tetramers, and the growth of aggregates through elongation/dissociation follows the formation of a thermodynamically unfavorable size of the nucleus. In the current study, we focused on the polymerization in the absence of *de novo* nucleation and the interaction between polymer and monomer (PM) interaction.

For stable growth, the lifespan of tubules is controlled by a guanidine triphosphate (GTP)-cap that forms at their ends [14]. The structure and motility of growing tubules is influenced by intrapolymeric Brownian motion and fluctuation; this provides elasticity to the microtubules [15]. Polymerization/de-polymerization is controlled by binding of adenosine triphosphate (ATP)/GTP, resulting in the assembly of monomeric proteins. The intermittent transition between slow growth and rapid shrinkage in polymeric assemblies of microtubules is termed *dynamic instability* [14]. Numerous models have been proposed to explain this instability; in particular, Zapperi and Mahadevan successfully identified two parameters: a structural mechanical parameter that characterizes the ratio of longitudinal to lateral interactions in an assembly, and a kinetic parameter that characterizes the ratio of timescales for growth and conformation change. These parameters serve to demarcate a region of uninterrupted growth from that of collapse [16].

In the current study, we consider a model assembly system that shows the unstable dynamics of assembly around the critical concentration of ATP/GTP. The present model utilizes CMT for describing the behavior of monomers in the solvent and polymer for simplification of analysis. We applied a kinetic model that unifies *de novo* nucleation and growth by considering the monomer-monomer interactions as a diffusion process. In addition, the diffusion process of the monomeric protein has been considered from the perspective of nonlinearity. According to Fick's law, the continuity of monomer concentration of c_i ($i = 1$, n) including chemical reaction items, may be described using diffusion coefficients D_i, kinetic coefficients k_i, and the concentrations of individual compounds c_i. Protein assembly is limited by the slow diffusion rate of monomer proteins,

which is a diffusion-rate limiting aggregation process. Therefore, diffusion items and reaction items cannot necessarily be separated; therefore, we described kinetic rate of c_i as follows [8]:

$$\frac{dc_i}{dt} = k_i D_i c_i + f(c_i) \tag{1}$$

Here, the first item on the right represents the diffusion rate. The second item, $f(c_i)$, denotes the function of kinetic rate of reactions other than the diffusion process. k_i represents a coefficient.

Methods
Numerical simulation Numerical calculations were performed using Mathematica 8 (Wolfram Research, Inc., Champaign, IL).

Results
General formulation of an assembly
The model consists of several steps: (i) the monomer achieves an interactive state by binding a cofactor (ATP/GTP) that provides the monomer with the ability to interact; (ii) the monomer itself possesses the ability to hydrolyze the cofactor and lose assembly activity; (iii) the monomer has the ability to exchange the inactive hydrolyzed cofactor (ADP/GDP) with an active non-hydrolyzed one; and (iv) ATP/GTP are supplied continuously from the external environment. The second requirement indicates a self-limiting property of the monomer that causes dynamic instability during monomer-monomer interaction. When examining protein interaction kinetics, analysis of the fluctuation in monomer concentrations was performed using Mathematica 9.

Protein interaction kinetics
The model scheme is shown in Fig. 1. There are three types of monomer: ATP/GTP-binding monomer X, ADP/GDP-binding monomer Y in the oligomer (W), and the released ADP/GDP-binding monomer Z. X has the higher assembly activity, and Y and Z have lower assembly activity. We set the oligomer concentration W to be a constant, as *de novo* assembly is considered much slower than monomer interaction in the steady state [11–14]. The individual steps are shown below:

First, X associates with the assembly nucleus W to be Y at the end of W.

$$X + W \rightarrow W + Y \quad (m_1; kinetic\,coefficients) \tag{2}$$

In the next step, the intermediate species Y is released to be Z:

Fig. 1 Scheme of monomer interaction. Individual globules or oblongs represent monomers X, Y, Z, and oligomer W. Kinetic coefficients, k_0, k_1, k_2, and k_3 are shown next to the arrows. Outside and inside signify the outside and inside of the cell, respectively. Y is located at the end of the oligomer W

$$Y \rightarrow Z \quad (m_2) \tag{3}$$

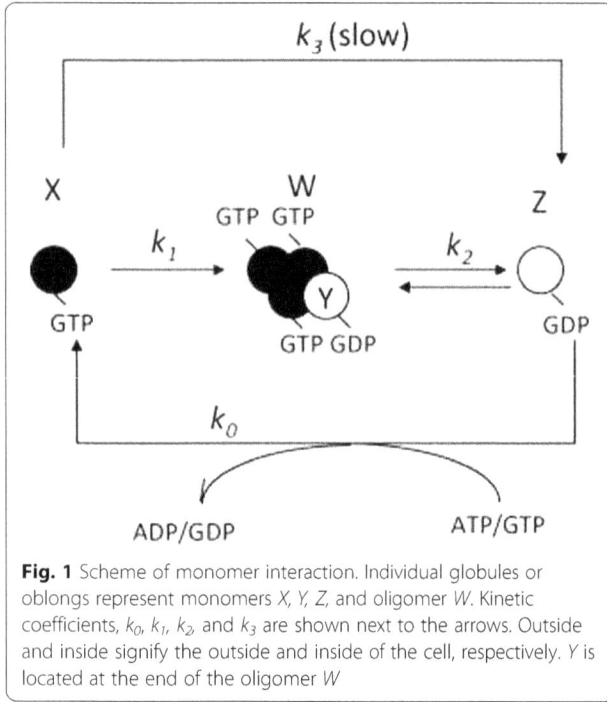

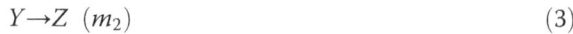

Z recovers its interaction activity by exchanging the active cofactor ATP/GTP (P) for the inactive cofactor ADP/GDP (P'), returning to X (see Fig. 1):

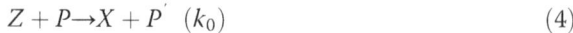

$$Z + P \rightarrow X + P' \quad (k_0) \tag{4}$$

In addition, direct slow conversion is supposed:

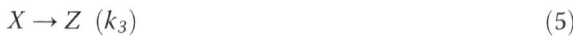

$$X \rightarrow Z \quad (k_3) \tag{5}$$

The kinetic equations were set according to the simple reaction cascade described above. We obtained equations for the protein interaction kinetics using the diffusion coefficient:

$$dX/dt = -m_1WX + k_0PZ - k_3X \tag{6}$$

$$dY/dt = m_1WX - m_2Y \tag{7}$$

$$dZ/dt = m_2Y - k_0PZ + k_3X \tag{8}$$

In addition, the total concentration of the monomer is maintained constant.

$$X + Y + Z = M \tag{9}$$

M, which represents the total concentration of the monomeric proteins, is maintained constant. A simple consideration of the diffusion-limited step implies that, when the kinetic rate can be described according to Fick's law using the diffusion coefficients D_X, D_Y and D_W then [17–19]:

$$m_1 \propto (D_X + D_W)/2 \approx D_X/2 \tag{10}$$

$$m_2 \propto (D_W + D_Y)/2 \approx D_Y/2 \tag{11}$$

As the oligomer diffusion rate is small, we set D_X, $D_Y >> D_W$. Therefore, m_1, and m_2 are substantially proportional to D_X and D_Y, respectively. Accordingly, kinetic coefficients k_1 and k_2 were defined as the proportional coefficients below:

$$m_1 \triangleq k_1 D_X$$
$$m_2 \triangleq k_2 D_Y$$

Rewriting (6), (7), and (8) using (10) and (11),

$$dX/dt = -k_1 D_X WX + k_0 PZ - k_3 X \tag{12}$$

$$dY/dt = k_1 D_X WX - k_2 D_Y Y \tag{13}$$

$$dZ/dt = k_2 D_Y Y - k_0 PZ + k_3 X \tag{14}$$

In the above equations, k_1 and k_2 represent the kinetic coefficients for the addition of the monomer to the oligomer and the release of the monomer from the oligomer, respectively.

In order to obtain the monomer concentration at the steady state of the reaction system, the right-hand side of Eqs. (12), (13), and (14) were set to be equal to zero and Eq. (9) were used to give:

$$X_e = \frac{D_Y k_2 Mp}{D_X k_3 k_2 + D_Y k_2 p + D_X D_Y k_1 k_2 W + D_X k_1 pW}$$

$$\sim \frac{D_Y k_2 M}{D_Y k_2 + D_X k_1 W} \tag{15}$$

$$Y_e = \frac{D_X k_1 MpW}{D_Y k_3 k_2 + D_Y k_2 p + D_X D_Y k_1 k_2 W + D_X k_1 pW}$$

$$\sim \frac{D_X k_1 MW}{D_Y k_2 + D_X k_1 W} \tag{16}$$

$$Z_e = \frac{D_Y k_2 M(k_3 + D_X k_1 W)}{D_Y k_3 k_2 + D_Y k_2 p + D_X D_Y k_1 k_2 W + D_X k_1 pW}$$

$$\sim \frac{D_Y k_2 M(D_X k_1 W)}{D_Y k_2 p + D_X k_1 pW} \sim 0 \tag{17}$$

In the above approximation, we omitted $D_X D_Y$ and k_3 as the diffusion coefficients and the direct conversion rate of X into Z is small.

Fluctuation of diffusion coefficient

Next, we considered the fluctuations of participant proteins using small letters x, y, and z:

$$X = X_e + x, Y = Y_e + y, Z = Z_e + z \qquad (18)$$

In Eq. (14), the subscript 'e' signifies values at the steady state.

In an assembly, monomers associate with other monomers. From Eq. (9),

$$x + y + z = 0 \qquad (19)$$

Therefore, the fluctuation y may be represented using $-x-z$. The fluctuation kinetics are thus provided by two parameters, namely x and z.

Given the nonlinearity during diffusion, we assume kinetic instability in the monomer-monomer interaction, and that the sensitivity of the assembly in response to environmental change may be evaluated. Indeed, the diffusion coefficient D_i of one macromolecule in the solution may generally be represented using the fluctuation concentration c_i:

$$\begin{aligned} D_i &= D_i{}^0 - \Sigma \alpha_{ij} c_i \\ &= D_i{}^0 - dD_i \text{ with } dD_i \equiv \Sigma \alpha_{ij} c_i (1 \le i \le 3, i \\ &= X, Y, Z) \end{aligned} \qquad (20)$$

c_j denotes the concentration of the solute, α_i is a coefficient, and D_i^0 is the diffusion coefficient when the fluctuation of monomeric protein is negligible. The dependence of the diffusion coefficient on the protein concentration has been reported [20, 21]. O'Learly reported that diffusion coefficients of proteins linearly decrease in proportion to the concentration, when the latter is sufficiently small. The fluctuation of the diffusion coefficient is obtained by considering the dependence of the coefficients on the concentration of the monomer from Eq. (20) [8]:

$$dD_X = \alpha x - \beta z \qquad (21)$$

$$dD_Z = \gamma x - \delta z \qquad (22)$$

Here, the fluctuation term αx ($\alpha > 0$) and γx ($\gamma > 0$) contributes to a decrease in D_X and D_Z, as higher assembly activity reduces diffusion. In contrast, an increase in the fluctuation terms βz ($\beta > 0$) and δz ($\delta > 0$) serves to increase the diffusion coefficients D_X and D_Z, as lower interaction or assembly activity increases diffusion. When the assembly activity of Z is lower, the fluctuation item δz is negligible, in accordance with the fluctuation kinetic equations given by (19), Eqs. (12), (14), (21), and (22):

$$\begin{aligned} dx/dt &= -k_1 W (D_X - \alpha x + \beta z)(X_e + x) \\ &\quad + k_0 P(Z_e + z) - k_3(X_e + x) \end{aligned} \qquad (23)$$

$$\begin{aligned} dz/dt &= k_3(X_e + x) - k_2(D_Y - \gamma x + \delta z)(Y_e + y) \\ &\quad - k_0 P(Z_e + z) \end{aligned} \qquad (24)$$

Here, y, fluctuation of intermediate species Y is negligible as the value is sufficiently small. In addition, we used the following equations to describe the balance in detail:

$$-k_1 D_X W X + k_0 P Z - k_3 X = 0 \qquad (25)$$

and

$$k_2 D_Y Y - k_0 P Z + k_3 X = 0 \qquad (26)$$

To simplify the notation in (23) and (24), we set:

$$\begin{aligned} k_1 D_X W &= D_1, k_1 D_X W \alpha = a, \ k_1 D_X W \beta = b \\ k_2 \gamma &= c, \ k_0 P = p, k_3 = k \end{aligned} \qquad (27)$$

and obtained:

$$\begin{aligned} dx/dt &= -(D_1 - aX_e - k)x + (-bX_e + p)z \\ &\quad + ax^2 - bxz \end{aligned} \qquad (28)$$

$$dz/dt = (k - cY_e)x - pz + cx^2 + cxz \qquad (29)$$

Eqs. (28) and (29) represent a master equation for the application of CMT.

Calculus simulation of concentration oscillations

For analysis of the behavior of the system, including multi-parameters, the examination of the linearization of behavior of the system near a steady state provides insights into the qualitative behavior of the system in the neighborhood of the point. In particular, the eigenvalues of the linear part of the governing kinetic equations enable determination of the stability of the system behavior. CMT is a rigorous formulation of this observation that enables the reduction of a large number of parameters [22].

Around the steady state $(x, z) = (0, 0)$ of Eqs. (28) and (29), the Jacobian matrix of $(dx/dt, dz/dt)$ is given by:

$$L = \begin{bmatrix} -D_1 + aX_e - k & -bX_e + p \\ k - cY_e & -p \end{bmatrix} \qquad (30)$$

Subsequently, the time-course of the monomer concentrations was simulated by substituting appropriate numerical values into Eqs. (28) and (29). The simulation results under the above conditions are shown in Fig. 2. A numerical calculation was performed over a sufficiently long period to evaluate the assembly trend. The

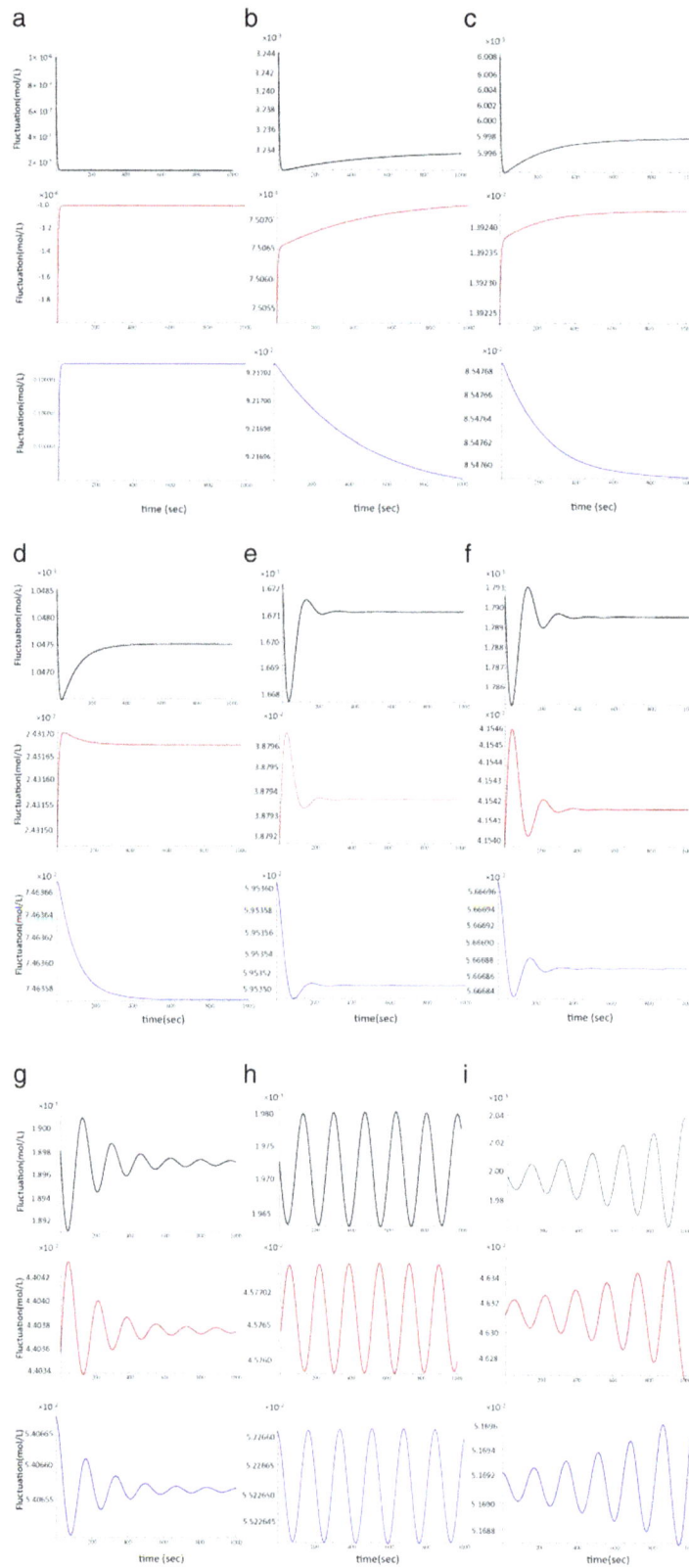

Fig. 2 (See legend on next page.)

(See figure on previous page.)

Fig. 2 Time-course of the fluctuation in monomer concentrations displays a oscillation. Diffusion of active cofactor binding monomer (X) and of inactive cofactor binding monomer (Z). p is (**a**) 0.000, (**b**) 0.001, (**c**) 0.002, (**d**) 0.004, (**e**) 0.008, (**f**) 0.009, (**g**) 0.01000, (**h**) 0.010705, (**i**) 0.011000. The graphs show plots of X (black), Y(red), and Z (blue). Lines represent the concentration of X and Z. The horizontal axis represents time ($0 \leq t \leq 1000$) and the vertical axis represents the concentration of X and Z. When p exceeds 0.01, oscillations are observed. The Mathematica (version 9, Wolfram Research, Inc., Champaign, IL) code for $p = 0.01$ is as follows: p = 0.01 X = ((D2 M p)/(D2 k + D2 p + D1 D2 W + D1 p W)) Y = ((D1 M p W)/(D2 k + D2 p + D1 D2 W + D1 p W)) Z = ((D2 M (k + D1 W))/(D2 k + D2 p + D1 D2 W + D1 p W)) M = 0.1 W = 1 D1 = 0.28 D2 = 0.012061855670103093` a = 150 b = 156 k = 0.005 c = 0.1 d = 0 NDSolve[{Derivative[1][x][t] == − (D1 - a X) x[t] + a x[t]^2 + (p - b X) z[t] - b x[t] z[t] - k x[t], Derivative[1][z][t] == k x[t] + c x[t]^2 + d x[t] z[t] - p z[t], x[0] == 1.`*^-6, z[0] == 1.`*^-6}, {x, z}, {t, 0, 3300}, MaxSteps ->50000] g001 = Plot[{X + x[t]} /. %, {t, 0, 1000}, PlotRange -> All, PlotStyle -> {RGBColor[0, 0, 0]}] g002 = Plot[{Y - x[t] - z[t]} /. %%, {t, 0, 1000}, PlotRange -> All, PlotStyle -> {RGBColor[1, 0, 0]}] g003 = Plot[{Z + z[t]} /. %%%, {t, 0, 1000}, PlotRange -> All, PlotStyle -> {RGBColor[0, 0, 1]}, PlotRange -> All] Show[g001, g002, g003]

steady-state concentrations of X and Z are given by Eqs. (15) and (17). The critical value of p_c is given by

$$det[\mathbf{L}] = (-D_1 + a\ X_e - k)(-p) - (k - cY_e)(-bX_e + p) = 0 \tag{31}$$

Here, the small affix c indicates the *critical point* of ATP/GTP concentration. Next, we conducted a simulation with values of $M = 0.1$, $X_e = 0.002$, $D_1 = 0.28$, $D_2 = 0.012$, $a = 150$, $b = 150$, $k = 0.005$, $c = 0.0$, and $d = 0$. Solving the above with respect to p with substitution of these values in Eq. (31), we find the critical value:

$$p_c = 0.011 \tag{32}$$

As a result, the fluctuations oscillate between decrease and increase in monomer concentrations, as shown in Fig. 2. When $p < p_c$, the fluctuation was found tobe attenuated (Fig. 2d) and the monomer concentration reached a plateau. However, when $p > p_c$, the fluctuation was found to diverge (Fig. 2f).

Evaluation of model stability using the center manifold around the equilibrium state

In order to demonstrate the Hopf-bifurcation around the critical state, in which $p = p_c$, we firstly defined the Jacobian matrix $\mathbf{L}_c$ according to (30) :

$$L_c = \begin{bmatrix} -D_1 + aX_e - k & -bX_e + p_c \\ k - cY_e & -p_c \end{bmatrix} \tag{33}$$

Using the eigenvectors of $\mathbf{L}_c$, $[\mathbf{l}_1\ \mathbf{l}_2]$, we performed the following coordinate transformation using novel parameters defined by following formulae:

$$\begin{bmatrix} u \\ v \end{bmatrix} = [\mathbf{l}_1\ \mathbf{l}_2]^{-1} \begin{bmatrix} x \\ z \end{bmatrix} \tag{34}$$

With reference to the numerical simulation (Fig. 2), when D_1, k, Y_e and p_c are sufficiently small,

$$\begin{bmatrix} l_1 & l_2 \end{bmatrix} = \begin{bmatrix} -a\ X_e & 1 \\ aX_e & 1 \end{bmatrix} \tag{35}$$

Eigenvalues λ of $\mathbf{L}_c$ are

$$\lambda^{\sim} aX_e, 0 \tag{36}$$

Using (34), we obtained:

$$du/dt = f_u(u,\ v)$$
$$= \left(D_1{}^2 p_c + k^2 v \left(b(u - v) + av \right) \right.$$
$$\left. + D_1 k \left(p_c(u - v) + buv \right) \right)/k(D_1 + k) \tag{37}$$

$$dv/dt = f_v(u,\ v) = (-k^2 v + D_1 u(-p_c + bv)$$

$$+ k(p_c(-u + v) + v(-D_1 + bu + (a - b)v))/(D_1 + k) \tag{38}$$

$$d\varepsilon/dt = 0 \tag{39}$$

The center manifold around the critical point ($p = p_c$) is then given as follows.

$$u = h(\varepsilon, v) = a_1 v^2 + a_2 v\varepsilon + a_3 \varepsilon^2 + a_4 v^3 + a_5 v^2 \varepsilon$$
$$+ a_6 v\varepsilon^2 + a_7 \varepsilon^3 + O(\varepsilon^4) \tag{40}$$

The eigenvalues of the Jacobian matrix, λ, in (33) are 0 and 2.9×10^{-4}. The given center manifold is an invariant manifold that is a tangent space of the center subspace, which is an eigenspace when the eigenvalue is equivalent to zero. The behavior of the fluctuation is complex when the real part of the eigenvalue is equivalent to zero. The above result in (36) shows that it is systematically necessary to analyze the behavior of the given system on the center manifold [22]. In order to analyze the behavior of the system, we investigated whether the change of the value in p around the critical value p_c gives u that satisfies $du/dt = 0$. When the two values of u are given, i.e., bifurcation of the system is shown, and oscillation and/ or other interesting behaviors may be predicted.

Using (40), we obtained:

$$u = (dv/dt)\partial h(u, \varepsilon)/\partial u + (d\varepsilon/dt)\partial h(u, \varepsilon)/\partial \varepsilon$$
$$= (2a_1 v + a_2 \varepsilon)f_u(u, v)$$

$$(41)$$

Using Eqs. (40) and (41), we then obtained:

$$(2a_1 v + a_2\varepsilon)f_u(u, v) = a_1 v^2 + a_2 v\varepsilon + a_3\varepsilon^2 + a_4 v^3$$
$$+ a_5 v^2\varepsilon + a_6 v\varepsilon^2 + a_7\varepsilon^3 + O(\varepsilon^4)$$

$$(42)$$

Solving Eq. (42) gives the coefficients of ai ($1 \le i \le 7$) in Eq. (40): $a3 = a7 = 0$. Substituting u in Eq. (40) given by v and ε into fv (u,v), we obtained the *kinetic stability equation* for fluctuation v using the coefficients n_i ($i = 1 \ldots, 7$) as follows:

$$dv/dt = n_1 v^2 + n_2 v\varepsilon + n_3\varepsilon^2 + n_4 v^3 + n_5 v^2\varepsilon + n_6 v\varepsilon^2$$
$$+ n_7\varepsilon^3 + O(\varepsilon^4)$$

$$(43)$$

Independent of the numerical values in Eq. (43),

$$n_3, n_6, n_7 = 0 \tag{44}$$

Then, we obtained:

$$dv/dt = n_1 v^2 + n_2 v\varepsilon + n_4 v^3 + n_5 v^2\varepsilon + O(\varepsilon^4) \tag{45}$$

By setting left-hand side equivalent to zero,

$$v = 0,$$
$$\left(-n_1 - n_5\varepsilon \pm \left((n_1 + n_5\varepsilon)^2 - 4n_2 n_4\varepsilon\right)^{1/2}\right)/2n_4 \tag{46}$$

We obtained an approximate solution to Eq. (46):

$$v = 0,$$
$$-2n_1 + (2n_2 n_4/n_1 - 2n_5)\varepsilon, -2n_1 n_2 \varepsilon/n_4 \tag{47}$$

From (40), we obtained the formulation of u using a constant coefficient c',

$$u \approx 0, c'(n_1/n_4)^2 \tag{48}$$

When D_1, k, p are sufficiently small, substituting $[l_1 l_2]$ in (35) into (33) approximately gives:

$$x = -(aX/k)u + v\tilde{\ }v \tag{49}$$

As a result, as we described v and x had two amplitudes in (47) demonstrating the oscillation of the fluctuation by bifurcation in v-ε plane (Fig. 2). Thus, stability analysis enables prediction of the behavior of the fluctuation around the critical point of the protein assembly system.

Discussion

In this work, we presented a model for protein assembly kinetics and analyzed the stability around the critical point using CMT. The nonlinear kinetic equations include three parameters (X, Y, and Z); however, only two are independent. In the simulations, ATP/GTP- or ADP/GDP-binding monomers periodically exhibit an oscillation between assembly and disassembly. This accurately reflects the microtubule kinetics showing unstable assembly [8].

To the best of our knowledge, this is one of the first reports on the application of CMT to the analysis of biological reaction systems [8]. The fluctuation of monomer concentrations was subjected to a perturbation expansion using a minimal increase in the supply of ATP/GTP near the concentration at the critical point. This mathematical method precisely treats nonlinear and multi-parameter systems around the critical point. The fluctuation kinetics is expected to change from convergence to divergence of the concentration fluctuation of the monomer, i.e., from stable to unstable around the critical point, as shown in Fig. 2. Because of this high sensitivity to the concentration of ATP/GTP, protein assembly is dynamically regulated by minimal changes in the supply of ATP/GTP, which in turn is subject to metabolic control. Via modeling of microtubule growth at the mesoscopic scale, Zapperi et al. showed the time course of transition between slow growth and rapid shrinkage during microtubule polymerization [16]. The present simulation may explain microscopic tubulin oligomerization oscillations during the initial steps of microtubule assembly. In addition, the present model may explain the transition from microscopic oligomerization and aggregation to mesoscopic scale assembly. The quantitative evaluation of the theoretical basis of protein assembly requires further investigation through experimental studies.

The present center manifold analysis enables elucidation of detailed behavior around the steady state and oscillatory dynamics of protein monomer concentration. In the current study, we further developed the mathematical framework using CMT and aimed to describe Hopf-bifurcation around the steady state, through the center manifold analysis, in a simple model. Coveney et al. have described a detailed model of protein assembly, including nucleation, its catalysis, and inhibition processes and performed a kinetic analysis of the initial nucleation process [23, 24]. The kinetic model of monomer-oligomerization or nucleation requires multiple concentrations that describe variable oligomer and nucleation. As shown by Coveney et al., it was challenging to predict the behavior of the system using a multiparametric (dimensional) center-manifold on the model. In the current study, we utilized a monomeric parameter and showed bifurcation of the system around the critical point. Therefore, CMT in a simple model serves to reduce the dimensions of the system to signal dimensions, as shown in this study. We expect that the theoretical framework in the current study provides a general theory of protein assembly kinetics and signal transduction [5, 25].

The analysis of growth kinetics of polymerization, according to Oosawa's model, has recently been reported by Michaels et al. [12]. The authors focused primarily on the dynamic phase of protein polymerization. As nucleation and polymerization to the nucleus proceeds in parallel, the analysis requires a detailed kinetic model of interaction between the nucleation and polymerization process [13, 14]. However, after the dynamic phase and before the plateau phase of polymerization, PM interactions are dominant during signal transduction. The present analysis illustrates the dynamics of cytoplasm in the stable state, and the corresponding influence on cell motility.

The present simulation was applied to such a quasi-statistic state, and the results revealed a possibility that oscillation of monomer concentration may occur when the ATP/GTP concentration exceeds the critical concentration. The calculated critical concentration of ATP/GTP, based on Hopf-bifurcation in (46) and amplitude of the fluctuation, coincided well with the amplitude obtained via the present simulation. The consistency in values in the simulation is important for verification. The periodic change in concentration may contribute to the coherently spatial-periodic viscosity and subsequently to contraction and elongation during cell movement. A recent study demonstrated the role of cytokeratin in determining keratinocyte motility and shape [26] and experimental method has greatly developed [27]. Structural components of cells determine non-linear cellular structural behavior and the contribution of various cell components to stability in response to mechanical stimuli. The cytoskeleton plays key roles in determining cellular stiffness. Our model captures non-linear structural behaviors including variable compliance along the cell surface and resistance to pull-out force [28]. The role of the microtubules in dynamic behavior may be investigated from the viewpoint of cell geometries. Measurement of the oscillation and determination of the critical concentration of ATP/GTP may reveal physical properties such as elasticity and compressibility.

Conclusion

Our model is expected to be useful for computing biophysical behavior in response to minute changes in GTP/ATP concentration using fluorescence intensity meter in two-dimensional cell geometries. In addition, the present model is expected to be suitable for use in algorithms for simulation of metabolic processes. Although further experimental studies are necessary for verification, our findings show that the current non-linear model of dynamic instability analysis captures the non-linear behaviors of cellular chemical and mechanical responses.

Abbreviations
ATP: Adenosine triphosphate ATP; CMT: Center manifold theory; GTP: Guanidine triphosphate; PM: Polymer and monomer (PM)

Acknowledgements
We are very thankful for Prof. Kenichi Yoshikawa, Dr. Masatoshi Ichikawa, and Prof. Masayuki Imai.

Funding
This work was supported by the Ministry of Education, Culture, Sport, Science and Technology, Japan, under the project name "Synergy of Fluctuation and Structure". Project number 2502. The funders had no role in the study design, data collection and analysis, decision to publish, or preparation of the manuscript.

Author' contributions
TT designed the study, implemented the final model, and wrote the manuscripts.

Competing interests
The authors declare that they have no competing interests.

References
1. Chang X, Wei J. Stability and Hopf bifurcation in a diffusive predator-prey system incorporating a prey refuge. Math Biosci Eng. 2013;10:979–96.
2. Zhang X, Zhao H. Bifurcation and optimal harvesting of a diffusive predator-prey system with delays and interval biological parameters. J Theor Biol. 2014;363:390–403.
3. Xiao M, Zheng WX, Cao J. Hopf bifurcation of an (n + 1) -neuron bidirectional associative memory neural network model with delays. IEEE Trans Neural Netw Learn Syst. 2013;24:118–32.
4. Yamaguchi I, Ogawa Y, Jimbo Y, Nakao H, Kotani K. Reduction theories elucidate the origins of complex biological rhythms generated by interacting delay-induced oscillations. PLoS One. 2011;6:e26497.
5. Tsuruyama T. A model of cell biological signaling predicts a phase transition of signaling and provides mathematical formulae. PLoS One. 2014; (in press).
6. Hazra P, Inoue K, Laan W, Hellingwerf KJ, Terazima M. Tetramer formation kinetics in the signaling state of AppA monitored by time-resolved diffusion. Biophys J. 2006;91:654–61.
7. Wu Z, Wang HW, Mu W, Ouyang Z, Nogales E, Xing J. Simulations of tubulin sheet polymers as possible structural intermediates in microtubule assembly. PLoS One. 2009;4:e7291.
8. VanBuren V, Cassimeris L, Odde DJ. Mechanochemical model of microtubule structure and self-assembly kinetics. Biophys J. 2005;89: 2911–26.
9. Symmons MF, Martin SR, Bayley PM. Dynamic properties of nucleated microtubules: GTP utilisation in the subcritical concentration regime. J Cell Sci. 1996;109:2755–66.

10. Voter WA, Erickson HP. The kinetics of microtubule assembly. Evidence for a two-stage nucleation mechanism. J Biol Chem. 1984;25:10430–8.

11. Zilberman M, Sofer M. A mathematical model for predicting controlled release of bioactive agents from composite fiber structures. J Biomed Mater Res A. 2007;80:679–86.

12. Oosawa F, Kasai M. A theory of linear and helical aggregations of macromolecules. J Mol Biol. 1962;4:10–21.

13. Michaels TC, Garcia GA, Knowles TP. Asymptotic solutions of the Oosawa model for the length distribution of biofilaments. J Chem Phys. 2014;140: 194906.

14. Chretien D, Jainosi I, Taveau JC, Flyvbjerg H. Microtubule's conformational cap. Cell Struct Funct. 1999;24:299–303.

15. Oosawa F, Asakura S. Thermodynamics of the Polymerisation of Proteins. New York and London: Acdemic Press; 1975. p. 204.

16. Zapperi S, Mahadevan L. Dynamic instability of a growing adsorbed polymorphic filament. Biophys J. 2011;101(2):267–75.

17. Wustner D, Solanko LM, Lund FW, Sage D, Schroll HJ, Lomholt MA. Quantitative fluorescence loss in photobleaching for analysis of protein transport and aggregation. BMC Bioinformatics. 2012;13:296.

18. Dorsaz N, De Michele C, Piazza F, Foffi G. Inertial effects in diffusion-limited reactions. J Phys Condens Matter. 2010;22:104116.

19. Kasche V, de Boer M, Lazo C, Gad M. Direct observation of intraparticle equilibration and the rate-limiting step in adsorption of proteins in chromatographic adsorbents with confocal laser scanning microscopy. J Chromatogr B Analyt Technol Biomed Life Sci. 2003;790:115–29.

20. O'Leary TJ. Concentration dependence of protein diffusion. Biophys J. 1987; 52:137–9.

21. Kenneth H. A diffusion model with a concentration-dependent diffusion coefficient for describing water movement in legumes during soaking. J Food Sci. 1983;48:618–23.

22. Guckenheimer J, Holmes PJ. Nonlinear oscillations, dynamical systems, and bifurcations of vector fields. 1st ed. New York: Springer; 1983. p. 1–459.

23. Wattis JAD, Coveney PV. Analysis of a generalized becker-doring model of self-reproducing micelles. Proc T Soc Lond A. 1996;452:2079–102.

24. Wattis JAD, Coveney PV. Mesoscopic models of nucleation and growth processes : a challenge to experiment. Phys Chem Chem Phys. 1999;1:2163–76.

25. Babu CVS, Song EJ, Yoo YS. Modeling and simulation in signal transduction pathways: a systems biology approach. Biochimie. 2006;88:277–83.

26. Nakata T, Okimura C, Mizuno T, Iwadate Y. The role of stress fibers in the shape determination mechanism of fish keratocytes. Biophys J. 2016;110: 481–92.

27. McGarry JG, Prendergast PJ. A three-dimensional finite element model of an adherent eukaryotic cell. Eur Cell Mater. 2004;7:27–33.

28. Burk AS, Monzel C, Yoshikawa HY, Wuchter P, Saffrich R, Eckstein V, et al. Quantifying adhesion mechanisms and dynamics of human hematopoietic stem and progenitor cells. Sci Rep. 2015;5:9370.

16

An efficient algorithm for identifying primary phenotype attractors of a large-scale Boolean network

Sang-Mok Choo[1] and Kwang-Hyun Cho[2*]

Abstract

Background: Boolean network modeling has been widely used to model large-scale biomolecular regulatory networks as it can describe the essential dynamical characteristics of complicated networks in a relatively simple way. When we analyze such Boolean network models, we often need to find out attractor states to investigate the converging state features that represent particular cell phenotypes. This is, however, very difficult (often impossible) for a large network due to computational complexity.

Results: There have been some attempts to resolve this problem by partitioning the original network into smaller subnetworks and reconstructing the attractor states by integrating the local attractors obtained from each subnetwork. But, in many cases, the partitioned subnetworks are still too large and such an approach is no longer useful. So, we have investigated the fundamental reason underlying this problem and proposed a novel efficient way of hierarchically partitioning a given large network into smaller subnetworks by focusing on some attractors corresponding to a particular phenotype of interest instead of considering all attractors at the same time. Using the definition of attractors, we can have a simplified update rule with fixed state values for some nodes. The resulting subnetworks were small enough to find out the corresponding local attractors which can be integrated for reconstruction of the global attractor states of the original large network.

Conclusions: The proposed approach can substantially extend the current limit of Boolean network modeling for converging state analysis of biological networks.

Keywords: Boolean network, Cell phenotypes, Attractors, Hierarchical partition, Systems biology

Background

In the realm of systems biology, mathematical modeling is essential to unravel the hidden principles underlying complex biological phenomena [1]. Among various mathematical modeling frameworks, the Boolean network is particularly useful for modeling large-scale biomolecular regulatory networks as it is a parameter-free logical model and thereby we can avoid parameter estimation which is often a critical limitation in mathematical modeling of such large-scale networks [2–5]. Once a Boolean network model is obtained, the converging state characteristics of the modeled network can be investigated by identifying attractor states which were known corresponding to cell phenotypes [6–11]. Finding attractors of interest is, however, an NP-hard problem [12–14] since we have to search the full state space and this is only possible for small networks with less than about 20 nodes [15, 16].

To tackle such a problem, there have been several attempts to reduce the original Boolean network model by eliminating some nodes or logically simplifying Boolean functions [17–27], or focusing only on point attractors [28, 29]. Another attempt was partitioning the original large Boolean network into smaller blocks and reconstructing the original attractors by integrating the local attractors of partitioned blocks [30, 31]. However, none of these could resolve the fundamental problem of computational complexity since both the reduced network and the partitioned subnetwork are still too large in most cases of biological networks. Even if the reduced network is small enough to search the full state

* Correspondence: ckh@kaist.ac.kr; http://sbie.kaist.ac.kr/
[2]Department of Bio and Brain Engineering, Korea Advanced Institute of Science and Technology (KAIST), Daejeon 34141, Republic of Korea
Full list of author information is available at the end of the article

space, the resulting attractor states of the reduced network can be different from the attractor states of the original network (this will be shown in the Results section). The existing partitioning approaches retain the complicated logic of the original network even after partitioning such that the whole set of attractors can still be found from the partitioned subnetworks. We found that the fundamental limitation of the previous partitioning methods lies in this point. To overcome such a limitation, we propose a different approach in this paper. The main idea is focusing only on particular phenotypic attractors of interest and simplifying the Boolean update rules of the original network by introducing some constraint equations such that the particular attractors are not affected. In this way, we can efficiently reduce the original large network. Then, we further partition the reduced network hierarchically and find out the local attractors of each partitioned subnetwork. We can finally obtain the global attractors for representing the particular phenotype in the original network by sequentially concatenating the local attractors. Our approach is based on the previous concept of strongly connected component (SCC) [30, 31], but the main difference is that our approach can efficiently find out the particular phenotypic attractors of interest by hierarchically partitioning the reduced network obtained by simplifying the state update rules using some constraint equations, whereas the previous approach attempts to partition the original network while retaining all the complicated state update rules and thereby results in still a large subnetwork even after partitioning.

We validated the usefulness of our approach by applying it to several large and complicated biological Boolean network models.

Methods

In this section, we describe the procedure of finding global attractors for the phenotype of interest from a large and strongly connected network and explain how to hierarchically partition the network into smaller-size subnetworks. We then describe a way of constructing the global attractors by sequential concatenation of local attractors of the subnetworks.

Procedure for construction of global attractors by concatenating local attractors

Let us consider that a large and complicated synchronous Boolean network is given and the states of nodes in the network are updated by logic functions or threshold (sign) functions. Our goal is to find attractors for a phenotype of interest (for example, apoptosis or proliferation) in the given network. The idea is to transform the original update rules into simplified update rules by fixing the state values of some nodes (Steps 1 and 2) and to convert the

original network into a simplified network using the simplified update rules. After hierarchically partitioning the simplified network into their SCCs (Step 3), the local attractors of each SCC can be found by using a full search algorithm (Step 4). Finally, we can find the global attractors by concatenating the local attractors (Step 5). Each step of the procedure is as follows:

Step 1-1. Determine the fixed state values of nodes for external environment. The environment is a stimulus, a tumour-promoting microenvironment or perturbation of a node as in [6, 9, 10]. Nodes for representing the environment are referred to as "external nodes".
Step 1-2. Insert the fixed values of the external nodes into the system of update equations. Applying Step1-2, we can obtain the fixed state values of some nodes, which are different from the external nodes and referred to as "secondary-external nodes". As a result, the original update rules are divided into two parts. The first is the set of external and secondary-external nodes (ESENs) with their fixed values. The other is the new update rules for those nodes except ESENs, which are called as "the semi-simplified update rules". We refer to the two parts as "the external condition" (Fig. 1 STEP1). Examples for Step 1 are given in S.Example 1 in Additional file 1 (see Additional files 2(c), 3(b), 4(c), 5(b) and 6(c) for details).
Step 2-1. Determine the nodes and their fixed state values for the definition of the phenotype. We consider networks with a node that is the phenotype as in [6, 9, 10]. For instance, in case that the network has the state update rule, Proliferation* = p70 & MYC & !p21, the network is said to have the node for proliferation, where the symbols * and (&,!) denote the next time step and the logic operators (and, or), respectively. Consistent activation of the phenotype is defined by both nodes and their fixed state values, which are referred to as "phenotype nodes and values". For the update rule, Proliferation* = p70 & MYC & !p21, the phenotype nodes are (p70, MYC, p21) with fixed values (1,1,0). Therefore the desired global attractor states for the proliferation phenotype must satisfy the constraints (p70, MYC, p21) = (1,1,0). The step is described in detail in Additional file 1.
Step 2-2. Insert the fixed values of the phenotype nodes into the semi-simplified update rules. After applying Step 2-2, we can obtain the fixed state values of some nodes, which are neither ESENs nor phenotype nodes and referred to as "secondary-phenotype nodes". Consequently, the semi-simplified update equations are divided into three parts. The first is the set of phenotype and secondary-phenotype nodes (PSPNs) with their fixed values. The second is constraint equations, which are introduced due to the constraints

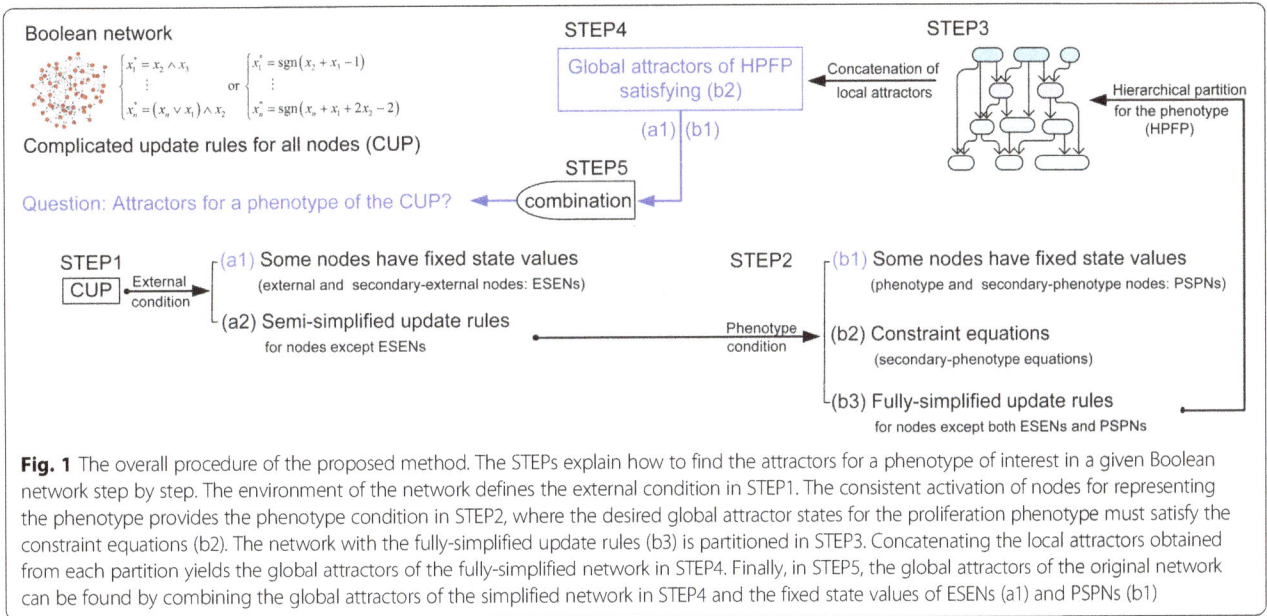

Fig. 1 The overall procedure of the proposed method. The STEPs explain how to find the attractors for a phenotype of interest in a given Boolean network step by step. The environment of the network defines the external condition in STEP1. The consistent activation of nodes for representing the phenotype provides the phenotype condition in STEP2, where the desired global attractor states for the proliferation phenotype must satisfy the constraint equations (b2). The network with the fully-simplified update rules (b3) is partitioned in STEP3. Concatenating the local attractors obtained from each partition yields the global attractors of the fully-simplified network in STEP4. Finally, in STEP5, the global attractors of the original network can be found by combining the global attractors of the simplified network in STEP4 and the fixed state values of ESENs (a1) and PSPNs (b1)

in Step 2-1 and explained in S.Example 3 in Additional file 1. The third is the new update rules for nodes except both ESENs and PSPNs, which are called as "the fully-simplified update rules". The three parts are referred to as "the phenotype condition" (Fig. 1 STEP2). Examples for Step 2 are given in Additional files 2(d), 3(c), 4(d), 5(c) and 6(d). Note that if update rules are defined by threshold (sign) functions, then constraint equations are changed to constraint inequalities as in the SCP network of the Result section and S.Remark 1 of Additional file 1.

Step 3. Construct the hierarchical partition of the network corresponding to the fully-simplified update rules.

Note that this partition is for nodes except both ESENs and PSPNs not for all nodes in the original network. The partition is referred to as "the hierarchical partition for the phenotype (HPFP)" as in Fig. 1 STEP3 and this step is explained in the Construction of HPFP section below. Examples for Step 3 are shown in Fig. 2 and the Result section.

Step 4-1. Find the local attractors of subnetworks in the HPFP.

Step 4-2. Construct the global attractors of the HPFP (Fig. 1 STEP4) by sequential concatenation of the local attractors satisfying the secondary-phenotype equations. Step 4 is explained in detail in the section

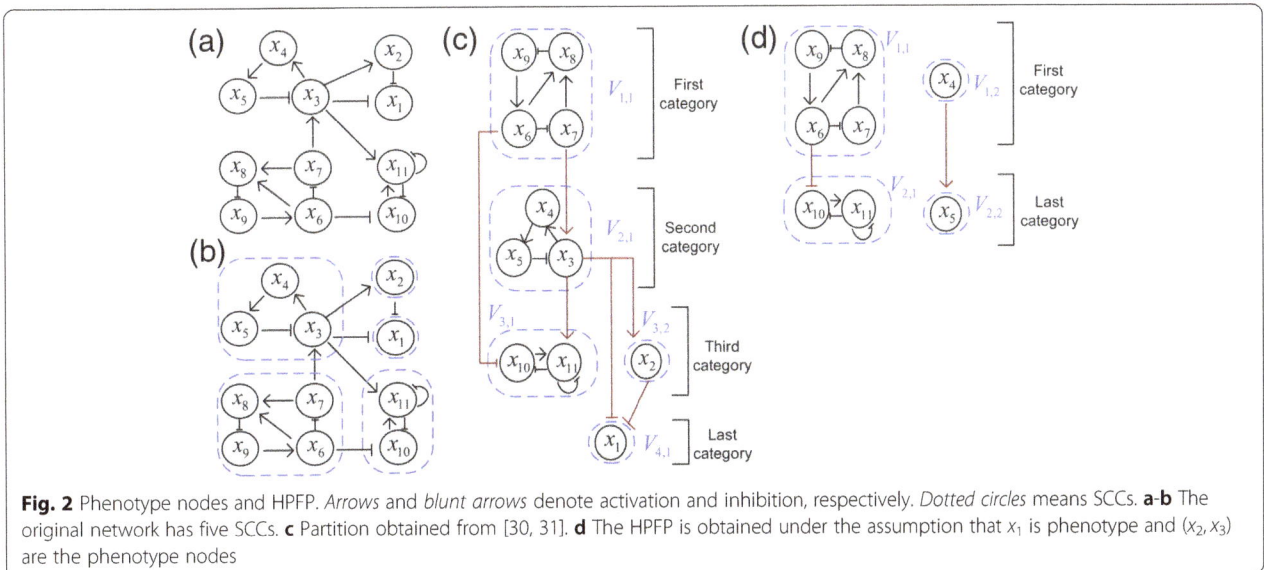

Fig. 2 Phenotype nodes and HPFP. *Arrows* and *blunt arrows* denote activation and inhibition, respectively. *Dotted circles* means SCCs. **a-b** The original network has five SCCs. **c** Partition obtained from [30, 31]. **d** The HPFP is obtained under the assumption that x_1 is phenotype and (x_2, x_3) are the phenotype nodes

of 'How to concatenate the local attractors of subnetworks?' and Additional files 7 and 8.

Step 5. Combine the global attractors of the HPFP with fixed values of ESENs and PSPNs as shown in STEP5 of Fig. 1. After applying Step 5, we can obtain all the desired phenotype attractors of the original network. Examples for such phenotype attractors are given in Additional files 2(g), 3(f), 4(f)-(g), 5(f)-(m) and 6(e).

Construction of HPFP

Let us explain how to construct the HPFP introduced in Step 3 above. The layers of the HPFP are referred to as categories. The first category consists of SCCs with zero indegree as the SCC $V_{1,1}$ in Fig. 2. The n-th category ($n \geq 2$) consists of SCCs satisfying two conditions: every link into SCCs in the n-th category comes from nodes in the k-th categories ($1 \leq k \leq n-1$) and SCCs in the n-th category have at least one input link coming from nodes in the (n-1)-th category. The HPFP is defined as the union of the hierarchical categories, each of which is also partitioned by SCCs. The HPFP has a simpler structure than the partition of the original network as the HPFP is not a partition of the original network but of the network simplified by the fully-simplified update rules.

For instance, let us assume that there exists a Boolean network in Fig. 2a, which has five SCCs in Fig. 2b. Applying the methods in [30, 31], the Boolean network is partitioned as in Fig. 2c. The goal of the partition in [30, 31] is to make a framework for finding all attractors of the original network without fixing a particular phenotype. However, we are interested in construction of other partition (HPFP) for finding particular attractors which represent a phenotype of interest. For simplicity, we assume that there exist no external, secondary-external and secondary-phenotype nodes. In addition, we assume that x_1 is a phenotype with the update equation

$$x_1^* = f_1(x_2, , x_3)$$

and that the phenotype nodes are (x_2, x_3) with one secondary-phenotype equation
$0 = x_3 = f_2(x_5, x_7)$.

Hence, the nodes $x_i (1 \leq i \leq 3)$ and the links connected to $x_i (1 \leq i \leq 3)$ are removed from Fig. 2c, thus changing Fig. 2c into Fig. 2d. Therefore, the first category of the HPFP in Fig. 2d comprises $V_{1,1} = \{x_6, x_7, x_8, x_9\}$ and $V_{1,2} = \{x_4\}$, which have a zero indegree. The SCCs $V_{2,1} = \{x_{10}, x_{11}\}$ and $V_{2,2} = \{x_5\}$ in the second category are the unique SCCs with links connected to the nodes in the first category. Comparing the two hierarchical partitions in Fig. 2c and d, we can find that the HPFP is simpler than the partition that can be obtained from previous methods [30, 31].

How to concatenate the local attractors of subnetworks in the HPFP?

Let us explain, through an example, how to sequentially concatenate the local attractors of subnetworks in categories of the HPFP in Fig. 3a for the construction of the global attractors of the HPFP. The algorithm for the concatenation is described in detail in Additional file 8. Our explanation is focused on concatenation from the HPFP, so that we do not take into account the external, phenotype, secondary-external and secondary-phenotype nodes, and secondary-phenotype equations. However, when applying our approach to biological networks in the Results section, these were taken into account. The HPFP is given with the fully-simplified update rules in Additional file 7(a). In the following, we describe the sequential concatenation step by step (the details of calculation in each step are given in Additional file 7).

Step 1. Find local attractors in the first category. There exists only one subnetwork $V_{1,1} = \{x_1, x_2\}$ in the first category as it is the unique subnetwork with no input links. Then $V_{1,1}$ has the update rules defined by x_1 and x_2 in Additional file 7(a), which yield three local attractors $a_{\langle 1 \rangle} = [\![10, 01]\!]$, $a_{\langle 2 \rangle} = [\![00]\!]$, $a_{\langle 3 \rangle} = [\![11]\!]$ in Additional file 7(b). Here the symbol $[\![10, 01]\!]$ denotes a cyclic attractor of length 2 and $[\![00]\!]$ a point attractor in Fig. 3b. In the next, we find global attractors containing the cyclic states $a_{\langle 1 \rangle}$ for x_1 and x_2. Similarly, the process for obtaining global attractors that include $a_{\langle 2 \rangle}$ and $a_{\langle 3 \rangle}$ is presented in detail in Additional file 8.

Step 2. Find local attractors in the second category. There exist two subnetworks $V_{2,1} = \{x_3, x_4\}$ and $V_{2,2} = \{x_5, x_6, x_7\}$ in the second category with input signals from the first category.

Step 2-1. Find local attractors of $V_{2,1}$. Due to $V_{2,1}^{in} = \{x_1\}$, the subnetwork $V_{2,1} = \{x_3, x_4\}$ gets the input signal $[\![1, 0]\!]$ generated from the state values of x_1 in the attractor $a_{\langle 1 \rangle}$. Note that change of the starting value of the input signal while preserving the order (1 and 0 with period 2) cannot affect the local attractors, which is proved in S.Theorem 1 in Additional file 8. Then we have a unique local attractor of $V_{2,1}^{in} \cup V_{2,1} = \{x_1, x_3, x_4\}$, which is $[\![101, 000, 100, 010]\!]$ in Fig. 3c and Additional file 7(c). Therefore, the local attractor of $V_{2,1} = \{x_3, x_4\}$ becomes $[\![01, 00, 00, 10]\!]$ in Fig. 3g.

Step 2-2. Find local attractors of $V_{2,2}$. As in Step 2-1, using the update rules for $V_{2,2} = \{x_5, x_6, x_7\}$ with the input signal $[\![0, 1]\!]$ generated from $V_{2,2}^{in} = \{x_2\}$, we have a unique local attractor of $V_{2,2}^{in} \cup V_{2,2}$, which is $[\![0000, 1010, 0101, 1000, 0001, 1001]\!]$ in Fig. 3d and Additional file 7(d). Therefore the local attractor of $V_{2,2} = \{x_5, x_6, x_7\}$ becomes $[\![000, 010, 101, 000, 001, 001]\!]$ in Fig. 3g.

Step 3. Find local attractors in the last category. There exist two subnetworks $V_{3,1} = \{x_8, x_9\}$ and $V_{3,2} = \{x_{10}, x_{11}\}$ in the last category with input signals from the first and second categories.

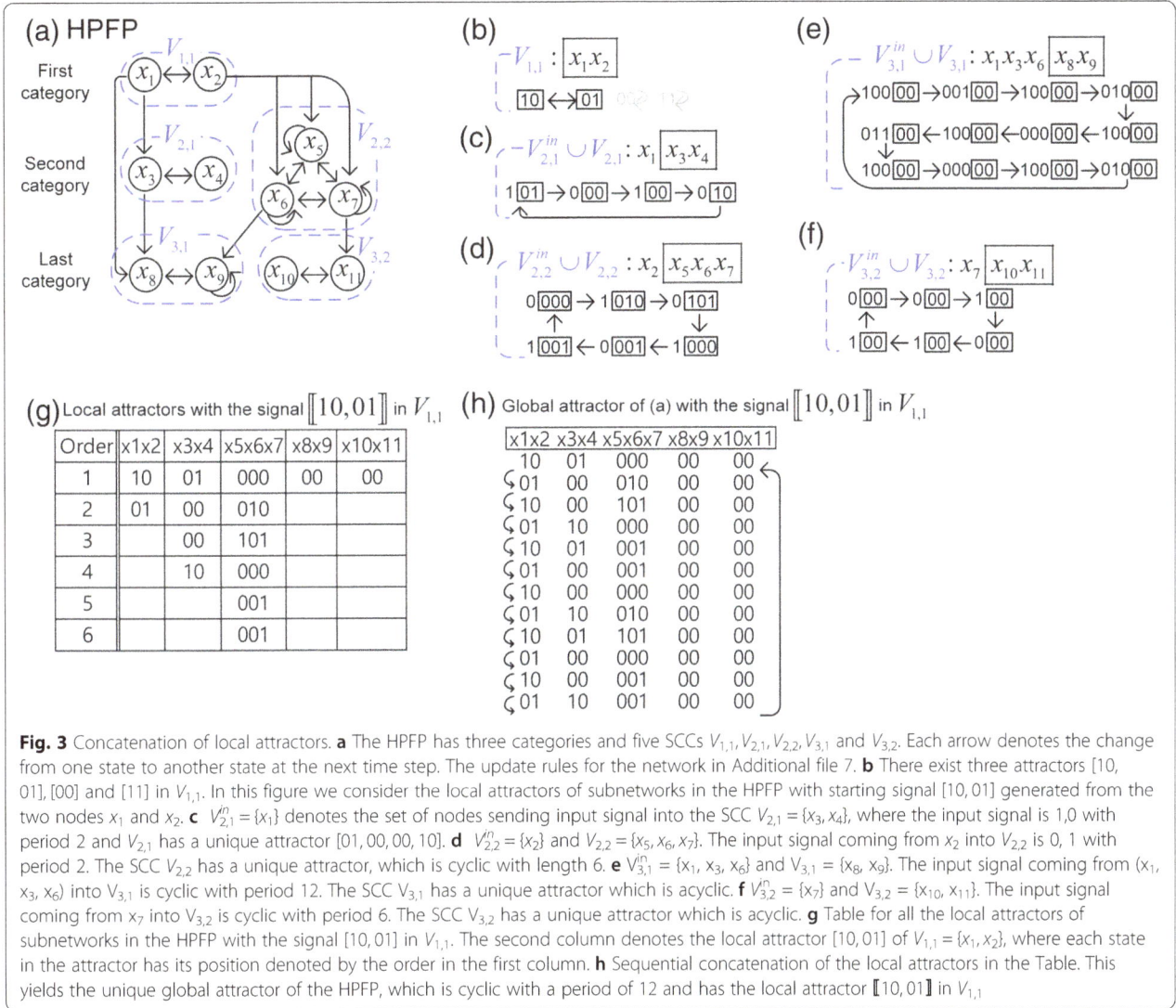

(a) HPFP

First category

Second category

Last category

(b)

$V_{1,1} : \boxed{x_1 x_2}$

$\boxed{10} \leftrightarrow \boxed{01}$

(c) $V_{2,1}^{in} \cup V_{2,1} : x_1 \boxed{x_3 x_4}$

$1\boxed{01} \to 0\boxed{00} \to 1\boxed{00} \to 0\boxed{10}$

(d) $V_{2,2}^{in} \cup V_{2,2} : x_2 \boxed{x_5 x_6 x_7}$

$0\boxed{000} \to 1\boxed{010} \to 0\boxed{101}$

$1\boxed{001} \leftarrow 0\boxed{001} \leftarrow 1\boxed{000}$

(e) $V_{3,1}^{in} \cup V_{3,1} : x_1 x_3 x_6 \boxed{x_8 x_9}$

$100\boxed{00} \to 001\boxed{00} \to 100\boxed{00} \to 010\boxed{00}$

$011\boxed{00} \leftarrow 100\boxed{00} \leftarrow 000\boxed{00} \leftarrow 100\boxed{00}$

$100\boxed{00} \to 000\boxed{00} \to 100\boxed{00} \to 010\boxed{00}$

(f) $V_{3,2}^{in} \cup V_{3,2} : x_7 \boxed{x_{10} x_{11}}$

$0\boxed{00} \to 0\boxed{00} \to 1\boxed{00}$

$1\boxed{00} \leftarrow 1\boxed{00} \leftarrow 0\boxed{00}$

(g) Local attractors with the signal $\llbracket 10,01 \rrbracket$ in $V_{1,1}$

Order	x1x2	x3x4	x5x6x7	x8x9	x10x11
1	10	01	000	00	00
2	01	00	010		
3		00	101		
4		10	000		
5			001		
6			001		

(h) Global attractor of (a) with the signal $\llbracket 10,01 \rrbracket$ in $V_{1,1}$

x1x2	x3x4	x5x6x7	x8x9	x10x11
10	01	000	00	00
01	00	010	00	00
10	00	101	00	00
01	10	000	00	00
10	01	001	00	00
01	00	001	00	00
10	00	000	00	00
01	10	010	00	00
10	01	101	00	00
01	00	000	00	00
10	00	001	00	00
01	10	001	00	00

Fig. 3 Concatenation of local attractors. **a** The HPFP has three categories and five SCCs $V_{1,1}, V_{2,1}, V_{2,2}, V_{3,1}$ and $V_{3,2}$. Each arrow denotes the change from one state to another state at the next time step. The update rules for the network in Additional file 7. **b** There exist three attractors [10, 01], [00] and [11] in $V_{1,1}$. In this figure we consider the local attractors of subnetworks in the HPFP with starting signal [10, 01] generated from the two nodes x_1 and x_2. **c** $V_{2,1}^{in} = \{x_1\}$ denotes the set of nodes sending input signal into the SCC $V_{2,1} = \{x_3, x_4\}$, where the input signal is 1,0 with period 2 and $V_{2,1}$ has a unique attractor [01, 00, 00, 10]. **d** $V_{2,2}^{in} = \{x_2\}$ and $V_{2,2} = \{x_5, x_6, x_7\}$. The input signal coming from x_2 into $V_{2,2}$ is 0, 1 with period 2. The SCC $V_{2,2}$ has a unique attractor, which is cyclic with length 6. **e** $V_{3,1}^{in} = \{x_1, x_3, x_6\}$ and $V_{3,1} = \{x_8, x_9\}$. The input signal coming from (x_1, x_3, x_6) into $V_{3,1}$ is cyclic with period 12. The SCC $V_{3,1}$ has a unique attractor which is acyclic. **f** $V_{3,2}^{in} = \{x_7\}$ and $V_{3,2} = \{x_{10}, x_{11}\}$. The input signal coming from x_7 into $V_{3,2}$ is cyclic with period 6. The SCC $V_{3,2}$ has a unique attractor which is acyclic. **g** Table for all the local attractors of subnetworks in the HPFP with the signal [10, 01] in $V_{1,1}$. The second column denotes the local attractor [10, 01] of $V_{1,1} = \{x_1, x_2\}$, where each state in the attractor has its position denoted by the order in the first column. **h** Sequential concatenation of the local attractors in the Table. This yields the unique global attractor of the HPFP, which is cyclic with a period of 12 and has the local attractor $\llbracket 10, 01 \rrbracket$ in $V_{1,1}$

Step 3-1. Find local attractors of $V_{3,1}$. Using the update rules for $V_{3,1} = \{x_8, x_9\}$ with the input signal $\llbracket 100, 001, 100, 010, 100, 000, 100, 011, 100, 000, 100, 010 \rrbracket$ from $V_{3,1}^{in} = \{x_1, x_3, x_6\}$, we have a unique attractor of $V_{3,1}^{in} \cup V_{3,1}$, which is cyclic with length 12

$$\begin{bmatrix} 10000, 00100, 10000, 01000, 10000, 00000, \\ 10000, 01100, 10000, 00000, 10000, 01000 \end{bmatrix}$$

in Fig. 3e and Additional file 7(e). Therefore the local attractor of $V_{3,1} = \{x_8, x_9\}$ becomes $\llbracket 00 \rrbracket$ in Fig. 3g.

Step 3-2. Local attractors of $V_{3,2}$. Using the update rules for $V_{3,2} = \{x_{10}, x_{11}\}$ with the input signal $\llbracket 0, 0, 1, 0, 1, 1 \rrbracket$ from $V_{3,2}^{in} = \{x_7\}$, we have a unique attractor of $V_{3,2}^{in} \cup V_{3,2}$, which is $\llbracket 000, 000, 100, 000, 100, 100 \rrbracket$ in Fig. 3f and Additional file 7(f). Therefore the local attractor of $V_{3,2} = \{x_{10}, x_{11}\}$ becomes $\llbracket 00 \rrbracket$ in Fig. 3g and Additional file 7(f).

Step 4. Construct the table with all the local attractors obtained from $a_{\langle 1 \rangle}$. We construct the table in Fig. 3g, in

which the first column denotes the order of states of each local attractor in the second column to the sixth column. The second column denotes the local attractor $\llbracket 10, 01 \rrbracket$ of $V_{1,1} = \{x_1, x_2\}$. Even though there are no states in the cells of the second column from the order 3, the second column is considered to be filled with states 10 and 01 with a period of 2. For instance, the states in the 3rd and 4th cells are 10 and 01, respectively. Similarly, the third column is filled with states 01, 00, 00 and 10 with a period of 4. Repeating this process completes the table.

Step 5. Construct the global attractor from the table. The concatenation of states in the i-th row of the table becomes the i-th state of the global attractor of the HPFP. Therefore, concatenating states from cells of the table in a row yields the unique global attractor of the HPFP with a period of 12, where the period is the least common multiple of periods of the five local attractors: 2, 4, 6, 1 and 1.

Finally, we obtain the unique global attractor by sequentially concatenating the local attractors that include the local attractor $a_{(1)}$ in Fig. 3h and Additional file 7(g), and confirm that the concatenated states become the global attractor by applying the original update rules in Additional file 7(g).

Results

To demonstrate the effectiveness of our framework in practice, we applied our method to three biological network models for finding attractors responsible for proliferation or apoptosis phenotypes: the first was a Mitogen-activated protein kinase (MAPK) model [6] with 53 nodes and 88 links, the second was a colitis-associated colon cancer (CACC) model [10] with 70 nodes and 152 links and the last was the simplified cancer pathways (SCP) model [9] with 96 nodes and 265 links. In general, Boolean update rules are classified into two types:

one is defined with logic functions and the other with threshold functions. The MAPK and CACC networks have update rules that correspond to the first type. The SCP network has update rules corresponding to the second type.

MAPK network

The MAPK model has four stimuli (DNA damage, TGFBR stimulus, EGFR stimulus and FGFR3 stimulus) and three phenotypes (proliferation, apoptosis and growth arrest) as in Fig. 4. The cascades in the MAPK network are strongly interconnected and the maximum number of nodes of the SCCs in the network is 37(69.8 %) as shown in Additional files 2(a) and (b). Therefore, due to computational complexity, the previous methods [30, 31] based on SCCs is not useful for this MAPK network.

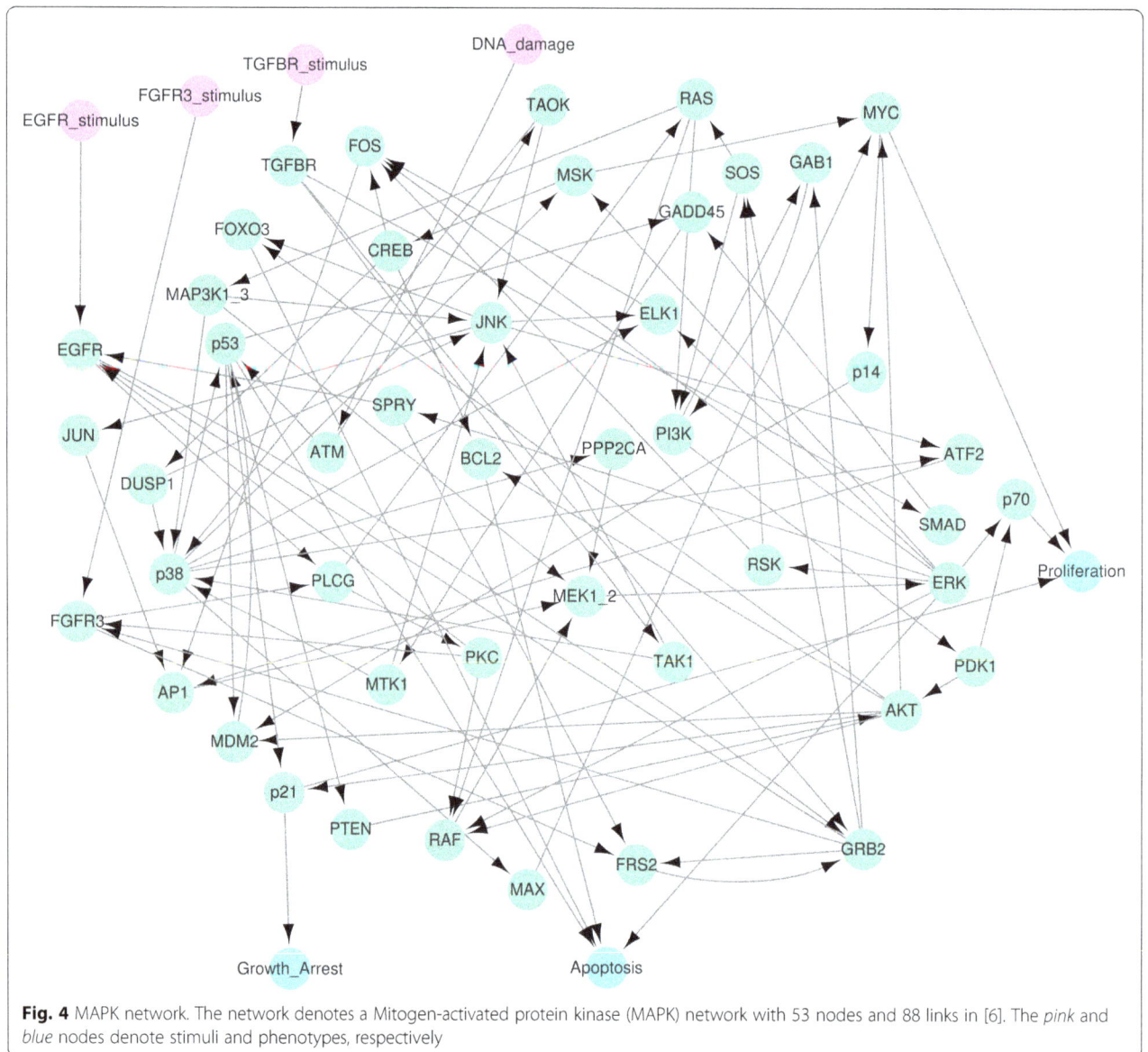

Fig. 4 MAPK network. The network denotes a Mitogen-activated protein kinase (MAPK) network with 53 nodes and 88 links in [6]. The *pink* and *blue* nodes denote stimuli and phenotypes, respectively

Proliferation attractors of the MAPK network

In order to find global attractors for proliferation of the MAPK network, we used the same simulation condition r30 in S3 Dataset of [6]: ERK perturbation with setting the values of the four stimuli to zero. Inserting the fixed values (ERK, TGFBR stimulus, EGFR stimulus, FGFR3 stimulus, DNA damage) = (1,0,0,0,0) of the external nodes into the update rules for the MAPK network in Additional file 2(a), we found the external condition: the external, secondary-external nodes (ESENs) and the semi-simplified update rules for nodes except ESENs in Additional file 2(c). Due to the update rule, Proliferation* = p70 & MYC & !p21, inserting the phenotype values (p70, MYC, p21) = (1,1,0) of the phenotype nodes into the semi-simplified update rules, we found the phenotype condition: the phenotype and secondary-phenotype nodes (PSPNs), the fully-simplified update rules for 21 nodes and the two secondary-phenotype equations

$$\text{MAX} \mid \text{AKT} = 1, \ !\text{AKT} \ p53 = 0$$

in Additional file 2(d). The fully-simplified update rules yield the HPFP for proliferation in Fig. 5 where the HPFP has 8 categories and the SCCs has four nodes at most whereas the SCC in the original network has 37 nodes. The yellow boxes on the three nodes p53, MAX, AKT in Fig. 5 denote the nodes included in the two secondary-phenotype equations, where the three nodes are referred to as "the equation nodes".

The SCCs with more than one node in the HPFP are $V_{1,1} = \{\text{GRB2, PKC, EGFR, PLCG}\}$ and $V_{4,1} = \{\text{p38, p53, GADD45, MTK1}\}$. The fully-simplified update rules for (GRB2, PKC, EGFR, PLCG) yield that $V_{1,1}$ has the unique attractor ⟦0100, 0000, 0010, 1011, 1101⟧ in Additional file 2(e), where the computing time was 0.028871 s by using a PC with 3.6GHz CPU and 32G RAM. The signal coming from {PLCG} in $V_{1,1}$ is transmitted to $V_{2,1} = \{\text{RAS}\}$ with the formula RAS* = PLCG and the signal from {RAS} becomes the input signal to $V_{3,1} = \{\text{MAP3K1\_3}\}$ with MAP3K1_3* = RAS. As a result, $V_{3,1}$ has a unique attractor ⟦1, 1, 0, 0, 0⟧. The unique input signal from {MAP3K1_2} to $V_{4,1}$ yields a unique attractor of $V_{4,1}$, which is the point attractor ⟦0000⟧ for (p38, p53, GADD45, MTK1) in Additional file 2(f). In this case, the computing time was 0.108352 s. Inserting the state values of the point attractor into the fully-simplified update rules in Additional file 2(d), we have

$$\text{PTEN}* = p53 = 0, \ \text{AKT}* = !\text{PTEN} = 1, \ \text{MAX}* = p38 = 0,$$

and then

$$\text{MAX}|\text{AKT} = 1, \ !\text{AKT} p53 = 0.$$

Hence the secondary-phenotype equations were satisfied for the local attractors of the seven subnetworks

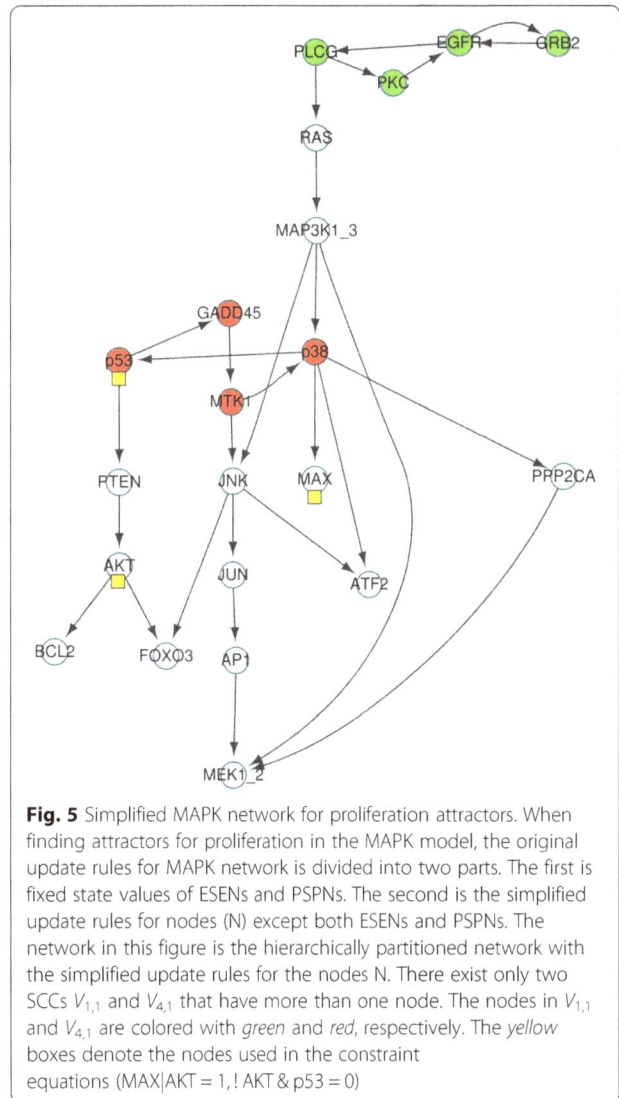

Fig. 5 Simplified MAPK network for proliferation attractors. When finding attractors for proliferation in the MAPK model, the original update rules for MAPK network is divided into two parts. The first is fixed state values of ESENs and PSPNs. The second is the simplified update rules for nodes (N) except both ESENs and PSPNs. The network in this figure is the hierarchically partitioned network with the simplified update rules for the nodes N. There exist only two SCCs $V_{1,1}$ and $V_{4,1}$ that have more than one node. The nodes in $V_{1,1}$ and $V_{4,1}$ are colored with green and red, respectively. The yellow boxes denote the nodes used in the constraint equations (MAX|AKT = 1, ! AKT & p53 = 0)

$V_{i,1}(1 \leq i \leq 4), V_{5,1} = \{\text{PTEN}\}, V_{5,3} = \{\text{MAX}\}, V_{6,1} = \{\text{AKT}\}.$

Note that the remaining eight subnetworks in Fig. 5 do not affect the states of the equation nodes (p53, MAX, AKT). Therefore, we can construct the table in Fig. 3g and found a unique global attractor for proliferation, which is cyclic with a period of 5 in Additional file 2(g). We confirmed that the set of cyclic states becomes a global attractor of the original network by applying the original update rules in Additional file 2(g).

Apoptosis attractors of the MAPK network

To find global attractors for apoptosis phenotype in the MAPK network, we used the simulation condition r4 in S3 Dataset of [6], which is FGFR3 perturbation with setting the values of the four stimuli to zero. The fully-simplified update rules yield the HPFP for apoptosis in Fig. 6, where the HPFP has 4 categories and each SCC has two nodes at most. We obtained a unique global

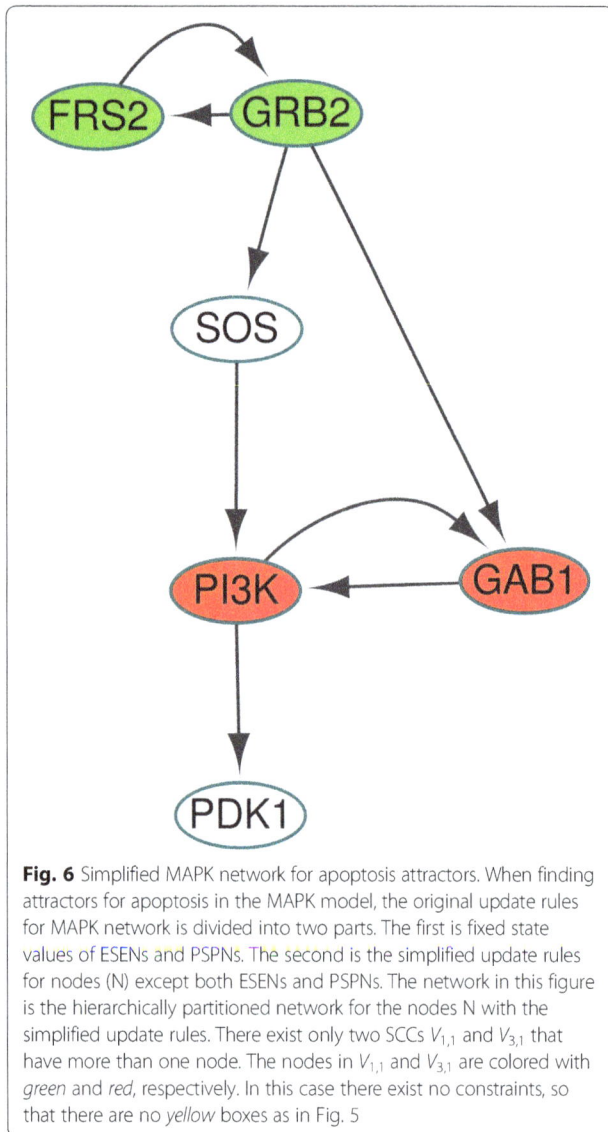

Fig. 6 Simplified MAPK network for apoptosis attractors. When finding attractors for apoptosis in the MAPK model, the original update rules for MAPK network is divided into two parts. The first is fixed state values of ESENs and PSPNs. The second is the simplified update rules for nodes (N) except both ESENs and PSPNs. The network in this figure is the hierarchically partitioned network for the nodes N with the simplified update rules. There exist only two SCCs $V_{1,1}$ and $V_{3,1}$ that have more than one node. The nodes in $V_{1,1}$ and $V_{3,1}$ are colored with *green* and *red*, respectively. In this case there exist no constraints, so that there are no *yellow* boxes as in Fig. 5

external nodes into the CACC update rules in Additional file 4(a), we found the external condition: the external and secondary-external nodes (ESENs) and the semi-simplified update rules for nodes except ESENs in Additional file 4(c). Due to Proliferation* = (FOS & CYCLIND1) & !(P21 | CASP3) with the secondary-external value FOS = 1, inserting the values (CYCLIND1, P21, CASP3) = (1,0,0) into the semi-simplified update rules, we found the phenotype condition: the phenotype, secondary-phenotype nodes (PSPNs), the fully-simplified update rules for three nodes (SOCS, STAT3, JAK) and no secondary-phenotype equation in Additional file 4(d). The fully-simplified update rules yield the HPFP for proliferation in Fig. 8 and the HPFP has one category with one SCC $V_{1,1}$, which has three nodes. The fully-simplified update rules for the three nodes (SOCS, STAT3, JAK) yield that $V_{1,1}$ has two attractors

010, 101, 111, 110, 100, 000, 001, 011

in Additional file 4(e). Therefore we found two global attractors for proliferation in Additional file 4(f) and (g), which are confirmed by applying the CACC update rules in Additional file 4(f) and (g).

Apoptosis attractors of the CACC network

Under the same condition for proliferation attractors of the CACC network above, the fully-simplified update rules yield the HPFP for apoptosis in Fig. 9. We applied our method to the CACC network and found that there exists no global attractor for apoptosis in the CACC network in Additional files 5 and 10. Since the simulation condition (DC, APC) = (1,1) denotes the strong tumour-promoting microenvironment (fixing DC at ON) for premalignant intestinal epithelial cells (fixing APC at ON), the nonexistence of global attractors for apoptosis can be expected.

SCP network

The simplified cancer pathway model has six inputs (Mutagen, GFs, Nutrients, TNfa, Hypoxia and Gli) and one phenotype (apoptosis) as shown in Fig. 10. The cascades in the original network are strongly interconnected with the maximum number of nodes of the SCC is 68 (70.8 %) as described in Additional file 6(a) and (b). The state vector (Mutagen, GFs, Nutrients, TNfa, Hypoxia) = (0,0,1,0,1) describes the normoxic microenvironment with the plenty of nutrients and growth factors [9].

We considered the apoptosis phenotype under the same microenvironment in [9]. Inserting the external values (Mutagen, GFs, Nutrients, TNfa, Hypoxia, Gli) = (0,0,1,0,1,0) into the SCP update rules in Additional file 6(a), we found the external condition: the external, secondary-external nodes (ESENs) and the semi-simplified update rules for nodes except

attractor for apoptosis, which is cyclic with a period of 4 in Additional files 3 and 9.

CACC network

The CACC model has one input (APC) and two phenotypes (proliferation and apoptosis) as in Fig. 7. The CACC network model is strongly interconnected with the maximum number of nodes of the SCC in the network is 65 (92.9 %) as described in Additional file 4(a) and (b).

Proliferation attractors of the CACC network

Under the strong tumour-promoting microenvironment (fixing DC at ON) for premalignant intestinal epithelial cells (fixing APC at ON) as in [10], we applied the proposed method to find global attractors for proliferation in the CACC network with the update rules in Additional file 4(a). Inserting the external values (DC, APC) = (1,1) of the

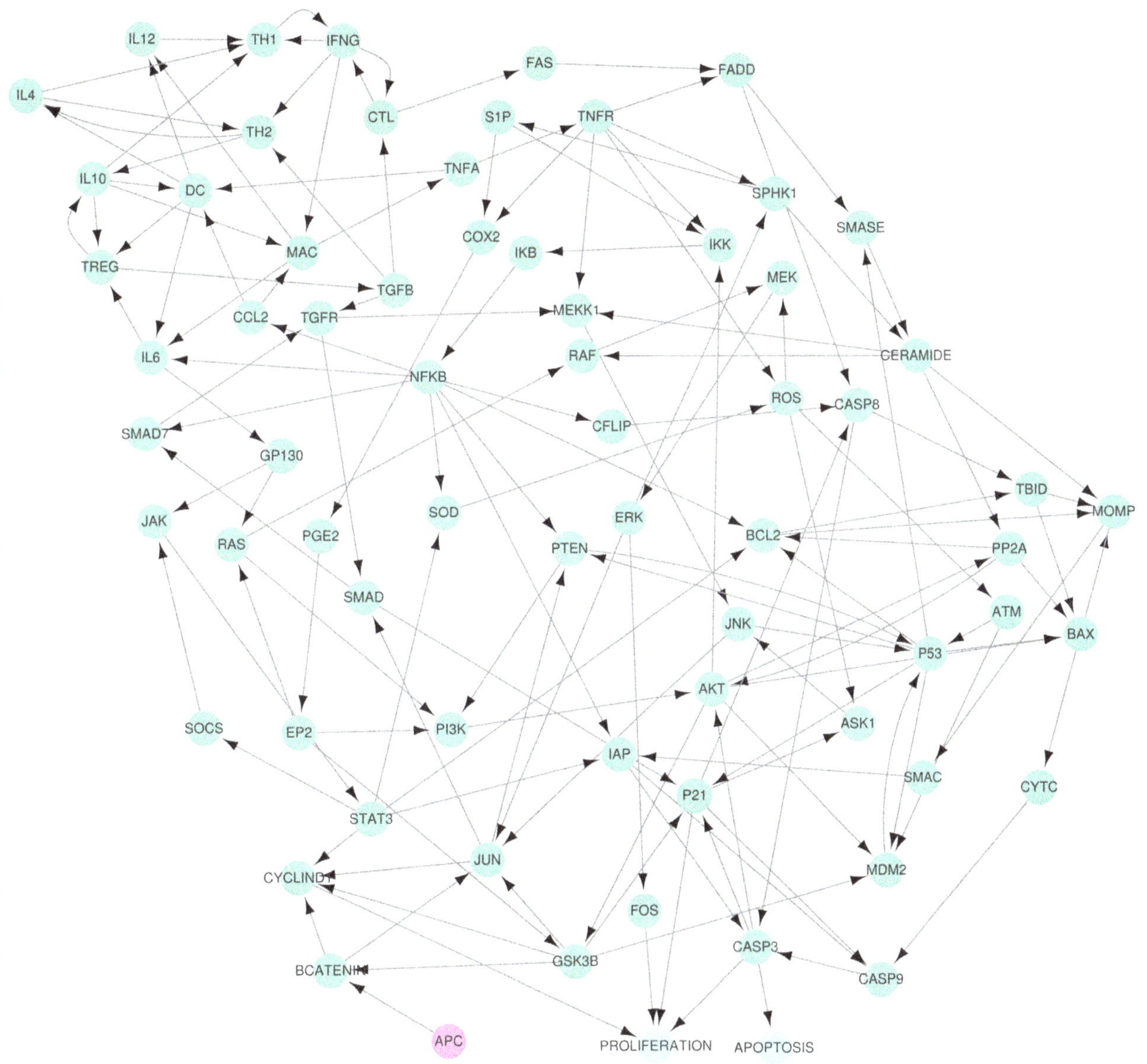

Fig. 7 CACC network. The network denotes a colitis-associated colon cancer (CACC) network with 70 nodes and 152 links in [10]. The *pink* circle denotes the input node, the adenomatous polyposis coli (APC) protein that represents premalignant intestinal epithelial cells when consistently activated. The *cyan* nodes denote phenotypes

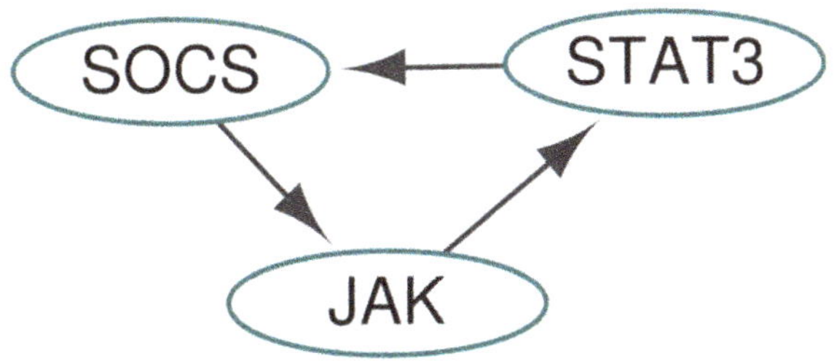

Fig. 8 Simplified CACC network for proliferation attractors. The network denotes the simplified CACC network for nodes except ESENs and PSPNs with the simplified update rules when finding attractors for proliferation in the CACC model

ESENs in Additional file 6(c). The secondary-external value Caspase8 = 0 yields

$$\text{Apoptosis}* = \text{sgn}[\text{Caspase8} + \text{Caspase9}]$$
$$= \text{sgn}[\text{Caspase9}].$$

Note that Apoptosis =1 if and only if Caspase9 = 1, which is also equivalent to (Caspase9, Cytoc/APAF1) = (1,1) and the constraint inequality 0 < -AKT + p53-BCL_2-Bcl_XL since Caspase9* = Cytoc/APAF1 and Cytoc/APAF1* = sgn(-AKT + p53-BCL_2-Bcl_XL). As a result of the external condition, the set of ESENs satisfies the constraint inequality and becomes the set of all nodes in the SCP network in

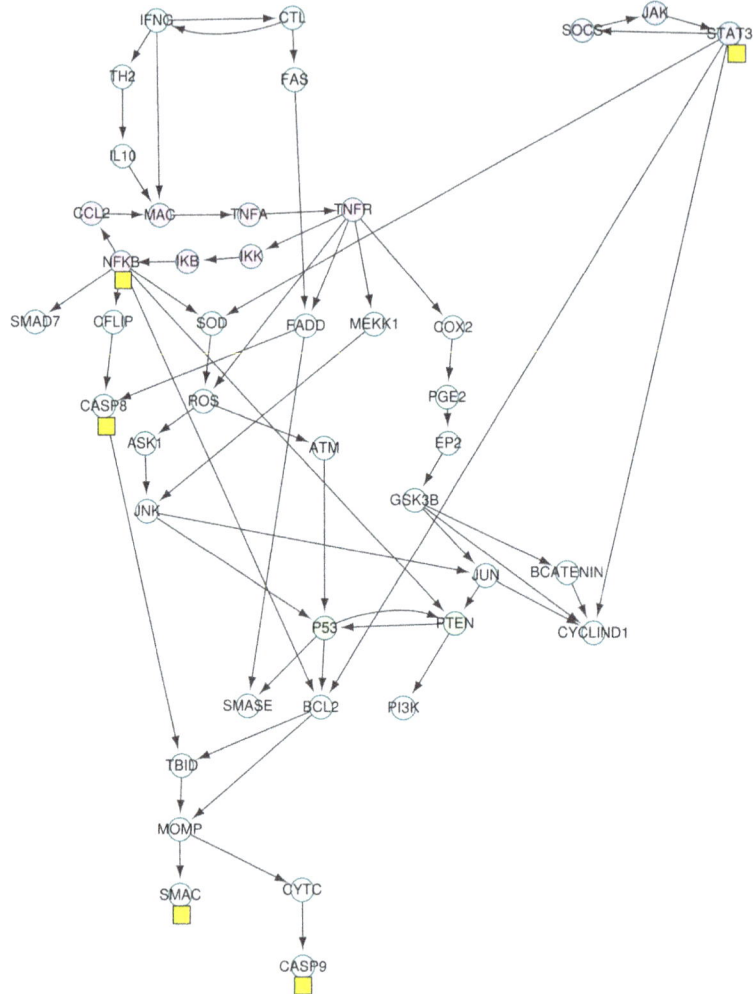

Fig. 9 Simplified CACC network for apoptosis attractors. When finding attractors for apoptosis in the CACC model, the network denotes the hierarchically partitioned CACC network for nodes except ESENs and PSPNs with the simplified update rules. There exist four SCCs with more than one node, which are represented with different colors: $V_{1,1}$ = {IFNG, CTL}, $V_{1,2}$ = {SOCS, JAK, STAT3}, $V_{4,1}$ = {CCL2, MAC, TNFA, TNFR, IKK, IKB, NFKB} and $V_{10,1}$ = {P53, PTEN}. The yellow boxes denote the nodes used in the constraint equations ((NFKB|STAT3)&(~ SMAC) = 0, CASP8|CASP9 = 1)

Additional file 6(c). Then there are no PSPNs, no secondary-phenotype inequalities and no fully-simplified update rules.

Therefore the vector of the fixed values of the 96 nodes (ESENs) is the unique attractor for apoptosis under the microenvironment. We confirmed that the concatenated state vector becomes the global attractor by applying the SCP update rules to Additional file 6(e).

Resolving the two problems: large size and strong interconnection

We summarized in Table 1 the aforementioned results obtained for the MAPK, CACC and SCP networks. The three biological networks have a large number of nodes that are strongly interconnected. The first two networks have nodes for representing proliferation and apoptosis phenotypes, but the last has a node for only apoptosis.

The column with the title #External and phenotype nodes in Table 1 shows that the number of nodes for a given external environment and a phenotype of interest are independent of the size and degree of interconnection. We found that the number of SCCs in the HPFP is approximately proportional to the number of categories as the number of categories gets increased. Even if the original networks are strongly interconnected and the number of categories and that of SCCs in the HPFP are high, the number of nodes in SCCs in the HPFP is small enough such that full search of the state space can be performed to find local attractors and therefore we can find global attractors for the phenotype in the original networks by concatenating the local attractors.

In particular, in the case of the SCP network, the set of ESENs becomes the set of all the 97 nodes in the SCP network with no secondary-phenotype inequalities, which

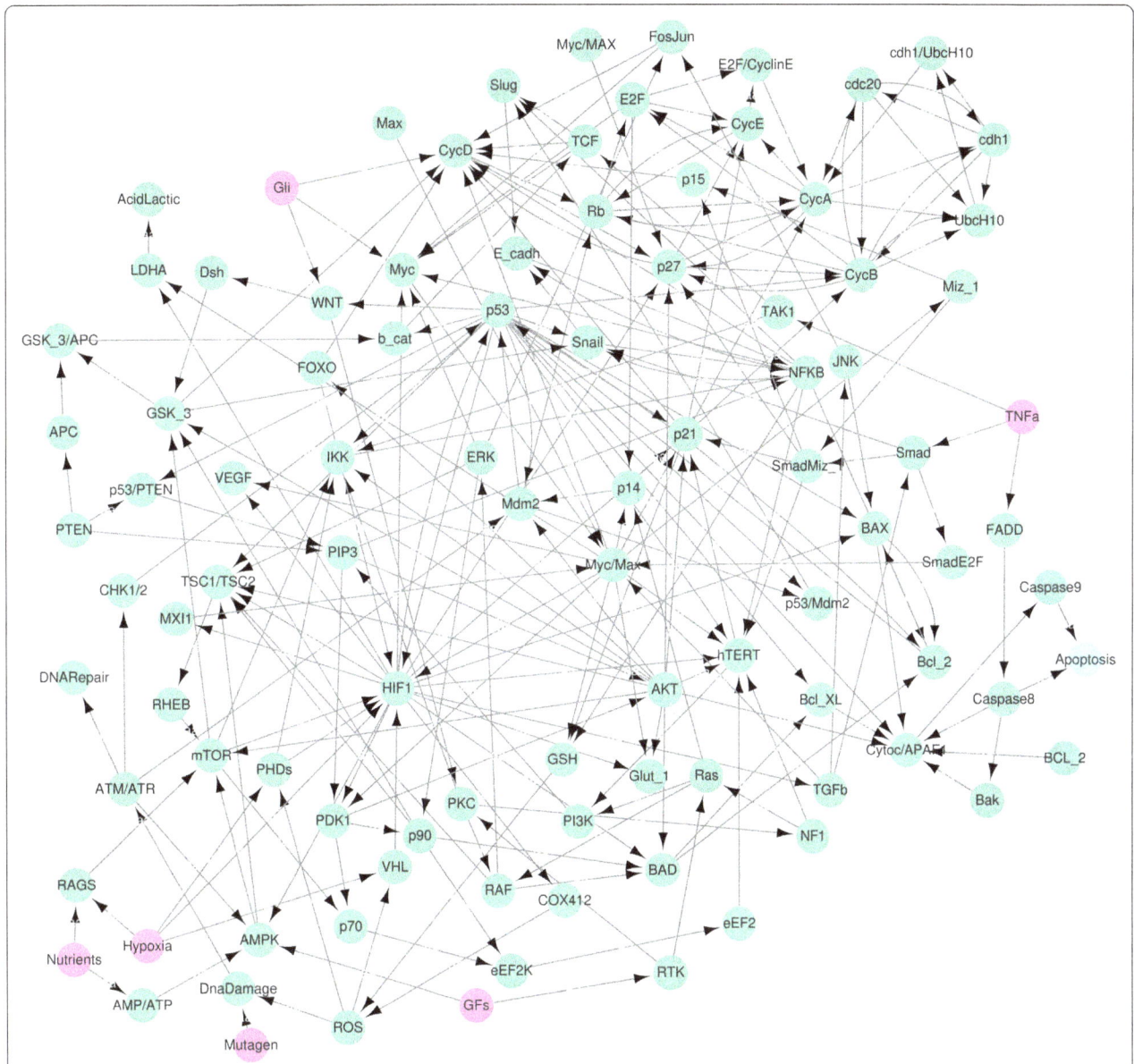

Fig. 10 SCP network. The network denotes a simplified cancer pathways (SCP) network with 96 nodes and 265 links in [9]. The *pink* and *cyan* nodes denote input and apoptosis phenotype, respectively

Table 1 The symbol #A denotes the number of A. And the symbol Max #Nodes in SCCs denotes the maximum of the numbers of nodes in strongly connected components

| | Original network | | HPFP for nodes except ESENs and PSPNs | | | | |
	#Nodes	Max #Nodes in SCCs	Phenotype	#External and phenotype nodes	#Categories	#SCCs	Max #Nodes in SCCs
MAPK	53	37 (69.8 %)	Proliferation	8	8	15	4
			Apoptosis	9	4	4	2
CACC	70	65 (92.9 %)	Proliferation	6	1	1	3
			Apoptosis	3	15	33	7
SCP	97	68 (70.8 %)	Apoptosis	8	0	0	0

explains why the number of categories, that of SCCs and the maximum are all zero in Table 1. The SCP network does not include a node for proliferation, thus we could not find attractors for proliferation phenotype.

Comparison of our method with the reduction and random sampling methods

Previous methods based on SCCs for finding all global attractors cannot be applied to the three networks in Table 1 due to the large size of networks. To deal with such a problem, two methods are usually used: reduction method and random sampling method, which were compared with our method for the MAPK, CACC and SCP networks.

Under the condition r30 in S3 Dataset of [6] used in the Proliferation attractors of the MAPK network section, the reduction method yields two attractors for proliferation phenotype, which are cyclic with a period of 6 in S3 Dataset of [6]. However, we found that the MAPK update rules under the same condition result in the unique attractor for proliferation, which is cyclic with a period of 5 in Additional file 2(g). Such different results show that the reduced network obtained by the reduction method does not preserve the attractors of the original network, even though the periodic property of attractors is preserved. In addition, under the condition r4 in S3 Dataset of [6], such difference was also found between S3 Dataset of [6] and Additional file 3(f).

Reduced update rules are given for a reduced CACC network in [10] and we used the update rules to find attractors for proliferation of the reduced CACC network in Additional file 11. We compared the proliferation attractors obtained by the reduced update rules with the proliferation attractors obtained from the original CACC network by using the proposed method. As a result, we did not find such a difference between S3 Dataset of [6] and Additional file 3(f). However, the reduced method has a disadvantage that the attractors obtained by reduced update rules for reduced networks do not have the information on states of all nodes in the original networks.

Under the condition (Mutagen, GFs, Nutrients, TNfa, Hypoxia) = (0,0,1,0,1) in S1 Dataset of [9], the random sampling method and our method yield a unique attractor for apoptosis phenotype, which is a fixed point attractor in S1 Dataset of [9] and Additional file 6(e). Even though a random sampling method (i.e. randomly sampling the initial states and tracing the converging state trajectories) can provide an estimate of global attractors of large-size networks while compromising the computational complexity, such an approximation cannot guarantee to find all global attractors for a phenotype of interest. However, our method can always guarantee the full search result even for a large network.

Validation of the proposed algorithm

To show that all attractors in a given network can be found by applying our method, we applied our method to two Boolean networks with known attractors. The first network is shown in Fig. 3a and it has 11 nodes with the update rules in Additional file 7(a), where the network has neither input nor output nodes. This network is suitable for the validation of concatenating local attractors obtained from HPFP without considering ESENs and PSPNs. By applying a full search algorithm to this network, we could find all attractors and, as a result, we confirmed that these attractors are exactly the same as we found by applying our method (see Additional file 12 for details).

For the validation of PSPNs as well as the concatenation, the second network is adopted from [10] and it has 21 nodes. This network has two cyclic attractors of length 2 and 6, both representing cell proliferation, as shown in Fig. 3c and Supplementary Table S6 in [10]. To find out all attractors representing cell proliferation in this network model, we have applied our method to this network and could obtain the same attractors (see Additional file 12 for details), as shown in Fig. 3c and Supplementary Table S6 in [10].

Discussion

Previous approaches of partitioning a large Boolean network model to resolve the computational complexity issue preserve the regulatory links of the original network to identify all the attractor states from the partitioned network. The primary point we noticed is that we have to simplify the state update rules and reduce the regulatory links to obtain small subnetworks of practically computable size. For this purpose, we focused only on a set of attractors for a particular phenotype of interest and developed a novel algorithm that can efficiently find out the attractor states from the hierarchically partitioned subnetworks obtained by simplifying the state update rules and replacing some regulatory links with constraint equations while preserving the particular attractor states. In contrast with the previous approaches, the proposed approach can result in small and simplified subnetworks of computable size. An important point is that we can always find out all the attractors corresponding to the phenotype of interest in the original large network by sequentially concatenating the local attractors that are obtained from the hierarchically partitioned subnetworks.

The limitation of our approach is that we cannot find out all attractors at the same time using this approach. However, in many practical case studies, only a few particular phenotypes are of interest and therefore, by applying the proposed approach to each particular phenotype, we can find out all attractor states of interest. Another limitation is that our approach is based on synchronous update rules, so it is currently not applicable to the Boolean network models based on asynchronous update rules. This remains as a future study.

From the case studies where we applied our approach to the three large biological network examples as well as small and medium size networks (Additional files 12 and 13), we found that the resulting subnetworks (i.e. SCCs) are composed of seven nodes at most. Of course, we cannot guarantee such a small size subnetwork in all cases, but we can always obtain much smaller subnetworks compared to previous approaches since our approach simplifies the state update rules in a practical way. Moreover, we can further reduce the subnetwork size if any other biological information on the molecular state of a node in the converging phenotypic feature is available. As there are many other biological networks that are different from the networks employed in this study, it remains as a future study to further investigate the power of our method by using extensive simulation-based analysis of synthetic networks with respect to various topological properties such as different size, level of interconnections, etc.

When we hierarchically partition a network, we simplify the state update rules by considering the fixed state values of marker nodes in the attractor states. As a result, the state space after the hierarchical partitioning becomes a subset of the state space of the original network. So, we cannot measure the basin of attraction to the particular phenotype of the original network in our framework. However, some modification of our framework might be able to resolve the problem. This also remains as a future study.

Conclusions

Although Boolean network modeling is becoming popular in modeling large-scale biological regulatory networks, looking for attractors for converging state analysis is still challenging for large networks due to computational complexity. There have been some attempts to resolve this problem by partitioning the large network into smaller subnetworks and reconstructing the global attractors by concatenating the local attractors obtained from each subnetwork, but the resultant subnetworks were still too large in most cases and therefore not much useful in practice. So, in this study, we have developed a novel approach of identifying a set of global attractors for a particular phenotype of interest by hierarchically partitioning the original large network such that the resulting subnetworks are small enough to guarantee that the full search of the local attractors of them is possible. We have applied the proposed method to several biological networks and confirmed its usefulness.

Throughout the hierarchical partitioning, we can obtain the hierarchical partitioned structure of the original network with the fixed state values of some nodes. Such structural information on the network might be also useful in identifying certain target nodes to control the phenotypic behavior of a biological network. For instance, we can use this information to find out control target nodes, the perturbation of which results in preventing the convergence of the dynamical network state to the particular attractor state of interest. This is an important subject for a future study.

Additional files

Additional file 1: Definition of external and phenotype nodes.

Additional file 2: Attractors for proliferation in the Mitogen-activated protein kinase (MAPK) model.

Additional file 3: Attractors for apoptosis in the Mitogen-activated protein kinase (MAPK) model.

Additional file 4: Attractors for proliferation in the colitis-associated colon cancer (CACC) model.

Additional file 5: Attractors for apoptosis in the colitis-associated colon cancer (CACC) model.

Additional file 6: Attractors for apoptosis in the simplified cancer pathways (SCP) model.

Additional file 7: Example of how to find the phenotype attractors in HPFP.

Additional file 8: Explanation of how to concatenate the local attractors of subnetworks in the HPFP.

Additional file 9: Attractors for apoptosis in the MAPK network.

Additional file 10: Attractors for apoptosis in the CACC network.

Additional file 11: Attractors for proliferation in the reduced CACC model.

Additional file 12: Validation of the proposed algorithm for two small and medium networks.

Additional file 13: Attractor in the logical T-cell signaling model.

Abbreviations

CACC: Colitis-associated colon cancer; ESENS: External and secondary-external nodes; HPFP: The hierarchical partition for the phenotype; MAPK: Mitogen-activated protein kinase; PSPNs: Phenotype and secondary-phenotype nodes; SCC: Strongly connected component; SCP: Simplified cancer pathways

Acknowledgments

We thank Sang-Min Park for his help in developing the programming code of finding attractors and also Je-Hoon Song for his help in preparing the figures.

Funding

This work was supported by the National Research Foundation of Korea (NRF) grants funded by the Korea Government, the Ministry of Science, and ICT & Future Planning (2015M3A9A7067220, 2014R1A2A1A10052404, and 2013M3A9A7046303). It was also supported by a grant of the Korean Health Technology R&D Project, Ministry of Health and Welfare, Republic of Korea (HI13C2162), the KUSTAR-KAIST Institute, KAIST, Korea, and the KAIST Future Systems Healthcare Project from the Ministry of Science, ICT and Future Planning.

Authors' contributions

Conceived and designed the experiments: S-MC and K-HC. Performed the experiments: S-MC Analyzed the data: S-MC and K-HC. Wrote the paper: S-MC and K-HC. Both authors read and approved the final manuscript.

Competing interests

The authors declare that they have no competing interests.

Author details

[1]Department of Mathematics, University of Ulsan, Ulsan 44610, Republic of Korea. [2]Department of Bio and Brain Engineering, Korea Advanced Institute of Science and Technology (KAIST), Daejeon 34141, Republic of Korea.

References

1. De Jong H. Modeling and simulation of genetic regulatory systems: a literature review. J Comput Biol. 2002;9(1):67–103.
2. Kauffman SA. Metabolic stability and epigenesis in randomly constructed genetic nets. J Theor Biol. 1969;22(3):437–67.
3. Kauffman S. Homeostasis and differentiation in random genetic control networks. Nature. 1969;224:177–8.
4. Glass L, Kauffman SA. The logical analysis of continuous, non-linear biochemical control networks. J Theor Biol. 1973;39(1):103–29.
5. Wang R, Saadatpour A, Albert R. Boolean modeling in systems biology: an overview of methodology and applications. Phys Biol. 2012;9(5):055001.
6. Grieco L, Calzone L, Bernard-Pierrot I, Radvanyi F, Kahn-Perles B, Thieffry D. Integrative modelling of the influence of MAPK network on cancer cell fate decision. PLoS Comput Biol. 2013;9(10):e1003286.
7. Hanahan D, Weinberg RA. The hallmarks of cancer. Cell. 2000;100(1):57–70.
8. Smith G, Carey FA, Beattie J, Wilkie MJ, Lightfoot TJ, Coxhead J, Garner RC, Steele RJ, Wolf CR. Mutations in APC, Kirsten-ras, and p53–alternative genetic pathways to colorectal cancer. Proc Natl Acad Sci U S A. 2002;99(14):9433–8.
9. Fumiã HF, Martins ML. Boolean network model for cancer pathways: predicting carcinogenesis and targeted therapy outcomes. PLoS One. 2013;8(7):e69008.
10. Lu J, Zeng H, Liang Z, Chen L, Zhang L, Zhang H, Liu H, Jiang H, Shen B, Huang M, Geng M, Spiegel S, Luo C. Network modelling reveals the mechanism underlying colitis-associated colon cancer and identifies novel combinatorial anti-cancer targets. Sci Rep. 2015;5:14739.
11. Li Q, Wennborg A, Aurell E, Dekel E, Zou JZ, Xu Y, Huang S, Ernberg I. Dynamics inside the cancer cell attractor reveal cell heterogeneity, limits of stability, and escape. Proc Natl Acad Sci U S A. 2016;113(10):2672–7.
12. Akutsu T, Kuhara S, Maruyama O, Miyano S. A system for identifying genetic networks from gene expression patterns produced by gene disruptions and overexpressions. Genome Inform. 1998;9:151–60.
13. Zhao Q. A remark on " Scalar equations for synchronous Boolean networks with biological Applications" by C. Farrow, J. Heidel, J. Maloney, and J. Rogers. IEEE Trans Neural Netw. 2005;16(6):1715–6.
14. Akutsu T, Kosub S, Melkman AA, Tamura T. Finding a periodic attractor of a Boolean network. IEEE/ACM Trans Comput Biol Bioinform. 2012;9(5):1410–21.
15. Berntenis N, Ebeling M. Detection of attractors of large Boolean networks via exhaustive enumeration of appropriate subspaces of the state space. BMC Bioinform. 2013;14(1):1.
16. Zañudo JG, Albert R. An effective network reduction approach to find the dynamical repertoire of discrete dynamic networks. Chaos. 2013;23(2):025111.
17. Bryant RE. Graph-based algorithms for boolean function manipulation. IEEE Trans Comput. 1986;100(8):677–91.
18. Bilke S, Sjunnesson F. Stability of the Kauffman model. Phys Rev E. 2001;65(1):016129.
19. Socolar JE, Kauffman SA. Scaling in ordered and critical random Boolean networks. Phys Rev Lett. 2003;90(6):068702.
20. Dubrova E, Teslenko M, Martinelli A. Kauffman networks: analysis and applications. In: Anonymous IEEE Computer Society, editor. Proceedings of the 2005 IEEE/ACM International conference on Computer-aided design. 2005. p. 479–84.
21. Mihaljev T, Drossel B. Scaling in a general class of critical random Boolean networks. Phys Rev E. 2006;74(4):046101.
22. Di Cara A, Garg A, De Micheli G, Xenarios I, Mendoza L. Dynamic simulation of regulatory networks using SQUAD. BMC Bioinform. 2007;8(1):1.
23. Garg A, Xenarios I, Mendoza L, DeMicheli G. An efficient method for dynamic analysis of gene regulatory networks and in silico gene perturbation experiments. In: Anonymous Springer, editor. Research in computational molecular biology. 2007. p. 62–76.
24. Saadatpour A, Albert I, Albert R. Attractor analysis of asynchronous Boolean models of signal transduction networks. J Theor Biol. 2010;266(4):641–56.
25. Naldi A, Remy E, Thieffry D, Chaouiya C. Dynamically consistent reduction of logical regulatory graphs. Theor Comput Sci. 2011;412(21):2207–18.
26. Veliz-Cuba A. Reduction of Boolean network models. J Theor Biol. 2011;289:167–72.
27. Zheng D, Yang G, Li X, Wang Z, Liu F, He L. An efficient algorithm for computing attractors of synchronous and asynchronous Boolean networks. PLoS One. 2013;8(4):e60593.
28. Hong C, Hwang J, Cho K, Shin I. An efficient steady-state analysis method for large boolean networks with high maximum node connectivity. PLoS One. 2015;10(12):e0145734.
29. Veliz-Cuba A, Aguilar B, Hinkelmann F, Laubenbacher R. Steady state analysis of Boolean molecular network models via model reduction and computational algebra. BMC Bioinform. 2014;15(1):1.
30. Zhao Y, Kim J, Filippone M. Aggregation algorithm towards large-scale Boolean network analysis. IEEE Trans Automatic Control. 2013;58(8):1976–85.
31. Guo W, Yang G, Wu W, He L, Sun M. A parallel attractor finding algorithm based on Boolean satisfiability for genetic regulatory networks. PLoS One. 2014;9(4):e94258.

Permissions

All chapters in this book were first published in SB, by BioMed Central; hereby published with permission under the Creative Commons Attribution License or equivalent. Every chapter published in this book has been scrutinized by our experts. Their significance has been extensively debated. The topics covered herein carry significant findings which will fuel the growth of the discipline. They may even be implemented as practical applications or may be referred to as a beginning point for another development.

The contributors of this book come from diverse backgrounds, making this book a truly international effort. This book will bring forth new frontiers with its revolutionizing research information and detailed analysis of the nascent developments around the world.

We would like to thank all the contributing authors for lending their expertise to make the book truly unique. They have played a crucial role in the development of this book. Without their invaluable contributions this book wouldn't have been possible. They have made vital efforts to compile up to date information on the varied aspects of this subject to make this book a valuable addition to the collection of many professionals and students.

This book was conceptualized with the vision of imparting up-to-date information and advanced data in this field. To ensure the same, a matchless editorial board was set up. Every individual on the board went through rigorous rounds of assessment to prove their worth. After which they invested a large part of their time researching and compiling the most relevant data for our readers.

The editorial board has been involved in producing this book since its inception. They have spent rigorous hours researching and exploring the diverse topics which have resulted in the successful publishing of this book. They have passed on their knowledge of decades through this book. To expedite this challenging task, the publisher supported the team at every step. A small team of assistant editors was also appointed to further simplify the editing procedure and attain best results for the readers.

Apart from the editorial board, the designing team has also invested a significant amount of their time in understanding the subject and creating the most relevant covers. They scrutinized every image to scout for the most suitable representation of the subject and create an appropriate cover for the book.

The publishing team has been an ardent support to the editorial, designing and production team. Their endless efforts to recruit the best for this project, has resulted in the accomplishment of this book. They are a veteran in the field of academics and their pool of knowledge is as vast as their experience in printing. Their expertise and guidance has proved useful at every step. Their uncompromising quality standards have made this book an exceptional effort. Their encouragement from time to time has been an inspiration for everyone.

The publisher and the editorial board hope that this book will prove to be a valuable piece of knowledge for researchers, students, practitioners and scholars across the globe.

List of Contributors

Oleg Lenive
ICR, SM2 5NG Sutton, UK

Paul D. W. Kirk
MRC Biostatistics Unit, Cambridge Institute of Public Health, Cambridge, UK

Michael P. H. Stumpf
Imperial College, London, Centre for Integrative Systems Biology and Bioinformatics, SW7 2AZ London, UK

Antje Jensch, Caterina Thomaseth and Nicole E. Radde
Institute for Systems Theory and Automatic Control, University of Stuttgart, Pfaffenwaldring 9, 70569 Stuttgart, Germany

Bram Thijssen and Lodewyk F. A. Wessels
Computational Cancer Biology, The Netherlands Cancer Institute, Plesmanlaan 121, 1066 CX Amsterdam, The Netherlands
Faculty of EEMCS, Delft University of Technology, Mekelweg 4, 2628CD Delft, The Netherlands

Tjeerd M. H. Dijkstra
Max Planck Institute for Developmental Biology, Spemannstrasse 35, 72076 Tübingen, Germany
Centre for Integrative Neuroscience, University Clinic Tübingen, Otfried-Müller-Strasse 25, 72076 Tübingen, Germany

Tom Heskes
Radboud University Nijmegen, Institute for Computing and Information Sciences, Heyendaalseweg 135, 6525 AJ Nijmegen, The Netherlands

Miriam Leon, Mae L. Woods and Alex J. H. Fedorec
Department of Cell and Developmental Biology, University College London, Gower Street, WC1E 6BT, London, UK

Chris P. Barnes
Department of Cell and Developmental Biology, University College London, Gower Street, WC1E 6BT, London, UK
Department of Genetics, Evolution and Environment, University College London, Gower Street, WC1E 6BT, London, UK

Daesik Choi, Byungkyu Park, Hanju Chae, Wook Lee and Kyungsook Han
Department of Computer Science and Engineering, Inha University, 22212 Incheon, South Korea

Stanislav Sokolenko, Marco Quattrociocchi and Marc G. Aucoin
Department of Chemical Engineering, University of Waterloo, 200 University Avenue West, N2L 3G1 Waterloo ON, Canada

Matthew J. Simpson
School of Mathematical Sciences, Queensland University of Technology (QUT), Brisbane, Australia

Kai-Yin Lo
Department of Agricultural Chemistry, National Taiwan University, 10617 Taipei, Taiwan

Yung-Shin Sun
Department of Physics, Fu-Jen Catholic University, 24205 New Taipei City, Taiwan

Lizhong Liu and Jian-Dong Huang
School of Biomedical Sciences, Li Ka Shing Faculty of Medicine, University of Hong Kong, Pok Fu Lam, Hong Kong, People's Republic of China
Shenzhen Institute of Research and Innovation, University of Hong Kong, Shenzhen 518057, People's Republic of China
The Centre for Synthetic Biology Engineering Research, Shenzhen Institutes of Advanced Technology, Shenzhen 518055, People's Republic of China

Wei Huang
Department of Biology, Shenzhen Key Laboratory of Cell Microenvironment, South University of Science and Technology of China, Shenzhen 518055, People's Republic of China

Ying Liu
Department of Biostatistics, Columbia University, New York, NY, USA

Wenfei Zhang, Mindy Zhang, Cheng Zhu and Yuefeng Lu
Sanofi, Framingham, MA, USA

Binh P. Nguyen and Lisa Tucker-Kellogg
Centre for Computational Biology and Program in Cancer and Stem Cell Biology, Duke-NUS Medical School, 169857 Singapore, Singapore
BioSystems and Micromechanics (BioSyM) Singapore – MIT Alliance for Research and Technology, 138602 Singapore, Singapore

Hans Heemskerk
BioSystems and Micromechanics (BioSyM) Singapore – MIT Alliance for Research and Technology, 138602 Singapore, Singapore
Centre for Computational Biology and Program in Cancer and Stem Cell Biology, Duke-NUS Medical School, 169857 Singapore, Singapore

Peter T. C. So
Department of Mechanical Engineering, Massachusetts Institute of Technology, 02139 Cambridge, MA, USA
BioSystems and Micromechanics (BioSyM) Singapore – MIT Alliance for Research and Technology, 138602 Singapore, Singapore

Philippe Nimmegeers, Dries Telen, Filip Logist and Jan Van Impe
KU Leuven, Department of Chemical Engineering, BioTeC+ and OPTEC, Gebroeders De Smetstraat 1, 9000 Ghent, Belgium

Mostafa Herajy
Department of Mathematics and Computer Science, Faculty of Science, Port Said University, 42521 Port Said, Egypt

Fei Liu
School of Software Engineering, South China University of Technology, 510006 Guangzhou, People's Republic of China

Christian Rohr and Monika Heiner
Computer Science Institute, Brandenburg University of Technology, 10 13 44 Cottbus, Germany

Minsu Lee and HyungJune Lee
Department of Computer Science and Engineering, Ewha Womans University, Seoul, South Korea

Roy Lau, Samuel H. Friedman, Ahmadreza Ghaffarizadeh, David B. Agus, Shannon M. Mumenthaler and Paul Macklin
Lawrence J. Ellison Institute for Transformative Medicine, University of Southern California, Los Angeles, California, USA

Edwin F. Juarez
Lawrence J. Ellison Institute for Transformative Medicine, University of Southern California, Los Angeles, California, USA
Department of Electrical Engineering, Viterbi School of Engineering, University of Southern California, Los Angeles, California, USA

Edmond Jonckheere
Department of Electrical Engineering, Viterbi School of Engineering, University of Southern California, Los Angeles, California, USA

Tatsuaki Tsuruyama
Department of Pathology, Kyoto University, Graduate School of Medicine, Yoshida-Konoe-Cho 1, Kyoto, Kyoto Prefecture 606-8501, Japan

Sang-Mok Choo
Department of Mathematics, University of Ulsan, Ulsan 44610, Republic of Korea

Kwang-Hyun Cho
Department of Bio and Brain Engineering, Korea Advanced Institute of Science and Technology (KAIST), Daejeon 34141, Republic of Korea

Index

www.ingramcontent.com/pod-product-compliance
Lightning Source LLC
Chambersburg PA
CBHW082024190326
41458CB00010B/3262